AF352769

Modelling, Test and Practice of Steel Structures

Modelling, Test and Practice of Steel Structures

Editors

Zhihua Chen
Hanbin Ge
Siu-lai Chan

MDPI • Basel • Beijing • Wuhan • Barcelona • Belgrade • Manchester • Tokyo • Cluj • Tianjin

Editors

Zhihua Chen
Tianjin University
China

Hanbin Ge
Meijo University
Japan

Siu-lai Chan
The Hong Kong Polytechnic
University
China

Editorial Office
MDPI
St. Alban-Anlage 66
4052 Basel, Switzerland

This is a reprint of articles from the Special Issue published online in the open access journal *Metals* (ISSN 2075-4701) (available at: https://www.mdpi.com/journal/metals/special_issues/simulation_steel).

For citation purposes, cite each article independently as indicated on the article page online and as indicated below:

LastName, A.A.; LastName, B.B.; LastName, C.C. Article Title. *Journal Name* **Year**, *Volume Number*, Page Range.

ISBN 978-3-0365-4883-8 (Hbk)
ISBN 978-3-0365-4884-5 (PDF)

Contents

Editorial

Modelling, Test and Practice of Steel Structures

Zhihua Chen [1,*], Hanbin Ge [2] and Siulai Chan [3]

[1] School of Civil Engineering, Tianjin University, Tianjin 300072, China
[2] Department of Civil Engineering, Meijo University, Nagoya 468-8502, Japan; gehanbin@meijo-u.ac.jp
[3] Department of Civil and Environmental Engineering, The Hong Kong Polytechnic University, Hong Kong 999077, China; siu-lai.chan@polyu.edu.hk
* Correspondence: zhchen@tju.edu.cn

Citation: Chen, Z.; Ge, H.; Chan, S. Modelling, Test and Practice of Steel Structures. *Metals* **2022**, *12*, 1212. https://doi.org/10.3390/met12071212

Received: 11 July 2022
Accepted: 15 July 2022
Published: 18 July 2022

Publisher's Note: MDPI stays neutral with regard to jurisdictional claims in published maps and institutional affiliations.

1. Introduction and Scope

Steel structures have been widely used in civil engineering in recent decades across applications such as large spatial structures, high-rise buildings, and bridges. With the development of techniques and economy, steel structures are increasingly popular in fabricated industry and residential buildings. Modelling and tests are the main methods for realizing the behaviors of steel structures, including the bearing capacity, the ductility, the seismic performance, etc. The behaviors of entire structures can be realized, and the construction method can be proposed in practical engineering.

This Special Issue on the Modelling, Testing, and Practice of Steel Structures provides an international forum for the presentation and discussion of the latest developments in structural steel research and their applications. The topics of this issue include the modelling, testing, and practice of steel structures and steel-based composite structures.

2. Contributions

In this Special Issue, 17 high-quality papers covering a wide range of steel structure research including modelling, testing, and construction research on material properties, components, assemblages, connection, and structural behaviors have been published.

Three papers focused on the material properties of structural steel, which presented investigations on the chemical composition of weld metals [1]; the fracture performances of SBHS500, SM570, and SM490 steel [2]; and the cyclic performance of structural steels after exposure to various heating–cooling treatments [3].

Nine papers focused on the mechanical properties of the components and joints of steel and steel-based composite structures, which presented investigations on shear square section steel tube dampers [4], diagonally stiffened steel plate walls [5], offshore platform deck structures [6], ship local structures [7], an L-shaped column composed of RAC-filled steel tubes [8], corrugated steel plate shear walls [9], partially connected steel plate shear walls [10], pipe-in-pipe systems [11], and bolted ball-cylinder joints [12].

Four papers focused on the structural behaviors, which presented investigations on the stability behaviors of a large-span spatial grid arch structure [13], the fatigue performance of a long-span steel truss arched bridge [14,15], and the structural safety performance of prestressed steel structures [16].

One paper focused on the planning method for a material-allocation path and the construction of prestressed steel structures [17].

3. Conclusions and Outlook

Topics such as material properties, the mechanical properties of components and joints, construction methods, and structural behaviors are covered by this Special Issue, presenting the latest developments in structural steel research and their applications. As Guest Editors of this Special Issue, we hope that the reported studies will be useful to researchers in advancing their respective research.

Funding: This research received no external funding.

Acknowledgments: As Guest Editors, we highly appreciate the valuable research works from the contributing authors, the professionalism of the reviewers and editors, and the efforts of the staff working on this Special Issue. Special gratitude goes to the Metals Editorial Office for its great support.

References

1. Kim, J.H.; Jung, C.J.; Park, Y.I.L.; Shin, Y.T. Development of Closed-Form Equations for Estimating Mechanical Properties of Weld Metals According to Chemical Composition. *Metals* **2022**, *12*, 528. [CrossRef]
2. Liu, Y.; Ikeda, S.; Liu, Y.; Kang, L.; Ge, H. Experimental Investigation of Fracture Performances of SBHS500, SM570 and SM490 Steel Specimens with Notches. *Metals* **2022**, *12*, 672. [CrossRef]
3. Du, P.; Liu, H.; Xu, X. Cyclic Performance of Structural Steels after Exposure to Various Heating–Cooling Treatments. *Metals* **2022**, *12*, 1146. [CrossRef]
4. Xiao, L.; Li, Y.; Hui, C.; Zhou, Z.; Deng, F. Experimental Study on Mechanical Properties of Shear Square Section Steel Tube Dampers. *Metals* **2022**, *12*, 418. [CrossRef]
5. Yang, Y.; Mu, Z.; Zhu, B. Numerical Study on Elastic Buckling Behavior of Diagonally Stiffened Steel Plate Walls under Combined Shear and Non-Uniform Compression. *Metals* **2022**, *12*, 600. [CrossRef]
6. Zhou, H.; Han, Y.; Zhang, Y.; Luo, W.; Liu, J.; Yu, R. Numerical and Experimental Research on Similarity Law of the Dynamic Responses of the Offshore Stiffened Plate Subjected to Low Velocity Impact Loading. *Metals* **2022**, *12*, 657. [CrossRef]
7. Xiao, S.; Han, Y.; Zhang, Y.; Wei, Q.; Wang, Y.; Wang, N.; Wang, H.; Liu, J.; Liu, Y. A Reliability Analysis Framework of Ship Local Structure Based on Efficient Probabilistic Simulation and Experimental Data Fusion. *Metals* **2022**, *12*, 805. [CrossRef]
8. Ma, T.; Chen, Z.; Du, Y.; Zhou, T.; Zhang, Y. Mechanical Properties of L-Shaped Column Composed of RAC-Filled Steel Tubes under Eccentric Compression. *Metals* **2022**, *12*, 953. [CrossRef]
9. Tan, Z.; Zhao, Q.; Zhao, Y.; Yu, C. Probabilistic Seismic Assessment of CoSPSW Structures Using Fragility Functions. *Metals* **2022**, *12*, 1045. [CrossRef]
10. Yang, Y.; Mu, Z.; Zhu, B. Study on Shear Strength of Partially Connected Steel Plate Shear Wall. *Metals* **2022**, *12*, 1060. [CrossRef]
11. Zhang, Z.; Chen, Z.; Liu, H. Lateral Buckling of Pipe-in-Pipe Systems under Sleeper-Distributed Buoyancy—A Numerical Investigation. *Metals* **2022**, *12*, 1094. [CrossRef]
12. He, J.; Wu, B.; Wu, N.; Chen, L.; Chen, A.; Li, L.; Xiong, Z.; Lin, J. Numerical and Theoretical Investigation on the Load-Carrying Capacity of Bolted Ball-Cylinder Joints with High-Strength Steel at Elevated Temperatures. *Metals* **2022**, *12*, 597. [CrossRef]
13. Chen, Z.; Lin, H.; Wang, X.; Liu, H.; Kawaguchi, K.; Matsui, M. Structural Stability Analysis of Eye of the Yellow Sea, a Large-Span Arched Pedestrian Bridge. *Metals* **2022**, *12*, 1138. [CrossRef]
14. Liu, P.; Chen, Y.; Lu, H.; Zhao, J.; An, L.; Wang, Y.; Liu, J. Fatigue Analysis of Long-Span Steel Truss Arched Bridge Part I: Experimental and Numerical Study of Orthotropic Steel Deck. *Metals* **2022**, *12*, 1117. [CrossRef]
15. Liu, P.; Lu, H.; Chen, Y.; Zhao, J.; An, L.; Wang, Y.; Liu, J. Fatigue Analysis of Long-Span Steel Truss Arched Bridge Part II: Fatigue Life Assessment of Suspenders Subjected to Dynamic Overloaded Moving Vehicles. *Metals* **2022**, *12*, 1035. [CrossRef]
16. Zhu, H.; Wang, Y. Intelligent Analysis for Safety-Influencing Factors of Prestressed Steel Structures Based on Digital Twins and Random Forest. *Metals* **2022**, *12*, 646. [CrossRef]
17. Liu, Z.; Shi, G.; Qin, J.; Wang, X.; Sun, J. Prestressed Steel Material-Allocation Path and Construction Using Intelligent Digital Twins. *Metals* **2022**, *12*, 631. [CrossRef]

Article

Structural Stability Analysis of Eye of the Yellow Sea, a Large-Span Arched Pedestrian Bridge

Zhihua Chen [1,2], Hao Lin [1], Xiaodun Wang [1,2,*], Hongbo Liu [2,3,*], Ken'ichi Kawaguchi [4] and Minoru Matsui [5]

[1] Department of Civil Engineering, Tianjin University, Tianjin 300072, China; zhchen@tju.edu.cn (Z.C.); lhao001@126.com (H.L.)
[2] State Key Laboratory of Hydraulic Engineering Simulation and Safety, Tianjin University, Tianjin 300072, China
[3] Department of Civil Engineering, Hebei University of Engineering, Handan 056000, China
[4] Institute of Industrial Science, The University of Tokyo, Tokyo 153-8505, Japan; kawaken@iis.u-tokyo.ac.jp
[5] Kawaguchi & Engineers Co., Ltd., Tokyo 102-0071, Japan; matui_m@kawa-struc.com
* Correspondence: maodun2004@126.com (X.W.); hb_liu2008@163.com (H.L.)

Abstract: To date, scholars' research on the stability behavior of the arch structure mainly focuses on solid–web section arches, steel tubular truss arches and concrete-filled steel tubular arches, but the stability behavior of the novel spatial grid arch structure, which integrates the characteristics of grid structure and arch structure, is not yet clear. Based on the Eye of the Yellow Sea pedestrian bridge project in Rizhao, China, the stability behavior of this large-span spatial grid arch structure was studied, in this paper, by the project's structure design team. The project is a glass covered steel arch pedestrian bridge with a span of 177 m, a height of 63.5 m, an elliptical section with a long axis of 18 m, and a short axis of 13.5 m. The elastic and the nonlinear elasto-plastic stability behavior considering different initial geometric imperfections, was analyzed by the ABAQUS finite element model. The buckling modes and the full-range load-displacement curve of the structure were analyzed, and the stress distribution, deformation mode and overall structural performance during the whole loading process were analyzed. The effects of initial imperfections, geometric nonlinearity and material nonlinearity on the ultimate load-carrying capacity of the structure were studied. The stability behavior of large-span spatial grid arch structure was studied in this paper, which provides an important reference for the design and analysis of such structures.

Keywords: arch structure; stability behavior; spatial grid arch structure; stability analysis; initial imperfections; nonlinear factors; ultimate load-carrying capacity

Citation: Chen, Z.; Lin, H.; Wang, X.; Liu, H.; Kawaguchi, K.; Matsui, M. Structural Stability Analysis of Eye of the Yellow Sea, a Large-Span Arched Pedestrian Bridge. *Metals* **2022**, *12*, 1138. https://doi.org/10.3390/met12071138

Academic Editor: Ulrich Prahl

Received: 9 May 2022
Accepted: 2 July 2022
Published: 3 July 2022

Publisher's Note: MDPI stays neutral with regard to jurisdictional claims in published maps and institutional affiliations.

1. Introduction

The arch structure is one of many structural systems, and has a wide range of applications in building and bridge projects. To date, a large amount of research has been conducted on the stability behavior of traditional arch structures commonly used in engineering. By using an analytical method, Timoshenko [1] derived the equation of the in-plane buckling load from the equilibrium differential equation of the pin-ended arch under uniform compression, which laid the foundation for the theory of the in-plane stability of the arch. Pi and Bradford [2–9] studied the in-plane nonlinear elastic stability behavior of circular arches with boundary conditions of hinged, fixed, one hinged and one fixed, and unequal rotational restraints under a centrally concentrated load. Hodges [10] studied the nonlinear in-plane deformation and bucking of rings and high arches under a hydrostatic pressure effect. Sakimoto and Namita [11] investigated the out-of-plane buckling of solid rib arches braced with transverse bars. By using the principal of stationary potential energy and Rayleigh–Ritz method, Pi and Bradford [12–23] proposed analytical solutions for the elastic bending and torsional buckling loads of circular arches with boundary conditions of hinged, fixed and in-plane elastic rotation constraints under concentrated forces, uniform compression and uniform bending. Dou et al. [24] first proposed a method for solving the shear and torsional stiffness of truss arches, and derived an analytical solution for the buckling load of lateral bending and torsional instability of truss arches with fixed restraints under uniform compression and bending based on the energy method. Guo et al. [25,26] derived elastic support stiffness thresholds for the out-of-plane instability of truss arches and solid web arches with lateral discrete bracing. By using the finite element method, Komatsu and Sakimoto [27,28] studied the out-of-plane elasto-plastic buckling load of a box cross-sectional arch with rise-to-span ratios of 0.1 to 0.2 under a uniform load. Pi et al. [29–31] studied the out-of-plane elasto-plastic bending–torsional buckling and post-buckling behavior of H-section steel arches under various loads, introducing the regularized slenderness ratio and the stability coefficients of axial compression columns or pure bending beams into the Australian code, and proposed formulas for calculating the out-of-plane elastic load-carrying capacity of uniformly compressed arches and uniformly bent arches. Dou et al. [32] investigated flexural–torsional buckling and ultimate resistance of parabolic steel arches subjected to a uniformly distributed vertical load. Dou et al. [33] conducted static stability tests of three groups of circular arches with the same span but different rise-to-span ratios, and the out-of-plane load-carrying capacity of the arch under 3-point symmetric and 2-point asymmetric loading was investigated. The stability behavior of the novel large-span spatial grid arch structure applied to the Eye of the Yellow Sea pedestrian bridge has not yet been studied.

1.1. Project Overview

The Eye of the Yellow Sea, which is a glass-covered steel arched pedestrian bridge with a span of 177 m, a height of 63.5 m, an elliptical section with a long axis of 18 m, and a short axis of 13.5 m, is located over an estuary in Rizhao, China. The highest part of the bridge is equipped with a semicircular viewing platform extending to both sides, with a radius of 9 m and an area of 398 square meters.

1.2. Structure Information

In order to satisfy the experience of visitors passing inside the arch, and the features of open eyesight and good perspective, a novel spatial grid arch structure was applied to the Eye of the Yellow Sea large-span arched pedestrian bridge. The structure integrates the characteristics of grid structure and arch structure, consisting of arch components, ring components and bracing components, with a grid size of about 4.5 m × 4.5 m. The basic cross section of the structure is an ellipse with a long axis of 18 m and a short axis of 13.5 m. The overall structure model is shown in Figure 1.

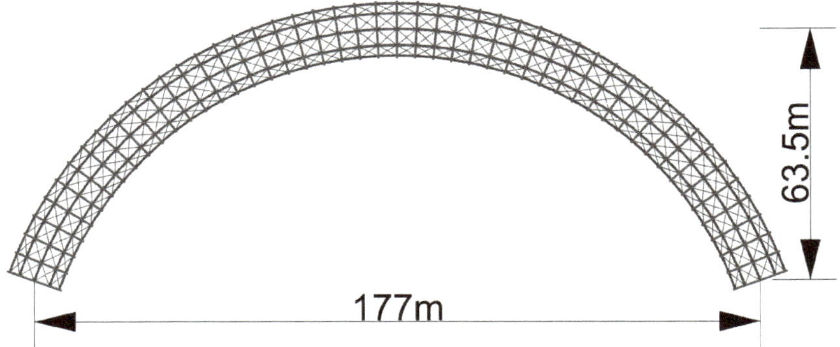

Figure 1. Overall structure model diagram.

As shown in Figure 2, the structure uses circular steel tubes for the arch components, H-shaped steel for the ring components, and steel tie rods for the bracing components.

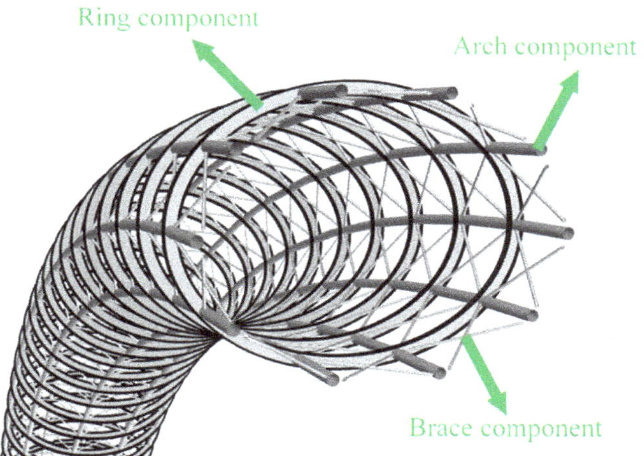

Figure 2. Basic cross section of the structure.

2. Finite Element Mode

According to the Eye of the Yellow Sea pedestrian bridge project, the structure model is divided into eight areas according to the different sizes of components' sections, and divided into three load regions according to the value of the vertical loads. Based on the above partition, the finite element model was established in ABAQUS. As shown in Figure 3, the component section partition and load partition are illustrated with the half structure.

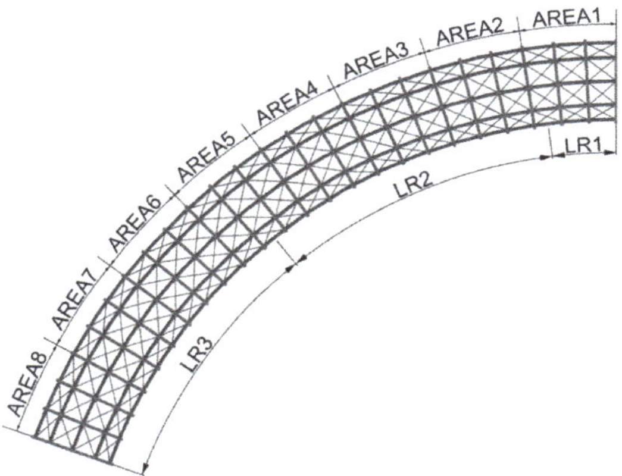

Figure 3. Regional division of the structure. Load region is notated as LR in the figure.

In the finite element model, the B31 element was used for the simulation of the ring and arch components, and the T3D2 truss element was used for the simulation of the bracing component, and were defined as tensile-only elements. The material's elastic modulus was set as 2.06×10^{11} Pa, Poisson's ratio as 0.3, density as 7.85×10^3 kg/m^3, yield strength as 345 MPa, and the material was considered as ideal elasto-plastic. The materials and major cross section sizes of the ring components, arch components and bracing components for the different areas are listed in Table 1.

Table 1. Summary of component information.

AREA	AREA1	AREA2	AREA3	AREA4	AREA5	AREA6	AREA7	AREA8
Ring (mm) Q345	B.Pipe-600 $\times$ 28	B.Pipe-600 $\times$ 28	B.Pipe-600 $\times$ 26	B.Pipe-600 $\times$ 26	B.Pipe-600 $\times$ 26	B.Pipe-600 $\times$ 26	B.Pipe-600 $\times$ 40	B.Pipe-600 $\times$ 40
Arch (mm) Q345	BH-700 $\times$ 300 $\times$ 22 $\times$ 22	BH-700 $\times$ 300 $\times$ 22 $\times$ 22	BH-700 $\times$ 300 $\times$ 20 $\times$ 20	BH-700 $\times$ 300 $\times$ 20 $\times$ 20	BH-700 $\times$ 300 $\times$ 20 $\times$ 20	BH-700 $\times$ 300 $\times$ 20 $\times$ 20	BH-700 $\times$ 300 $\times$ 36 $\times$ 22	BH-700 $\times$ 300 $\times$ 36 $\times$ 22
Brace (mm) Q460	Rod Φ185	Rod Φ185	Rod Φ185	Rod Φ185	Rod Φ185	Rod Φ185	Rod Φ185	Rod Φ185

According to the actual engineering situation, the dead loads (including 3 kN/m^2 for platform, deck and stairs, 9 kN/m^2 for escalator, and 10 kN/m^2 for elevator), and live loads (including 3.5 kN/m^2 for platform, deck and stairs, and 0.5 kN/m^2 for roof maintenance), were counted separately by different load regions, converted into nodal loads for each region, and added to the finite element model. The loads' direction was vertically downward, and the acceleration of gravity was 9.8 m/s^2. Model loads are shown as Figure 4.

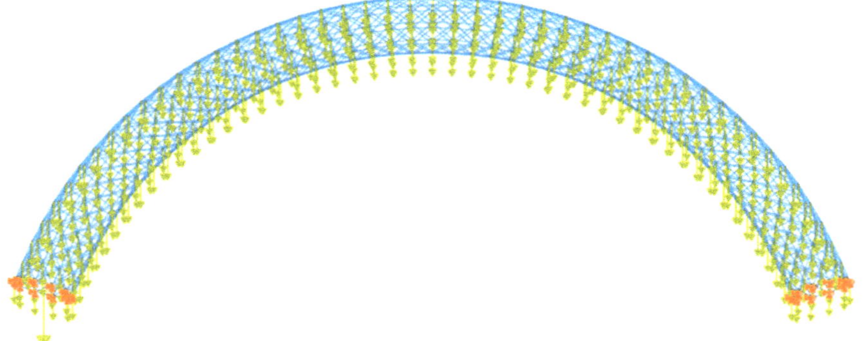

Figure 4. Nodal loads and gravity load in the finite element model.

The Merge function was used in the finite element model to create the welded connection between the components, the MPC Pin was used to simulate the hinge joint constraint, and the hinged constraint was set at the end of each arch component at the arch feet, as shown in Figure 5.

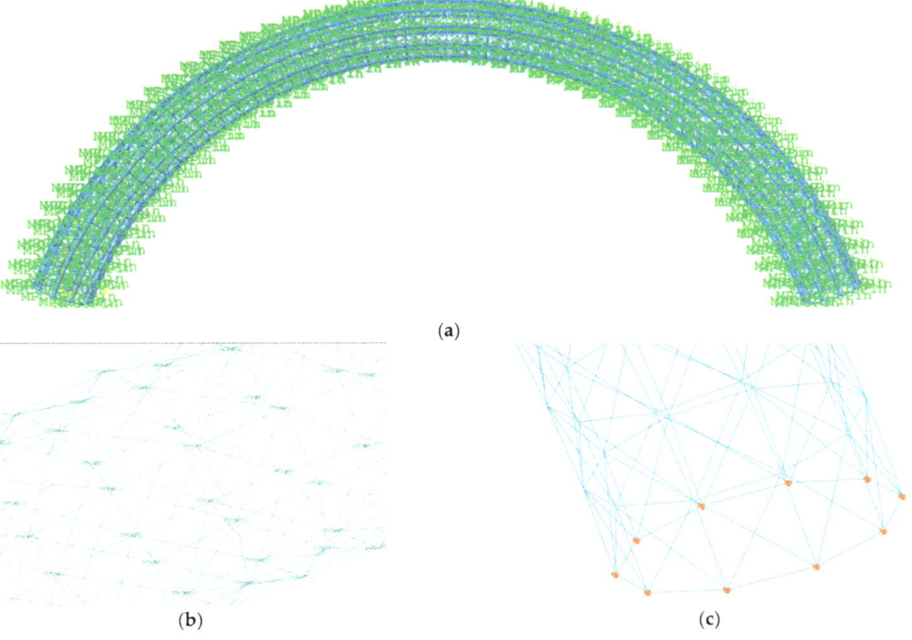

(a)

(b)   (c)

Figure 5. Finite element model constraint information: (**a**) Overview; (**b**) MPC Pin; (**c**) Hinged constraint.

3. Stability Behavior Analysis

3.1. Linear Buckling Analysis

Through linear buckling analysis, the elastic buckling load of the structure and the corresponding buckling modes was obtained to determine the weak regions of the structure. At the same time, consistent initial geometric imperfections were applied to the structure according to the buckling modes, which were used as the basis for the nonlinear elasto-plastic stability analysis of the structure [34]. The first six buckling modes of the structure under "1.0Dead Load + 1.0Live Load" load combination are shown in Figure 6.

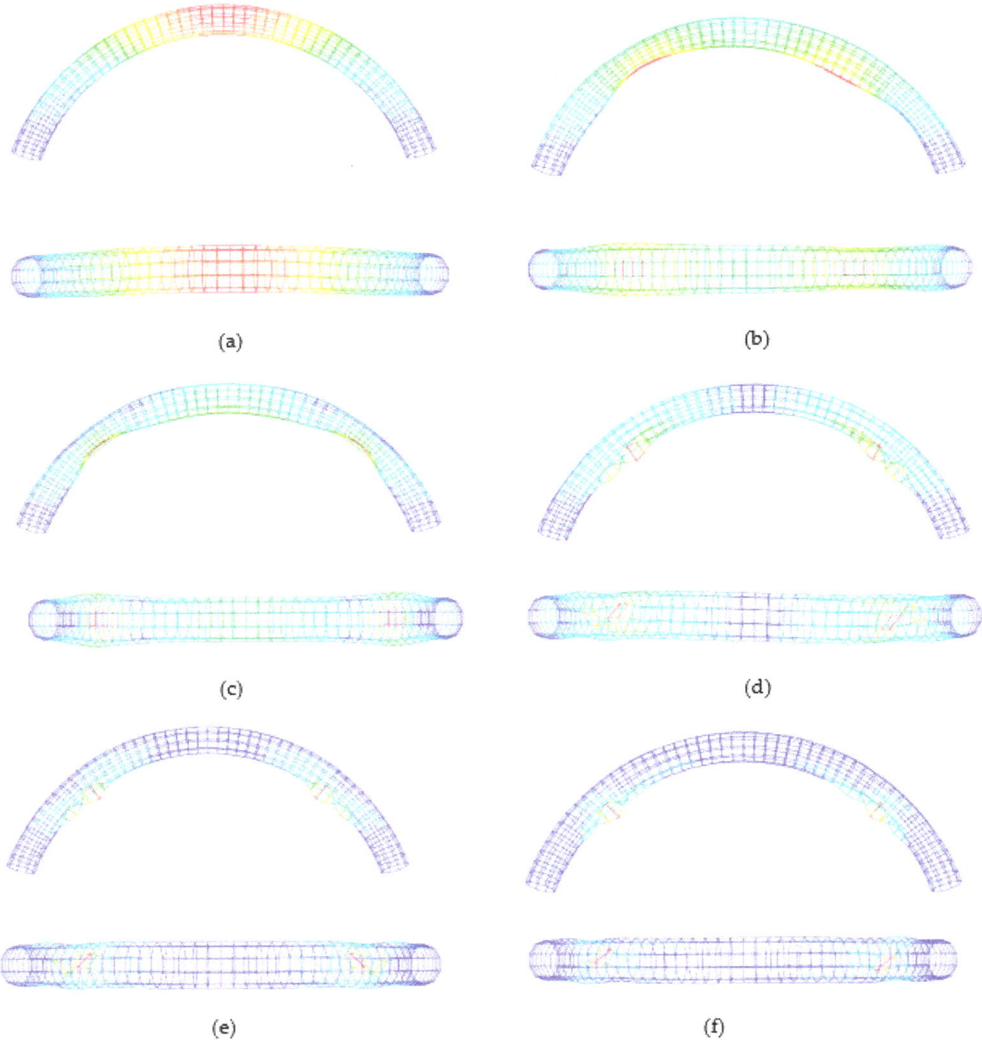

Figure 6. Buckling modes of the structure: (a) Mode 1; (b) Mode 2; (c) Mode 3; (d) Mode 4; (e) Mode 5; (f) Mode 6.

The elastic buckling loads for each mode are shown in Table 2.

Table 2. Summary of the linear buckling loads.

Modes	Mode 1	Mode 2	Mode 3	Mode 4	Mode 5	Mode 6
Linear buckling load factor *	56.57	75.75	82.89	83.76	84.03	84.19
Symmetry of deformation	symmetric	anti-symmetric	symmetric	anti-symmetric	symmetric	anti-symmetric

* Linear buckling load factor is the ratio of the elastic buckling load to the initial load (1.0Dead Load + 1.0Live Load).

In order to show the buckling mode more clearly, the deformation visualization scale factor was expanded to three times the original. In the first bucking mode, an overall lateral displacement out of the arch plane occurred, and the value of the lateral displacement decreased from the middle of the span to the feet of the arch. In the second bucking mode, the elliptical arch section of the left 1/4 span was compressed in the direction of the short axis, and the section of the right 1/4 span was elongated in the direction of the short axis. In the third bucking mode, the arch sections of both the left and right 1/4 span were compressed in the direction of the short axis, and the section at each foot of the arch was slightly bulged. In the fourth bucking mode, the grids at the left and right 1/4 span underwent large deformation. The fifth and sixth buckling modes were similar to the fourth, with larger grid deformation in the local area and change of deformation symmetry.

From the third buckling mode, the buckling load gradually became closer in value, and the buckling mode of the structure transited from the overall buckling to the buckling of the local region. In the third to sixth buckling mode, the location where buckling occurred was AREA5&6 of the structure. Compared with other regions, the ring component cross-section was BH-700 × 300 × 20 × 20, and the arch component cross-section was B.Pipe-600 × 26 in AREA5&6, both of which were the smallest in cross-sectional dimensions. At the same time, the maximum axial pressure occurred in these regions as Figure 7 shows, making them the most likely locations for buckling to occur.

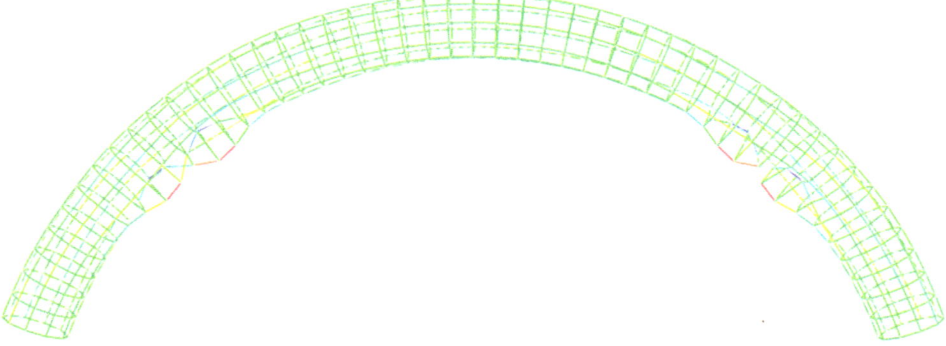

Figure 7. Axial force diagram of Mode 4.

3.2. Nonlinear Elasto-Plastic Stability Analysis

In order to analyze the ultimate load-bearing capacity of this structure more accurately, the initial geometric imperfections were introduced by the consistent mode imperfection method in the nonlinear elasto-plastic stability analysis. The initial geometric imperfection distribution was adopted from the first buckling mode of the structure, and the amplitude was taken as 1/300 of the span, respectively [35]. The full-range elasto-plastic analysis of the structure was carried out considering both geometric and material nonlinearities.

In the first buckling mode of linear buckling analysis, the point with maximum displacement was located in the middle of the structure, and the vertical displacement at this point was extracted as representative of the structural deflection in the elasto-plastic full-range analysis. The vertical axis was the load proportionality factor (LPF) of the structure, which is the ratio of the actual load to the initial load (1.0Dead Load + 1.0Live Load). The load-displacement curve of elasto-plastic full-range analysis is shown as Figure 8.

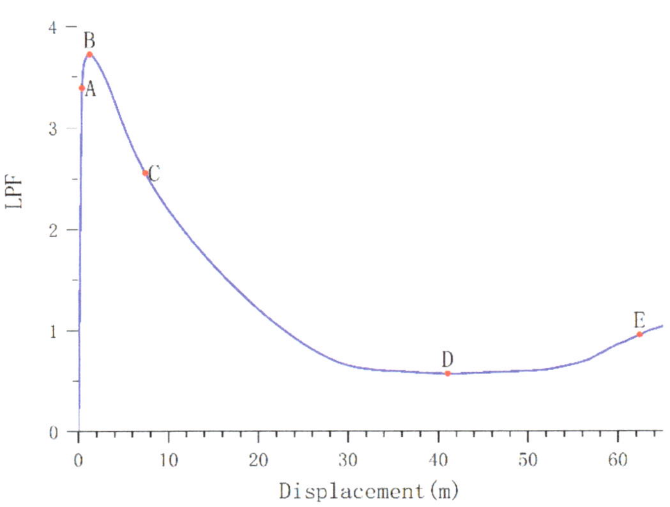

Figure 8. The load-displacement curve of elasto-plastic full-range analysis.

The stress distribution of the structure is shown in Figure 9. When the curve reached point A, the load reached 3.39 times the initial load (LPF = 3.39) and the vertical deflection of the structure was 0.479 m. The arch components in the upper part of the structure's middle section, the lower part of the 1/4 span section, and some at the feet of the structure, started to reach plasticity. As the load-displacement curve was linear, the overall structure was in the elastic stage in the stage O–A, and the structural stiffness was large.

Subsequently, the structure could still continue to carry the load, and the plastic range developed further with the increase in load, weakening the stiffness of the overall structure. The arch components and a few ring components at the middle and 1/4 span of the structure entered plasticity, and four more arch components at each foot of the structure reached plasticity. When the load reached 3.72 times the initial load, the structure reached the elasto-plastic ultimate load-bearing capacity and buckled, and the vertical deflection of the structure was 1.363 m. At this stage, most of the arch and ring components at the middle and 1/4 span of the structure reached plasticity, and all the arch components at the feet of the arch entered plasticity.

After buckling, the load decreased, with the displacement increasing rapidly, the structure as a whole was still in the compression and bending state, and the plastic range continued to increase, such as the stress and displacement pattern at point C. At this stage, most of the arch components reached plasticity, and the number of ring components that reached plasticity in the middle of the span, 1/4 span, and at the feet, continued to increase, and the plastic range became larger.

When the load continued to be applied, the vertical displacement of the structure increased rapidly. The structure as a whole gradually changed from compression and bending state, to tension and bending state. The tension edge of the structure reached the yield stress while the compression yield region gradually decreased, and the structure reached the reverse equilibrium state at point D. At this stage, the load was 0.57 times the initial load, and the structural deflection was 40.74 m. After point D, the bearing capacity of the structure was increasing. As the structure, as a whole, was in the tension and bending state, the compression yield region continued to decrease and the tensile yield region continued to increase, as shown in the Figure 9e. Although the simulation results show that the structure still has load carrying capacity, the state is no longer meaningful in engineering.

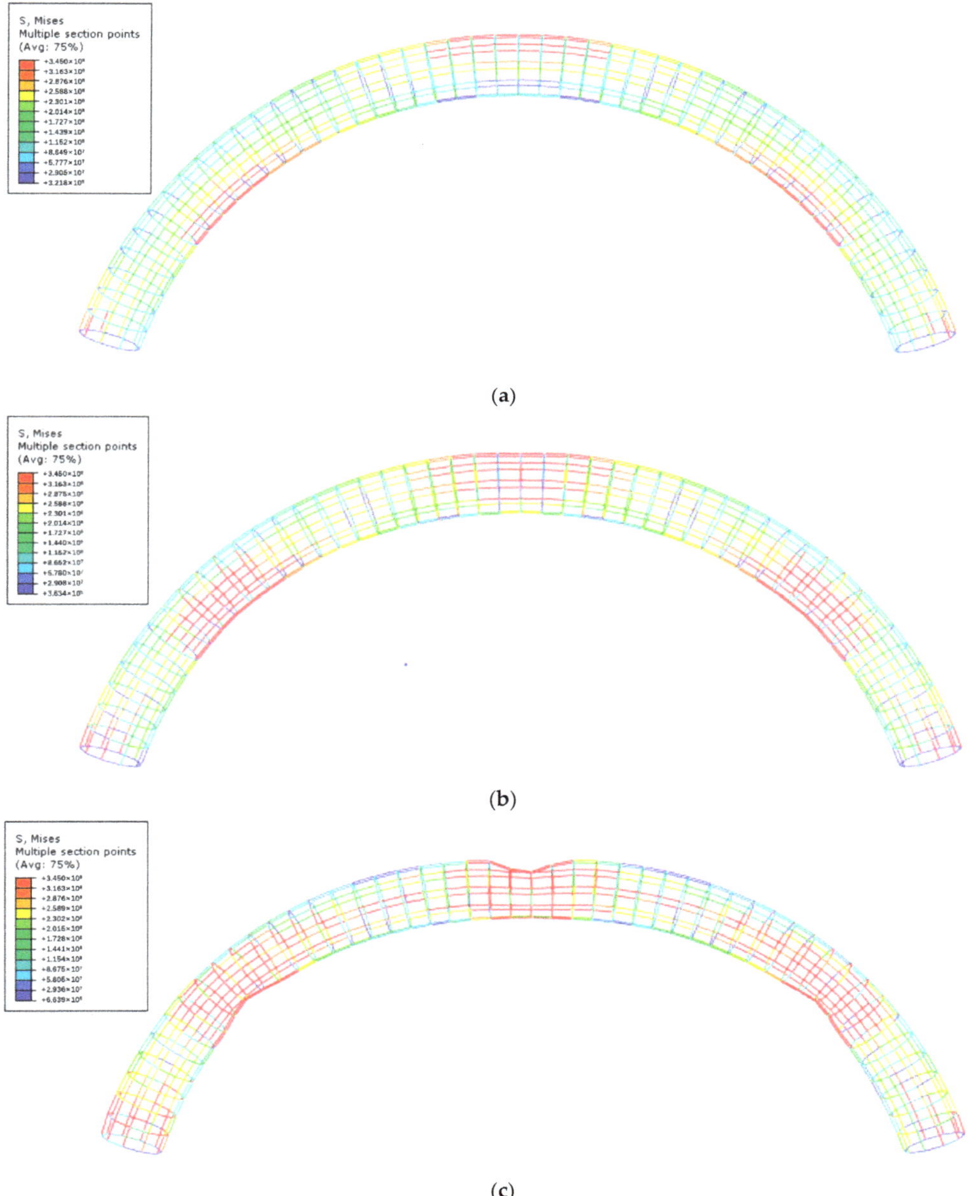

(a)

(b)

(c)

Figure 9. *Cont.*

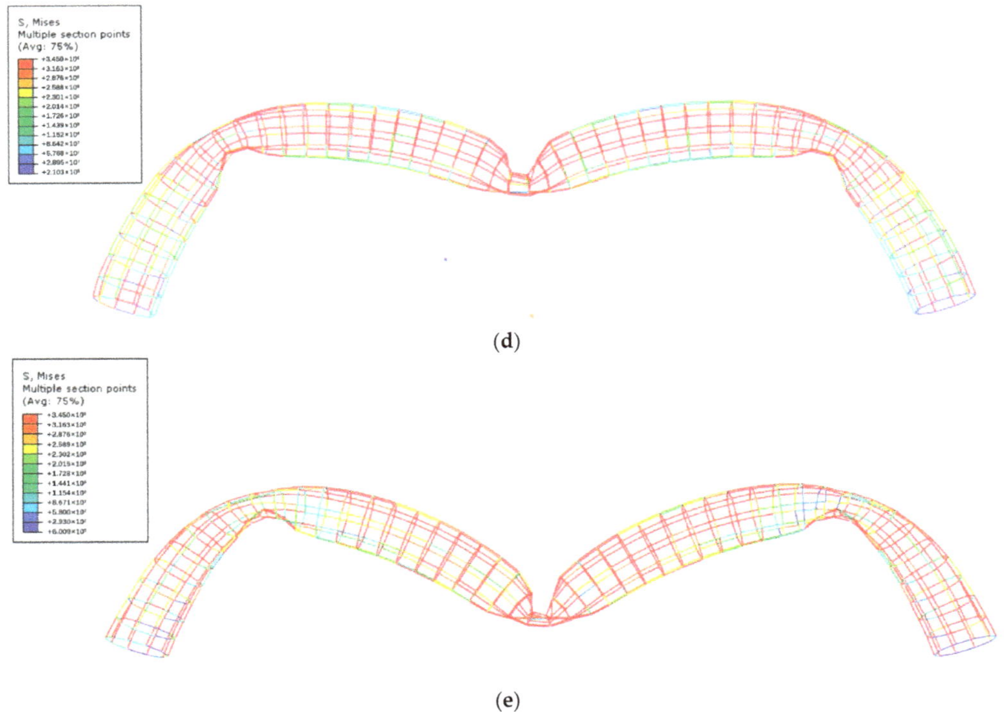

(d)

(e)

Figure 9. Stress distribution of the structure: (**a–e**) corresponds to the stage at point A–E in the load-displacement curve.

4. Analysis of the Factors Influencing Stability Bearing Capacity

4.1. Influence of Initial Geometric Imperfections

The actual engineering structure inevitably had various initial imperfections, including production and installation deviations and initial eccentricity of members to nodes, initial stresses caused by various reasons, etc. In order to study the influence of initial geometric imperfections on the structure's stability bearing capacity, four finite element models with initial geometric imperfection amplitudes of 1/100, 1/300, 1/500, and 1/1000 of the structure span, and one without geometric imperfection, were established. Both geometric nonlinearity and material nonlinearity were considered. The first buckling mode was chosen as the geometric initial imperfection distribution of the structure. Because the structural deformation pattern, according to this mode, was in the minimum potential energy state, there was a tendency for deformation along this mode at the initial load applying stage for the structure [35]. If the imperfection distribution of the structure was in the same form as the first buckling mode, it would have the most adverse effect on the structural load-bearing performance.

In the first buckling mode of linear buckling analysis, the point with maximum displacement was located in the middle of the structure, and the vertical displacement at this point was extracted as a representative of the structural deflection in the elasto-plastic full-range analysis. The vertical axis was the LPF of the structure. Figure 10 shows the load-displacement curves of full-range analysis with different initial imperfections.

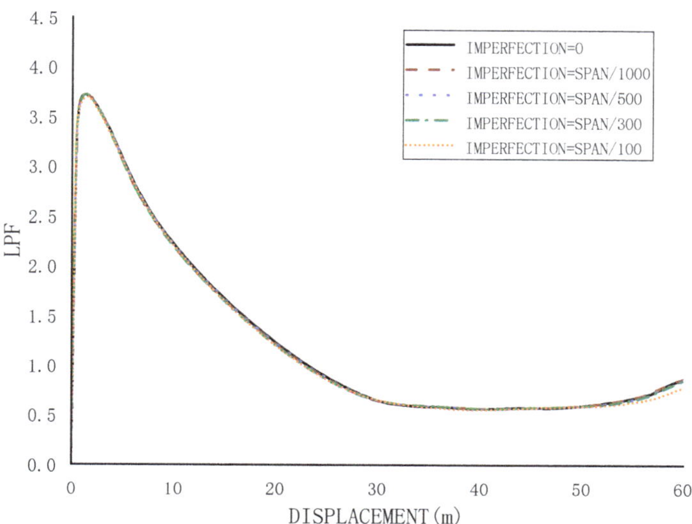

Figure 10. Load-displacement curves of full-range analysis with different initial imperfections.

It can be seen from Figure 10 that the load-displacement curves of the models with initial imperfections of 1/1000, 1/500 and 1/300 almost coincided with that of the model without imperfections. Only the ultimate load of the model with initial imperfections of 1/100 (LPF = 3.70) was slightly different from that of the model without imperfections (LPF = 3.72), while the imperfection maximum value reached 1.77 m, which is far beyond the actual construction-allowed range. It can be concluded that the construction error allowed in the project has a very small effect on the stability bearing capacity of the structure. At the same time, as the value of imperfection increases, the ultimate load-bearing capacity decreases, and deviations should be minimized during the construction of the structure.

4.2. Influence of Nonlinear Factors

In order to study the effects of material nonlinearity and geometric nonlinearity on the ultimate load-carrying capacity of the structure, the full-range analysis of the structure considering only geometric nonlinearity and only material nonlinearity was carried out separately and compared with the results of elasto-plastic stability analysis of the structure. The load-displacement curves of the structure are shown as Figure 11.

The load proportionality factor (LPF) gradually increased from 0 to point a (LPF = 3.39), during which the A, B and C curves coincide, illustrating that the structure is in the elastic stage.

The elastic stability behavior of the structure with large deformation can be obtained from load-displacement curve A, considering only the geometric nonlinearity, that the structure does not have an elastic ultimate load. By analyzing the stress and deformation of the structure, it was found that although the mid span area of the structure was most likely to buckle, the area was dominated by bending deformation. With the load increasing, the structure continued to bend downward, and its deformation mode and equilibrium mode remained unchanged. Therefore, it will continue to bear the load without buckling, and the cross-sectional stresses of the member in some areas have long exceeded the elastic limit.

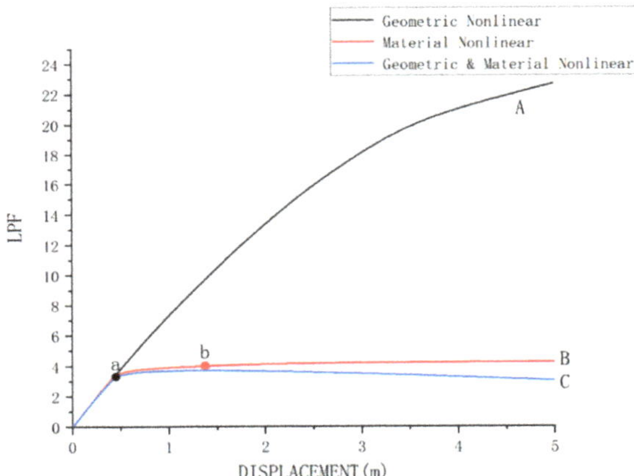

Figure 11. Load-displacement curves of full-range analysis with different nonlinear factors.

The elasto-plastic stability behavior of the structure with small deformation can be obtained from load-displacement curve B, considering only material nonlinearity, that the arch and ring components at the middle and 1/4 span, as well as the arch components at the feet, gradually reach plasticity with LPF increasing from 3.39 (point a) to 4.0 (point b). At this stage, the structural behavior is similar to the one considering both material nonlinearity and geometric nonlinearity, and curve B is slightly higher than curve C. The plasticity range gradually increases, and the LPF of curve B does not decrease. While the displacement keeps increasing, the load-displacement curve tends to be horizontal, and the behavior of the overall structure is nearly plastic at this stage. Since the geometric nonlinearity was not considered, the structure was still calculated according to the small deformation assumption, and its deformation mode did not change. Unlike the load-displacement curve C which considered both material nonlinearity and geometric nonlinearity, there is no descending portion in curve B.

The LPF corresponding to the ultimate load, considering only material nonlinearity, was 4.22, and the one considering material nonlinearity and geometric nonlinearity was 3.72. The influence of geometric nonlinearity on the ultimate bearing capacity of the structure was about 11.8%. The ultimate load-carrying capacity of the structure is sensitive to material nonlinearity, and a full-range analysis of the structure considering only geometric nonlinearity will seriously overestimate the ultimate load-carrying capacity of the structure, which would be unsafe for the engineering project. In summary, to evaluate the ultimate load-carrying capacity of the structure, both material nonlinearity and geometric nonlinearity should be considered.

5. Conclusions

In this paper, the elastic and elasto-plastic stability behavior considering the initial geometric imperfections of the large-span spatial grid arch structure based on Eye of the Yellow Sea project were analyzed, and the buckling mode as well as the full-range load-displacement curve of the structure were obtained. The full-range stress distribution, deformation mode and overall performance of the structure were studied. The influence of initial geometric imperfections and nonlinearities on the structural ultimate load-carrying capacity were investigated. The following main conclusions were obtained:

(1) In the first bucking mode, an overall out-of-plane lateral displacement occurred in the middle area of the span. The locations of maximum stress and deformation of the bucking modes 2~6 were at the left and right 1/4 span of the structure. The linear buckling load became progressively closer from bucking mode three, and the locations where buckling occurred were located in the 1/4 span local area, where

    the component cross-sectional dimensions were the smallest. The maximum axial pressure occurred in these areas, making them the most likely locations for buckling to occur;

(2) The full-range analysis considering geometric nonlinearity, material nonlinearity and initial imperfections showed that with the change of load, the structure went through the elastic stage, plastic development stage, the stage of buckling when the ultimate load was reached, the stage of displacement increasing rapidly while the load-carrying capacity was decreasing rapidly, and the stage of reverse equilibrium;

(3) When the initial geometric imperfection amplitude was set as 1/1000, 1/500, and 1/300 of the span, the stability behavior of the structure was almost unaffected. Only when the initial imperfection was 1/100 of the span, the ultimate load of the structure was reduced by 0.54%;

(4) The influence of geometric nonlinearity on the ultimate bearing capacity of the structure was 11.8%;

(5) The ultimate load-carrying capacity of the structure is sensitive to material nonlinearity. Considering only geometric nonlinearity will seriously overestimate the ultimate bearing capacity of the structure, which would be unsafe for the project. To evaluate the ultimate load-carrying capacity of the structure, both material nonlinearity and geometric nonlinearity should be considered.

Author Contributions: Conceptualization, Z.C. and X.W.; methodology, Z.C., X.W., H.L. (Hongbo Liu) and H.L. (Hao Lin); software, H.L. (Hao Lin); formal analysis, H.L. (Hao Lin); validation, H.L. (Hao Lin), K.K. and M.M.; investigation, H.L. (Hao Lin), K.K. and M.M.; writing—original draft preparation, H.L. (Hao Lin); writing—review and editing, Z.C., X.W., H.L. (Hongbo Liu) and H.L. (Hao Lin); project administration, Z.C. and X.W.; funding acquisition, H.L. (Hongbo Liu). All authors have read and agreed to the published version of the manuscript.

Funding: This research was funded by Hebei Province Full-time Top-level Talents Introduction Project, grant number [2020HBQZYC013].

Institutional Review Board Statement: Not applicable.

Informed Consent Statement: Not applicable.

Data Availability Statement: Not applicable.

Acknowledgments: Special thanks to the Institute of Steel Structures, Department of Civil Engineering, Tianjin University, for the help and support.

Conflicts of Interest: The authors declare no conflict of interest.

References

1. Timoshenko, S.P.; Gere, J.M. *Theory of Elastic Stability*, 2nd ed.; Dover Publications, Inc.: New York, NY, USA, 2009; pp. 297–302.
2. Pi, Y.L.; Bradford, M.A.; Uy, B. In-Plane Stability of Arches. *Int. J. Solids Struct.* **2002**, *39*, 105–125. [CrossRef]
3. Bradford, M.A.; Uy, B.; Pi, Y.-L. In-Plane Elastic Stability of Arches under a Central Concentrated Load. *J. Eng. Mech.* **2002**, *128*, 710–719.
4. Pi, Y.L.; Bradford, M.A.; Tin-Loi, F. Nonlinear Analysis and Buckling of Elastically Supported Circular Shallow Arches. *Int. J. Solids Struct.* **2007**, *44*, 2401–2425. [CrossRef]
5. Pi, Y.L.; Bradford, M.A.; Tin-Loi, F. Non-Linear in-Plane Buckling of Rotationally Restrained Shallow Arches under a Central Concentrated Load. *Int. J. Non-Linear Mech.* **2008**, *43*, 1–17. [CrossRef]
6. Pi, Y.L.; Bradford, M.A. Non-Linear in-Plane Postbuckling of Arches with Rotational End Restraints under Uniform Radial Loading. *Int. J. Non-Linear Mech.* **2009**, *44*, 975–989. [CrossRef]
7. Pi, Y.L.; Bradford, M.A. Non-Linear in-Plane Analysis and Buckling of Pinned-Fixed Shallow Arches Subjected to a Central Concentrated Load. *Int. J. Non-Linear Mech.* **2012**, *47*, 118–131. [CrossRef]
8. Pi, Y.L.; Bradford, M.A. Nonlinear Elastic Analysis and Buckling of Pinned-Fixed Arches. *Int. J. Mech. Sci.* **2013**, *68*, 212–223. [CrossRef]
9. Pi, Y.L.; Bradford, M.A. Nonlinear Analysis and Buckling of Shallow Arches with Unequal Rotational End Restraints. *Eng. Struct.* **2013**, *46*, 615–630. [CrossRef]
10. Hodges, D.H. Non-linear inplane deformation and buckling of rings and high arches. *Int. J. Non-Linear Mech.* **1999**, *34*, 723–737. [CrossRef]

11. Sakimoto, T.; Namita, Y. Out-Of-Plane Buckling of Solid Rib Arches Braced with Transverse Bars. *Proc. Jpn. Soc. Civ. Eng.* **1971**, *191*, 109–116. [CrossRef]
12. Pi, Y.L.; Bradford, M.A. Elastic flexural–torsional buckling of fixed arches. *Q. J. Mech. Appl. Math.* **2004**, *57*, 551–569. [CrossRef]
13. Pi, Y.L.; Bradford, M.A. Effects of Prebuckling Deformations on the Elastic Flexural-Torsional Buckling of Laterally Fixed Arches. *Int. J. Mech. Sci.* **2004**, *46*, 321–342. [CrossRef]
14. Pi, Y.L.; Bradford, M.A.; Trahair, N.S.; Chen, Y.Y. A further study of flexural-torsional buckling of elastic arches. *Int. J. Struct. Stab. Dyn.* **2005**, *5*, 163–183. [CrossRef]
15. Pi, Y.L.; Bradford, M.A.; Chen, Y.Y. An equilibrium approach for flexural-torsional buckling of elastic arches. *Adv. Steel Constr.* **2005**, *1*, 47–66.
16. Bradford, M.A.; Pi, Y.L. Flexural-Torsional Buckling of Fixed Steel Arches under Uniform Bending. *Proc. J. Constr. Steel Res.* **2006**, *62*, 20–26. [CrossRef]
17. Bradford, M.A.; Pi, Y.L. Elastic Flexural-Torsional Buckling of Circular Arches under Uniform Compression and Effects of Load Height. *J. Mech. Mater. Struct.* **2006**, *1*, 1235–1255. [CrossRef]
18. Pi, Y.L.; Bradford, M.A.; Tin-Loi, F. Flexural-Torsional Buckling of Shallow Arches with Open Thin-Walled Section under Uniform Radial Loads. *Thin-Walled Struct.* **2007**, *45*, 352–362. [CrossRef]
19. Bradford, M.A.; Pi, Y.L. Elastic Flexural-Torsional Instability of Structural Arches under Hydrostatic Pressure. *Int. J. Mech. Sci.* **2008**, *50*, 143–151. [CrossRef]
20. Pi, Y.L.; Bradford, M.A.; Tong, G.S. Elastic Lateral-Torsional Buckling of Circular Arches Subjected to a Central Concentrated Load. *Int. J. Mech. Sci.* **2010**, *52*, 847–862. [CrossRef]
21. Bradford, M.A.; Pi, Y.L. A New Analytical Solution for Lateral-Torsional Buckling of Arches under Axial Uniform Compression. *Eng. Struct.* **2012**, *41*, 14–23. [CrossRef]
22. Pi, Y.L.; Bradford, M.A. Lateral-Torsional Elastic Buckling of Rotationally Restrained Arches with a Thin-Walled Section under a Central Concentrated Load. *Thin-Walled Struct.* **2013**, *73*, 18–26. [CrossRef]
23. Pi, Y.L.; Bradford, M.A. Lateral-Torsional Buckling Analysis of Arches Having In-Plane Rotational End Restraints under Uniform Radial Loading. *J. Eng. Mech.* **2013**, *139*, 1602–1609. [CrossRef]
24. Dou, C.; Guo, Y.L.; Zhao, S.Y.; Pi, Y.L.; Bradford, M.A. Elastic Out-of-Plane Buckling Load of Circular Steel Tubular Truss Arches Incorporating Shearing Effects. *Eng. Struct.* **2013**, *52*, 697–706. [CrossRef]
25. Guo, Y.-L.; Zhao, S.-Y.; Bradford, M.A.; Pi, Y.-L. Threshold Stiffness of Discrete Lateral Bracing for Out-of-Plane Buckling of Steel Arches. *J. Struct. Eng.* **2015**, *141*, 04015004. [CrossRef]
26. Guo, Y.L.; Zhao, S.Y.; Dou, C. Out-of-Plane Elastic Buckling Behavior of Hinged Planar Truss Arch with Lateral Bracings. *J. Constr. Steel Res.* **2014**, *95*, 290–299. [CrossRef]
27. Komatsu, S.; Sakimoto, T. Ultimate Load Carrying Capacity of Steel Arches. *J. Struct. Div.* **1977**, *103*, 2323–2336. [CrossRef]
28. Sakimoto, T.; Komatsu, S. Ultimate Strength Formula for Steel Arches. *J. Struct. Eng.* **1983**, *109*, 613–627. [CrossRef]
29. Pi, Y.L.; Trahair, N.S. Out-of-Plane Inelastic Buckling and Strength of Steel Arches. *J. Struct. Eng.* **1998**, *124*, 174–183. [CrossRef]
30. Pi, Y.L.; Trahair, N.S. Inelastic lateral buckling strength and design of steel arches. *Eng. Struct.* **2000**, *22*, 993–1005. [CrossRef]
31. Pi, Y.L.; Bradford, M.A. Out-of-Plane Strength Design of Fixed Steel I-Section Arches. *J. Struct. Eng.* **2005**, *131*, 560–568. [CrossRef]
32. Dou, C.; Guo, Y.-L.; Pi, Y.-L.; Zhao, S.-Y. Flexural-Torsional Buckling and Ultimate Resistance of Parabolic Steel Arches Subjected to Uniformly Distributed Vertical Load. *J. Struct. Eng.* **2014**, *140*, 04014075. [CrossRef]
33. Dou, C.; Guo, Y.-L.; Zhao, S.-Y.; Pi, Y.-L. Experimental Investigation into Flexural-Torsional Ultimate Resistance of Steel Circular Arches. *J. Struct. Eng.* **2015**, *141*, 04015006. [CrossRef]
34. Chen, Z.; Xu, H.; Zhao, Z.; Yan, X.; Zhao, B. Investigations on the Mechanical Behavior of Suspend-Dome with Semirigid Joints. *J. Constr. Steel Res.* **2016**, *122*, 14–24. [CrossRef]
35. Liu, H.; Ding, Y.; Chen, Z. Static Stability Behavior of Aluminum Alloy Single-Layer Spherical Latticed Shell Structure with Temcor Joints. *Thin-Walled Struct.* **2017**, *120*, 355–365. [CrossRef]

metals

Article

Fatigue Analysis of Long-Span Steel Truss Arched Bridge Part I: Experimental and Numerical Study of Orthotropic Steel Deck

Peng Liu [1], Yixuan Chen [1], Hongping Lu [1], Jian Zhao [2], Luming An [2], Yuanqing Wang [3] and Jianping Liu [1,*]

[1] School of Architecture and Civil Engineering, Shenyang University of Technology, Shenyang 110870, China; pliu@sut.edu.cn (P.L.); cyx980201@163.com (Y.C.); lhp19971008@163.com (H.L.)
[2] China Railway Bridge Engineering Group Co., Ltd., Beijing 100039, China; zhaojianll@126.com (J.Z.); anluming@stdu.edu.cn (L.A.)
[3] School of Civil Engineering, Tsinghua University, Beijing 100084, China; wang-yq@mail.tsinghua.edu.cn
* Correspondence: liujianping024@163.com

Abstract: The orthotropic steel deck is sensitive to fatigue, and a number of cracks have been found in existing bridges. Based on the long-span Guangzhou Mingzhu Bay steel arched bridge, this paper focus on the cracking process, fatigue mechanism, and fatigue performance evaluation of an orthotropic steel bridge deck under traffic load. A finite element model of a three-U-rib and three-span bridge deck was first established to investigate the stress state and the most unfavorable wheel loading position under the longitudinal wheel load. Then, four full-scale single-U-rib specimens were fabricated with high-strength lower alloy structural steel Q370qD in compliance with construction standards. High-cycle loading was subsequently implemented according to the Specification for Design of Highway steel bridge (JTG D64-2015), and the crack initiation, propagation process, and fatigue failure modes were studied. The results showed the stress at structural concern points is larger than in other locations, which was located around 35 mm from the welding seam of the U-rib and the lower end of the diaphragm plate. The Mingzhu Bay steel bridge deck meets the fatigue design requirements. However, the bottom of the welding seam between the U-rib and diaphragm plate is a dangerous fatigue position, and attention should be paid to the welding quality at this position during construction.

Keywords: orthotropic steel deck; crack propagation; fatigue performance; finite element

Citation: Liu, P.; Chen, Y.; Lu, H.; Zhao, J.; An, L.; Wang, Y.; Liu, J. Fatigue Analysis of Long-Span Steel Truss Arched Bridge Part I: Experimental and Numerical Study of Orthotropic Steel Deck. *Metals* **2022**, *12*, 1117. https://doi.org/10.3390/met12071117

Academic Editors: Zhihua Chen, Hanbin Ge, Siu-lai Chan and Manuel José Moreira de Freitas

Received: 30 April 2022
Accepted: 27 June 2022
Published: 29 June 2022

Publisher's Note: MDPI stays neutral with regard to jurisdictional claims in published maps and institutional affiliations.

1. Introduction

Orthotropic steel decks (OSD) are composed of orthotopically deck plates stiffened by longitudinal ribs and transverse diaphragms, which are widely used in various types of large- and medium-sized long-span steel bridges due to the benefits of a high bearing capacity, light weight, and short construction period [1–6]. In the past decades, this type of steel deck has been improved in terms of design, fabrication, and maintenance, and the structural behavior has been enhanced. However, as the orthotropic steel decks withstand high-level cycles of traffic load, fatigue cracks may occur at the weld connection or other points of stress concentration points, which will seriously affect traffic safety [7–10].

In a literature review, it was found that many studies have been conducted on fatigue performance assessment and prediction models [11–16]. These models were generally based on fracture mechanics and cumulative damage theories. For example, Wu et al. [17] considered the influence of the crack closure effect on metals' mechanical properties and established a mathematical model for fatigue life prediction based on the law of metal fatigue characteristics. Macek et al. [18] proposed a mixed-mechanics measurement method to determine the three-dimensional port damage of specimens with fatigue bending loading history. Furthermore, the fatigue performance evaluation methods of a steel bridge deck mainly include the nominal stress method, hot spot stress method, and notch stress method [19–23]. The nominal stress method is convenient for calculation and engineering applications, which are widely adopted in bridge specifications [24–27].

Since the design, fabrication, and construction details of orthotropic steel decks are varied from the bridge, the fatigue performance analysis of bridges is generally based on the specific bridge [28]. Many efforts have been made to investigate the fatigue performance of orthotropic steel decks. Zeng et al. [29] conducted 1:2 scale fatigue experiments on the orthotropic steel bridge panel; the design life cycle loading was applied. It was concluded that the bridge met the design requirements and had a certain safety reserve during the service. Huang et al. [30] carried out fatigue tests and theoretical studies on the steel bridge deck of the Yangtze River Bridge under heavy traffic loads to evaluate the fatigue life. The results showed that the initial crack depth was sensitive to fatigue life evaluation. Deck plates, U-ribs, and transverse diaphragms are the load-bearing components and are the main sources of bridge deck stiffness and stability. However, the joints of longitudinal ribs, the deck plate, the transverse diaphragm, and the structural details of the transverse diaphragm are prone to fatigue cracking [31]. Cao et al. [32] analyzed the fatigue performance of the Jiangyin Yangtze River Bridge, the finite element model was established, and the influence of residual welding stress and vehicle load stress on the fatigue performance of the bridge deck was studied. The results showed that the steel deck thickness is relevant to the fatigue life of orthotropic steel decks. At the same time, Zhong et al. [33] analyzed the fatigue life of the Jiangyin Yangtze River Bridge. A coupling stress analysis FE model was established that considered the residual welding stress and vehicle load. The results showed that the residual tensile stress at the weld position was superimposed on the cyclic tensile stress of the main vehicle load, and the longitudinal stress relaxation exceeded the peak vehicle load stress. Ji et al. [34] employed traffic monitoring data to establish a load model to evaluate the root-deck fatigue durability of the Taizhou Bridge. Cheng et al. [35] analyzed the FE model of the Balinghe Bridge under traffic load, and the results showed that fatigue cracks initiated at the arc opening and extended to the welded seam of the diaphragm plate. A health detection plan was proposed. Yang et al. [36] established the finite element model of the Taizhou Yangtze River Bridge. The stress amplitude and fatigue damage of the steel bridge deck were studied. The stress state of the diaphragm under the lateral distribution of wheel load was obtained. It was shown that the stress amplitude of the diaphragm increased with the increase in the diaphragm spacing. Zeng et al. [37] established a 1:2 scale experiment of an orthotropic steel bridge deck under fatigue loading; it was found that with good welding quality and standard maintenance, the orthotropic steel bridge deck would not develop fatigue cracks during its life service.

Although scholars have completed a lot of studies on the fatigue properties of metal materials and orthotropic steel bridge decks, due to the differences in materials, designs, and construction details, it is still necessary to conduct numerical and experimental studies on the fatigue performance of specific long-span steel bridge decks in order to ensure safety during their service. Therefore, this study is based on the Mingzhu Bay steel truss arched bridge. A finite element model of a bridge deck was first established to investigate the stress state and the most unfavorable loading position. Four single U-rib specimens were used to investigate the crack initiation, propagation process, and fatigue failure modes. Finally, the fatigue performance of the steel deck was evaluated according to the specifications. Figure 1 shows the flow chart of this paper.

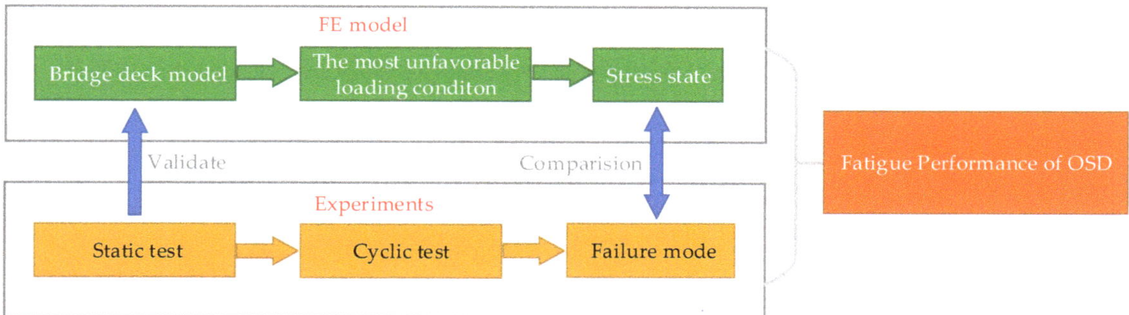

Figure 1. Flow chart.

2. Project Summary

The Mingzhu Bay Bridge main bridge adopts (96 + 164 + 436 + 164 + 96 + 60) m = 1016 m span continuous steel truss bridge [38]. The main span of the main bridge is three arched truss structures; the truss spacing is 18.1 m, the side truss height is 10.369 m, the middle truss height is 10.685 m, and the side span and the second side span are flat truss structures. The side span and the main arch rib are "N"-type trusses. The layout of the Mingzhu Bay Bridge is shown in Figure 2.

Figure 2. General layout of Guangzhou Mingzhu Bay Bridge.

The steel deck of the Mingzhu bay Bridge adopts a double-deck arrangement, as illustrated in Figure 3. The upper deck is a two-way, eight-lane highway with sidewalks on both sides. The total width of the main bridge deck is 43.2 m. An orthotropic steel deck with a thickness of 16 mm and a u-shaped closed rib with a spacing of 600 mm are adopted. A diaphragm is provided every 3 m longitudinally and bolted with the upper chord.

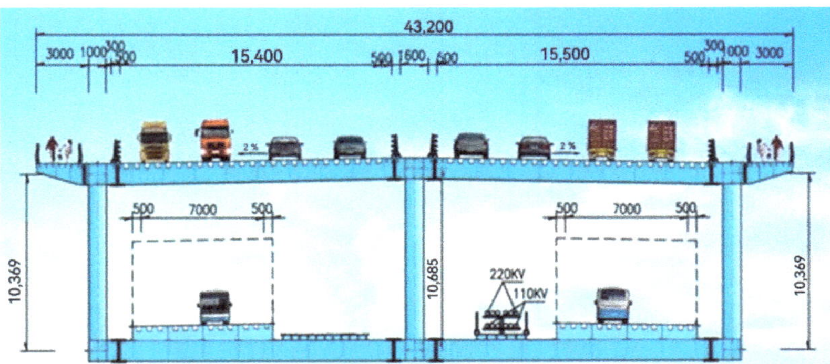

Figure 3. Deck arrangement (unit: mm).

3. Finite Element Model

3.1. Materials

A finite element model of the 3 spans and 3 U-ribs was established using Abaqus software (2016, Dassault SIMULIA, Paris, France). The FE model was 9 m long from the longitudinal direction and 1.5 m wide from the transverse direction. Solid element C3D8 was employed. The elastic modulus of the plate was 206 GPa, and the Poisson's ratio was 0.3. The fixed constraint was applied at the bottom of the diaphragm plate. The steel deck of Mingzhu Bay Bridge is made of Q370qD. The chemical composition and mechanical properties are shown in Tables 1 and 2.

Table 1. The chemical composition of Q370qD (%) [39].

C	Si	Mn	P	S	Als	Nb	V	Ti	N
$\leq$0.14	$\leq$0.55	1.00–1.60	$\leq$0.020	$\leq$0.010	0.010–0.045	0.10–0.090	0.010–0.080	0.006–0.030	$\leq$0.0080

Table 2. The mechanical properties of Q370qD.

Thickness (mm)	Yield Strength (MPa)	Tensile Strength (MPa)	Elongation after Fracture (%)	T ($^\circ$C)
$\leq$50	$\geq$370			−20
50–100	$\geq$360	$\geq$510	20	−40

3.2. Loading

To investigate the stress state and the most unfavorable wheel loading position, three loading conditions were considered, and the stress state of the bridge deck was analyzed. The details were as follows: (1) from the transverse direction, three wheel-loading positions were across the U-ribs, on the U-ribs, and between the U-ribs; (2) from the longitudinal direction, two loading positions were on middle span and OSD diaphragm. The details are shown in Figure 4.

The vehicle load index was obtained from specification (JTG D64-2015) [25]. The pavement layer was 55 mm, and taking the projected area from angle of 45°, the wheel loading area is (7.1 × 3.1) m^2, and the single wheel load is 120 kN. Hence, a single wheel's loading pressure on the bridge deck is 0.2726 GPa.

3.3. FE Analysis

The results of FE model under mid-span and diaphragm wheel loading are shown in Figure 5. It was observed that the maximum stress points were always located at the welding joint of the U-ribs and transverse diaphragm. Since this area is prone to fatigue cracks under cyclic load, this location is taken as the area of concern.

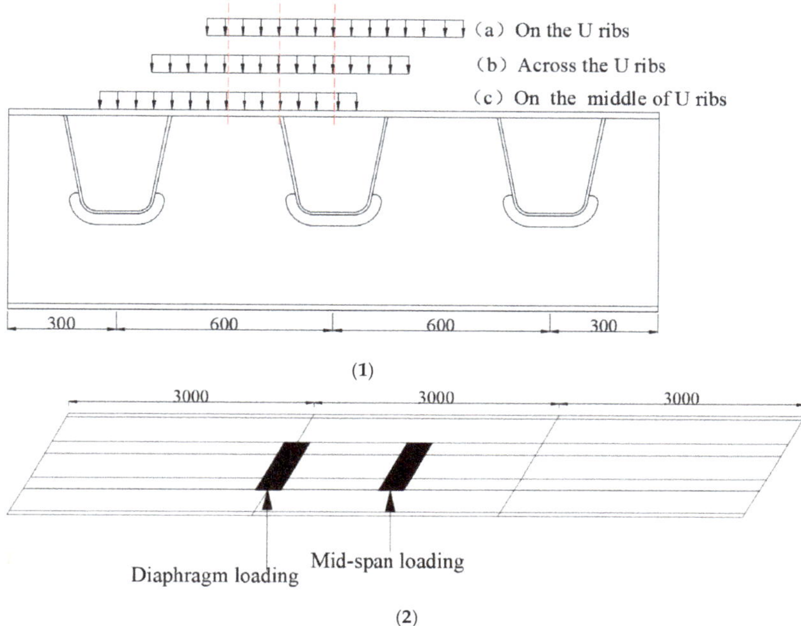

Figure 4. Wheel loading positions. (**1**) Wheel load in the transverse direction (unit: mm); (**2**) wheel load in the longitudinal direction (unit: mm).

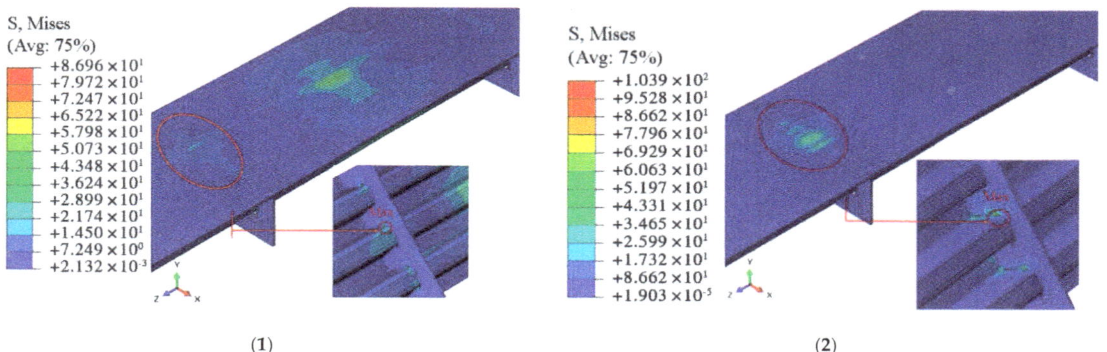

Figure 5. Stress state of the bridge deck (unit: MPa). (**1**) Mid-span loading; (**2**) diaphragm loading.

The stress state results of FE model are listed in Table 3.

Table 3. Stress results of FE model.

Transverse Loading Positions	Longitudinal Loading Positions	Stress Output	S_x (MPa)	S_z (Mpa)	Mises (Mpa)
Across the U−ribs	Mid−span	Min	−32.28	−24.03	2.132
		Max	35.84	44.41	86.96
	Diaphragm	Min	−17.17	−13.73	1.90
		Max	30.06	10.28	103.9
On the U−ribs	Mid−span	Min	−27.93	−29.38	1.902
		Max	48.57	39.87	104.6
	Diaphragm	Min	−27.62	−17.76	6.73
		Max	41.23	14.67	101.4
Between the U−ribs	Mid−span	Min	−31.10	−24.46	2.049
		Max	37.45	46.31	92.29
	Diaphragm	Min	−28.58	−16.09	5.201
		Max	36.87	11.22	101.1

Using stress-state data in the Table 3, we can see that the maximum stress is 104.6 MPa, while the most unfavorable wheel loading position was on the U-ribs of mid-span loading.

4. Fatigue Tests

4.1. Specimen Design

Based on the FE analysis, a single U-rib was selected for the fatigue specimen, and the loading was on the U-rib. The fatigue specimen design was fabricated according to the design of the Mingzhu Bay bridge; the height and length were 560 mm and 600 mm, respectively. The thickness of the bridge deck was 16 mm; the welded seam was 8 mm in width. An extra steel pad and stiffener were welded to the bottom to maintain the stability of the loading. The steel grade was Q370qD, and the welding wire and process were the same as the Mingzhu Bay Bridge. The dimension details are shown in Figure 6.

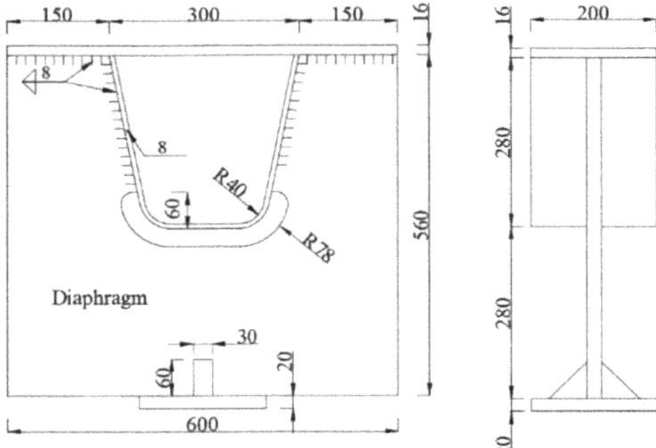

Figure 6. Schematic diagram of specimen (unit: mm).

4.2. Static Loading

The objective of static loading was to validate the FE model, which was a three-span, three-U-rib OSD structure. Since the fatigue wheel load was in the elastic range, the static loading and unloading procedure involved 50 kN increments to 350 kN and decreases to 50 kN. The strain gauges were distributed at the concerning points. The strain gauge arrangement is shown in Figure 7.

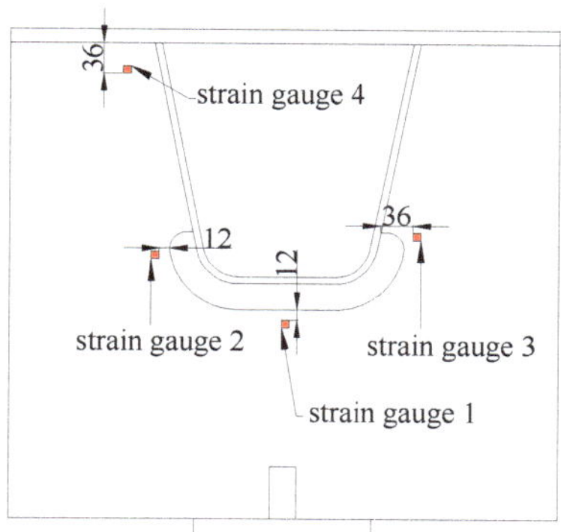

Figure 7. Strain gauge arrangement.

The strain was converted, and the comparisons between tests and the FE model are shown in Figure 8.

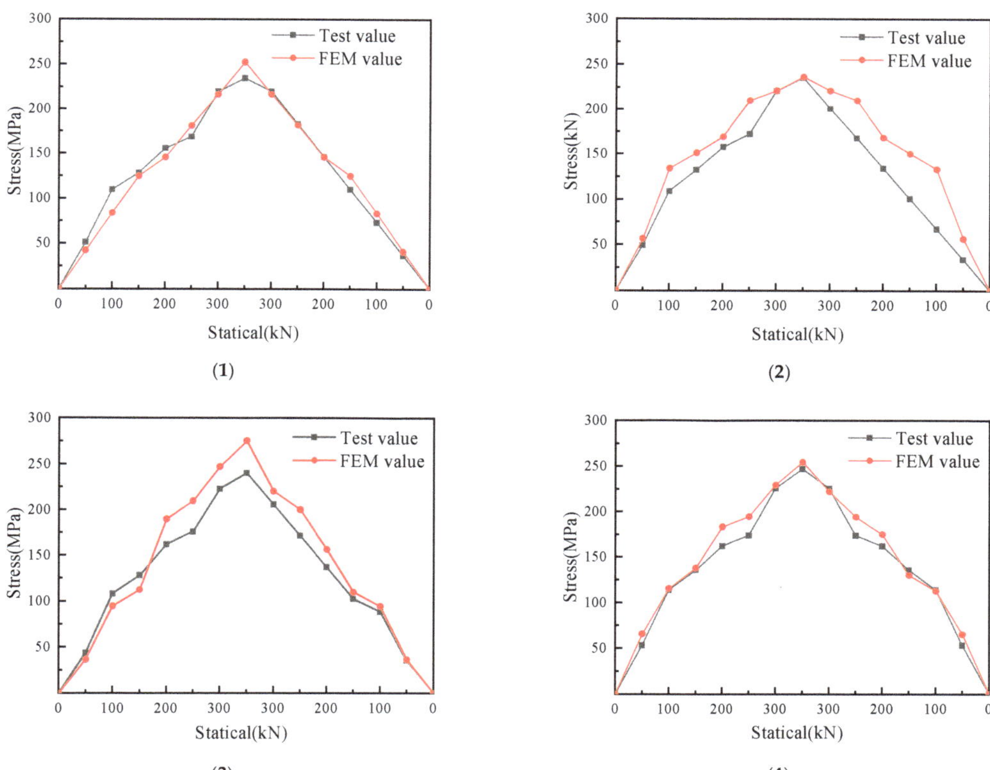

Figure 8. Stress comparison between test and FE model. (**1**) Strain gauge 1; (**2**) strain gauge 2; (**3**) strain gauge 3; (**4**) strain gauge 4.

It can be seen from the Figure 8 that the stress state and distribution between the FE model show great agreement with the test data.

4.3. Fatigue Loading

An MTS hydraulic fatigue test machine (Mechanical Testing & Simulation, Eden, MN, USA) was used for the fatigue test loading, as shown in Figure 9. The cyclic load was 120 kN with an amplitude of 80 kN and 100 kN, and the cyclic frequency was 5 Hz.

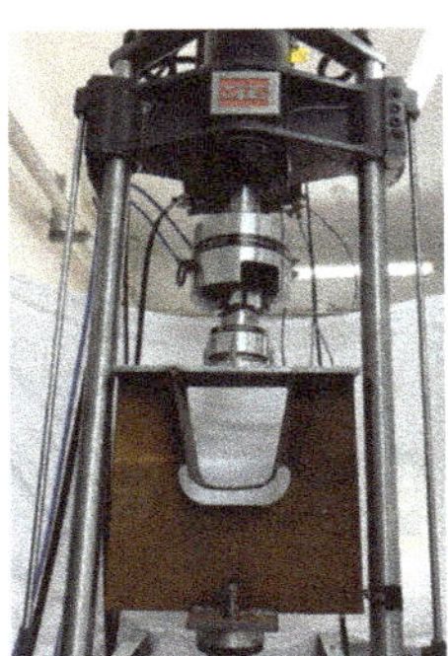

Figure 9. Fatigue loading of specimens.

The fatigue test results and failure mode of the specimen are listed in Table 4 and Figure 10.

Table 4. Fatigue test results.

Specimen Number	Load Amplitude (kN)	Cycle-Index	Crack Location
1	80	3,534,047	The bottom welding seam between U-rib and diaphragm
2	80	3,634,208	The welding seam between U-rib and diaphragm
3	100	3,065,567	The welding seam between U-rib and diaphragm
4	100	3,062,353	The welding seam between U-rib and diaphragm

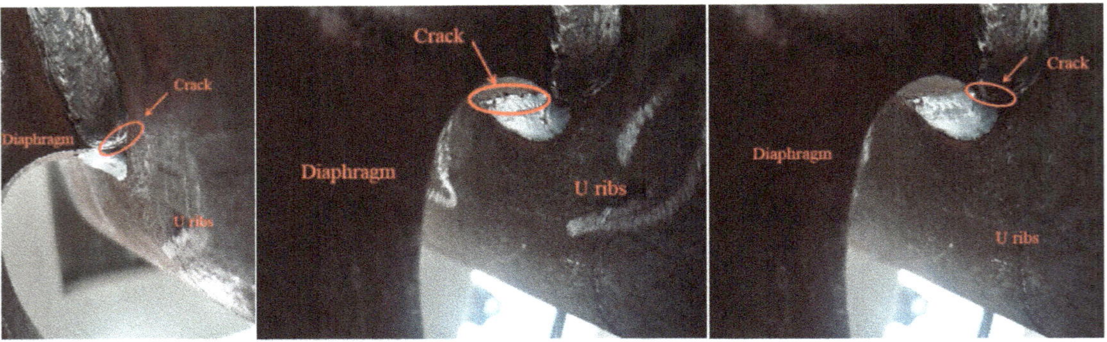

Figure 10. Failure mode of specimens.

It can be seen from Table 4 and Figure 10 that the cracks initialed at the concerned area of welded toe 1 and extended to both sides of the weld seam at three million cycles. The crack was about 2 mm wide and 20 mm long.

5. Fatigue Performance and Discussion

The fatigue performance of OSD is dependent on a variety of different parameters, such as the steel grade, welding quality, traffic load conditions, and design. It has been generally accepted that stress amplitudes under certain values, and the fatigue life could be treated as infinite. A high stress state or concentration increases the potential of fatigue cracks. Hence, the fatigue cracks were usually found at a higher stress state position around welding seams in the real construction. In other words, a higher stress state under the most unfavorable wheel load condition should receive more attention. For the FE model of the steel deck, the stress was extracted from A to B, as shown in Figure 11a. Stress values at toe 1, toe 2, and the welding root are shown in Figure 11b.

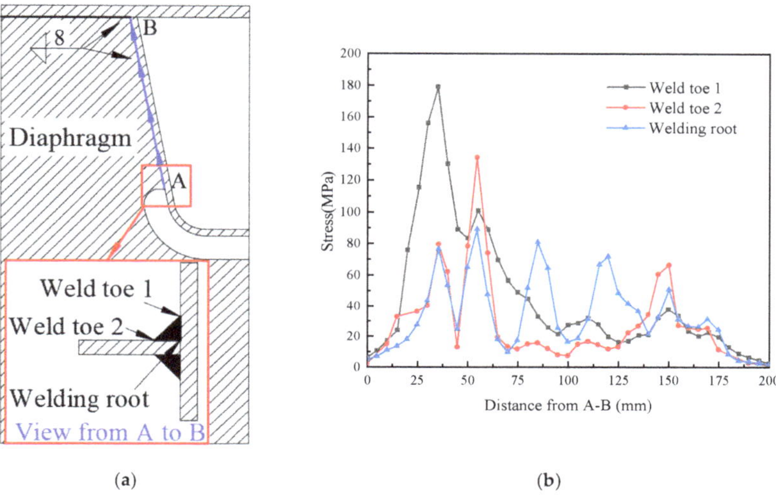

(**a**)   (**b**)

Figure 11. Area of concern and welded seam stress. (**a**) Schematic diagram of welded seams; (**b**) seam stress.

From A to B, the stress of the welding seam decreased gradually, and the stress at toe 1 and toe 2 was 180 MPa and 130 MPa, which was located 35 mm and 55 mm at the lower end of the U-rib and diaphragm. The stress of the welding root was around 80 MPa and was distributed evenly from A to B, which indicates that the potential fatigue crack area was at weld toe 1 35–55 mm from the arc cut-out location.

As seen in the experiment's phenomena, fatigue cracks also occurred at the welding seam between the U-rib and diaphragm, which were around the arc cut-out area of OSD. The fatigue cycles of all specimens were over two million in number and met the specification requirement.

6. Conclusions

In this paper, based on the Mingzhu Bay steel arch bridge, the fatigue performance of an orthotropic steel deck was studied through numerical and experimental analysis. The following conclusions were obtained: the fatigue crack was initialed at the point of concern, which was located 35–55 mm from the welding position at the lower end of the U-rib and diaphragm and then expanded along the weld to both sides. The crack was about 2 mm wide and 20 mm long at three million load cycles. The fatigue cycles of the specimens exceeded two million in number and satisfied the Specification for Design of Highway Steel Bridge (JTG D64-2015). The bridge deck of the Mingzhu Bay Bridge meets the fatigue requirements, but the welding seam of the U-rib and diaphragm is in a dangerous location of fatigue, so attention should be paid to the welding quality. The limitations of this paper are that the number of tests was inadequate, and the conclusions were based on the most unfavorable scenario.

Author Contributions: P.L., Y.C. and H.L. wrote the manuscript together, J.Z. and L.A. conducted data processing and drawings, Y.W. and J.L. revised the manuscript. All authors have read and agreed to the published version of the manuscript.

Funding: Funding was received from the Science and Technology Research and Development Project of China Railway Construction Bridge Engineering Bureau Group Co., Ltd. (DQJ-2018-A01); the Tianjin Science and Technology Development Plan Project (19YDLZSF00030); the National Natural Science Foundation of China (51038006).

Data Availability Statement: The data presented in this study are available on request from the corresponding author.

Acknowledgments: The author would like to thank the China Railway Construction and Bridge Bureau Group Co., Ltd. for the experiment support.

Conflicts of Interest: The authors declare no conflict of interest.

Nomenclature

$\varDelta$—Welded seam, unit: mm. Weld toe—the junction of weld surface and base metal. Weld root—the junction between the back of the weld and the base metal

References

1. Huang, Y.; Zhang, Q.; Bao, Y.; Bu, Y. Fatigue assessment of longitudinal rib-to-crossbeam welded joints in orthotropic steel bridge decks. *J. Constr. Steel Res.* **2019**, *159*, 53–66. [CrossRef]
2. Fisher, J.W.; Barsom, J.M. Evaluation of cracking in the rib-to-deck welds of the Bronx Whitestone Bridge. *J. Bridge Eng.* **2015**, *21*, 4015065. [CrossRef]
3. Dung, C.V.; Sasaki, E.; Tajima, K.; Suzuki, T. Investigations on the effect of Weld Penetration on Fatigue Strength of Rib-to-Deck Welded Joints in Orthotropic Steel Decks. *Int. J. Steel Struct.* **2015**, *15*, 299–310. [CrossRef]
4. Kainuma, S.; Jeong, Y.S.; Yang, M.; Inokuchi, S. Welding residual stress in roots between deck plate and U-rib in orthotropic steel decks. *Measurement* **2016**, *92*, 475–482. [CrossRef]
5. Connor, R.; Fisher, J.; Gatti, W.; Gopalartnam, V.; Kozy, B. *Manual for Design, Construction, and Maintenance of Orthotropic Steel Deck Bridges*; Federal Highway Adminition, U.S. Department of Transportation: Washington, DC, USA, 2012.
6. Liao, P.; Qu, B.; Huang, Y.L.; Jia, Y.; Wang, Y.B.; Zhu, H.F. Fatigue Life Assessment of Revised Cope-Hole Details in Steel Truss Bridges. *Metals* **2020**, *9*, 1092. [CrossRef]
7. Gou, H.; Shi, X.; Zhou, W.; Cui, K.; Pu, Q. Dynamic performance of continuous railway bridges: Numerical analyses and field tests. *Proc. Inst. Mech. Eng. Part F J. Rail Rapid Transit* **2018**, *232*, 936–955. [CrossRef]
8. Li, M.; Suzuki, Y.; Hashimoto, K.; Sugiura, K. Experimental study on fatigue resistance of rib-to-deck joint in orthotropic steel bridge deck. *J. Bridge Eng.* **2017**, *23*, 157–167. [CrossRef]
9. Hobbacher, A. *Recommendations for Fatigue Design of Welded Joints and Components*, 2nd ed.; IIW Doc. IIW 2259-15; Springer: Berlin/Heidelberg, Germany, 2016.
10. Chen, Y.X.; Lv, P.M.; Li, D.T. Research on Fatigue Strength for Weld Structure Details of Deck with U-rib and Diaphragm inOrthotropic Steel Bridge Deck. *Metals* **2019**, *9*, 484. [CrossRef]
11. Pineau, A.; Antolovich, S.D. High temperature fatigue of nickel-base superalloys–A review with special emphasis on deformation modes and oxidation. *Eng. Fail. Anal.* **2009**, *16*, 2668–2697. [CrossRef]
12. Yan, X.L.; Zhang, X.C.; Tu, S.T.; Mannan, S.L.; Xuan, F.Z.; Lin, Y.C. Review of creep–fatigue endurance and life prediction of 316 stainless steels. *Int. J. Press. Vessel. Pip.* **2015**, *126–127*, 17–28. [CrossRef]
13. Santecchia, E.; Hamouda, A.M.S.; Musharavati, F.; Zalnezhad, E.; Cabibbo, M.; El Mehtedi, M.; Spigarelli, S. A Review on Fatigue Life Prediction Methods for Metals. *Adv. Mater. Sci. Eng.* **2016**, *2016*, 9573524. [CrossRef]
14. Kamal, M.; Rahman, M.M. Advances in fatigue life modeling: A review. *Renew. Sustain. Energy Rev.* **2018**, *82*, 940–949. [CrossRef]
15. Cao, Y.; Nie, W.; Yu, J.; Tanaka, K. A Novel Method for Failure Analysis Based on Three-dimensional Analysis of Fracture Surfaces. *Eng. Fail. Anal.* **2014**, *44*, 74–84. [CrossRef]
16. Wang, X.W.; Zhang, W.; Zhang, T.Y.; Gong, J.M.; Abdel Wahab, M. A New Empirical Life Prediction Model for 9–12% Cr Steels under Low Cycle Fatigue and Creep Fatigue Interaction Loadings. *Metals* **2019**, *9*, 183. [CrossRef]
17. Wu, Q.; Liu, X.T.; Liang, Z.Q.; Wang, Y.S.; Wang, X.L. Fatigue life prediction model of metallic materials considering crack propagation and closure effect. *J. Braz. Soc. Mech. Sci. Eng.* **2020**, *42*, 424. [CrossRef]
18. Maceka, W.; Owsińskib, R.; Trembaczc, J.; Brancod, R. Three-dimensional fractographic analysis of total fracture areas in 6082 aluminium alloy specimens under fatigue bending with controlled damage degree. *Mech. Mater.* **2020**, *147*, 103410. [CrossRef]

19. Ye, X.W.; Su, Y.H.; Han, J.P. A State-of-the-Art Review on Fatigue Life Assessment of Steel Bridge. *Math. Probl. Eng.* **2014**, *2014*, 956473. [CrossRef]
20. Xiao, Z.G.; Yamada, K.; Ya, S.; Zhao, X.L. Stress analyses and fatigue evaluation of rib-to-deck joints in steel orthotropic deck. *Int. J. Fatigue* **2008**, *30*, 1387–1397. [CrossRef]
21. *BS 5400*; Part 10: Code of Practice for Fatigue. British Stangard Institution: London, UK, 1980.
22. *AASHTO LRFD Bridge Design Specifications*; American Association of State and Transportation Officials: Washington, DC, USA, 2007.
23. Marques, J.M.; Benasciutti, D.; Niesłony, A.; Slavič, J. An Overview of Fatigue Testing Systems for Metals under Uniaxial and Multiaxial Random Loadings. *Metals* **2021**, *11*, 447. [CrossRef]
24. *ENV 1991-2-6:1997*; Eurocode 1. Basis of Design and Actions on Structures. European Committee for Stangardization: Brussels, Belgium, 1997.
25. *JTG D64-2015*; Code for Design of Highway Steel Structure Bridges. National Standards of the People's Republic of China: Beijing, China, 2015.
26. Wang, Z.; Wang, Y. Fatigue crack propagation law of orthotropic steel bridge deck. *J. Cent. South Univ.* **2020**, *51*, 1873–1882. (In Chinese) [CrossRef]
27. Li, J.; Zhang, Q.; Bao, Y.; Zhu, J.; Chen, L.; Bu, Y. An equivalent Structural stress-based fatigue evaluation framework for rib-to-deck welded joints in orthotropic steel deck. *Eng. Struct.* **2019**, *196*, 109304. [CrossRef]
28. Wang, Q.; Ji, B.; Li, C.; Fu, Z. Fatigue evaluation of rib-deck welds: Crack-propagation-life predictive model and parametric analysis. *J. Constr. Steel Res.* **2020**, *173*, 106248. [CrossRef]
29. Huang, Y.; Zhang, Q.H.; Guo, Y.W.; Bu, Y.Z. Study on welding Detail Surface Defects and Fatigue effect of longitudinal Rib and Diaphragm plate of steel Bridge Deck. *Eng. Mech.* **2019**, *36*, 203–223. (In Chinese) [CrossRef]
30. Lv, P.M.; Wang, L.F.; Li, D.T.; Lei, J.; Zhang, X.L. Fatigue behavior of u-rib and diaphragm plate of orthotropic steel bridge deck. *J. Chang. Univ. Nat. Sci. Ed.* **2015**, *35*, 63–70. (In Chinese) [CrossRef]
31. Cao, B.Y.; Ding, Y.L.; Song, Y.S.; Zhong, W. Fatigue Life Evaluation for Deck-Rib Welding Details of Orthotropic Steel Deck Integrating Mean Stress Effects. *J. Bridge Eng.* **2019**, *24*, 04018114. [CrossRef]
32. Zhong, W.; Ding, Y.L.; Song, Y.S.; Geng, F.F. Relaxation Behavior of Residual Stress on Deck-to-Rib welded Joints by Fatigue Loading in an Orthotropic Bridge Deck. *Period. Polytech.-Civ. Eng.* **2020**, *63*, 1125–1138. [CrossRef]
33. Ji, B.; Liu, R.; Chen, C.; Maeno, H.; Chen, X. Evaluation on root-deck fatigue of orthotropic steel bridge deck. *J. Constr. Steel Res.* **2013**, *90*, 174–183. [CrossRef]
34. Cheng, J.H.; Xiong, J.M.; Zhou, J.Z. Finite Element Anaiysis on mechanical behavior of orthotropic steel bridge deck in Balinghe Bridge. *Adv. Mater. Res.* **2013**, *639–640*, 239–242. [CrossRef]
35. Yang, M.Y.; Liu, R.; Ji, B.H.; Xu, H.J.; Chen, C.; Zhao, D.D. Fatigue Stress Analysis of Disphragm Cap Hole in Orthotropic Steel Bridge Decks. *Appl. Mech. Mater.* **2012**, *204–208*, 3265–3269. [CrossRef]
36. Zeng, Y.; Tan, H.; Zhong, H.; Yu, Q. Experimental study of fatigue properties of orthotropic steel deck with U-shaped stiffeners. *Proc. Inst. Civ. Eng.-Struct. Build.* **2020**, *173*, 128–140. [CrossRef]
37. Liu, G.; Liu, Y.; Huang, Y. A novel structural stress approach for multiaxial fatigue strength assessment of welded joints. *Int. J. Fatigue* **2014**, *63*, 171–182. [CrossRef]
38. Hu, H.Y.; Zhao, J.; Ren, Y.L.; An, L.M.; Liu, Y.T. Overall design of main bridge of guangzhou mingzhu bay bridge. *Bridge Constr.* **2021**, *51*, 93–99. (In Chinese) [CrossRef]
39. *GB/T 714-2015*; Structural Steel for Bridges. National Standard of the People's Republic of China: Beijing, China, 2015.

Article

Cyclic Performance of Structural Steels after Exposure to Various Heating–Cooling Treatments

Peng Du [1], Hongbo Liu [1,2,3,*] and Xuchen Xu [2,3]

[1] School of Civil Engineering, Hebei University of Engineering, Handan 056038, China; dupeng@hebeu.edu.cn
[2] State Key Laboratory of Hydraulic Engineering Simulation and Safety, Tianjin University, Tianjin 300072, China; xcxu@tju.edu.cn
[3] School of Civil Engineering, Tianjin University, Tianjin 300072, China
* Correspondence: hbliu@tju.edu.cn

Abstract: The cyclic performance of structural steels after exposure to various elevated temperatures and cooling-down methods was experimentally investigated in this paper. Four types of frequently used structural steels were tested including Chinese mild steel Grade Q235, Chinese high-strength steel Grade Q345, and Chinese stainless steel Grade S304 and S316. A total of eighty specimens were prepared using three different heating–cooling processes before being subjected to cyclic loads. The post-fire basic features and hysteretic performances of the four types of structural steels exposed to various target temperatures (100–1000 °C), heat soak times (30 min or 180 min) and cooling-down methods (natural air or water) were recorded and discussed. The results show that all the tested structural steels prepared using different heating–cooling treatments exhibited proper ductility and energy dissipation capacity, while the heat soak times and cooling-down methods had a definite effect on their energy dissipation capacity; no Masing phenomenon was found in the tested structural steels. Finally, a set of skeleton curves were proposed for the four types of structural steels under cyclic loading based on the Ramberg–Osgood model, which could serve as the foundation for the seismic capacity evaluation of steel structures after a fire.

Keywords: structural steel; elevated temperatures; cooling-down method; cyclic performance; skeleton curve

Citation: Du, P.; Liu, H.; Xu, X. Cyclic Performance of Structural Steels after Exposure to Various Heating–Cooling Treatments. *Metals* **2022**, *12*, 1146. https://doi.org/10.3390/met12071146

Academic Editors: Andrey Belyakov and Ulrich Prahl

Received: 9 May 2022
Accepted: 26 June 2022
Published: 5 July 2022

Publisher's Note: MDPI stays neutral with regard to jurisdictional claims in published maps and institutional affiliations.

1. Introduction

Structural steels have become one of the most widely used materials in civil engineering owing to their light weight, high strength, excellent ductility, and good energy dissipation capacity. However, it has been demonstrated that the mechanical properties of structural steels are changed in a fire. Therefore, the performance of structural steels after exposure to elevated temperatures is important for the post-fire reliability evaluation of steel structures.

Extensive studies were conducted on the post-fire mechanical properties of structural steels, which found that the post-fire mechanical properties of structural steels are dependent on steel grades, cold forming procedures, exposure temperatures, heating methods, and cooling methods [1,2]. Corresponding predictive equations for the mechanical properties of structural steels were proposed to incorporate the aforementioned factors into those studies.

The post-fire mechanical properties of high-strength steels were found to be quite different from those of mild steels [3,4]. For Chinese hot-rolled Q235, Q345, and Q420 steels and ASTM A572 Gr.50 steels, their post-fire mechanical properties changed significantly after exposure to temperatures above approximately 700 °C [1,5]. The post-fire mechanical properties of European standard high-strength steels, including S460, S690, and S960, are not obviously affected until the temperature reaches above 600 °C, and they begin to degrade after cooling down from the temperatures beyond 600 °C up to 1000 °C [3,6,7].

Furthermore, the critical temperatures turn into approximately 400 °C–700 °C for the Chinese high-strength steels Grade Q460 and Q690 when cooled by air, water, and fire-fighting foam [4,8–12]. For cold-formed steels, the post-fire mechanical properties are reduced after they have been exposed to temperatures exceeding 300 °C [1,13].

The heating rate and repeated heating/cooling were found to affect the post-fire elastic modulus and yield strength of high-strength steel Q690; however, the effect on tensile strength and fracture strain was negligible [9]. Conversely, no obvious effect of cyclic heating–cooling was observed in hot-rolled Q235, Q345, Q420, and cold-formed Q235 steel [1]. Due to the microstructure transformation, the cooling method has been found to have an obvious influence on the residual strength and ductility of structural steel after exposure to sufficiently elevated temperatures, whereas there is no observable influence on elastic modulus [1,8–10]. The strength enhancement of structural steels due to cold forming or heat treatment is weakened with increasing heat treatment target temperatures, but their ductility is observed to be recovered [1,7,14]. Stainless steels show superior post-fire performance and a higher strength retention capacity than carbon steels [14–16]. According to some studies [17,18], the heat soak time was found to have a negligible effect on the post-fire mechanical properties of carbon steels and austenitic stainless steels, whereas other studies have shown that the ductility of lean duplex stainless steels and high-strength quenched and tempered steels are improved with heat soak time [16,19].

Most spatial structures can avoid collapse after a fire event, owing to their sufficient redundancy. For economic reasons, these fire-suffered spatial structures are considered to remain in service after a post-fire reliability evaluation. The post-fire seismic capacity is one of the most important components in the post-fire reliability evaluation. The mechanical properties of structural steels under cyclic loading are quite different from those under monotonic loading, and the hysteretic properties of structural steels play a pivotal role in the seismic performance of structures [20,21]. According to the review of state-of-the-art literature, most of the included studies focused on the post-fire mechanical properties of structural steels under monotonic loading conditions, and the post-fire cyclic performance of structural steels was rarely discussed. This paper therefore presents a detailed experimental study on the post-fire mechanical properties of commonly used structural steels under cyclic loading condition, including Chinese mild steel Grade Q235, Chinese high-strength steel Grade Q345, and Chinese stainless steel Grade S304 and S316. There were three heating–cooling treatments considered in the preparation of the specimens. Research funding can serve as the foundation for the seismic capacity evaluation of steel structures after a fire.

2. Experimental Programs

2.1. Specimen Design

Chinese mild steel Grade Q235, Chinese high-strength steel Grade Q345, and Chinese stainless steel Grade S304 and S316 were tested in this paper, and were in accordance with GB/T700-2006, GB/T1591-2008, GB/T1220-2007, and GB/T3280-2015, respectively [22–25]. According to GB/T 15248-2008 [26], eighty-four specimens were designed and classified into four groups according to their steel categories. Each specimen group had twenty-one specimens and each included one room-temperature specimen and twenty heating–cooling-treated specimens exposed to different heating treatments and cooling-down methods. Figure 1a,b show the dimensions of specimens for Q235/Q345 and S304/S316, respectively.

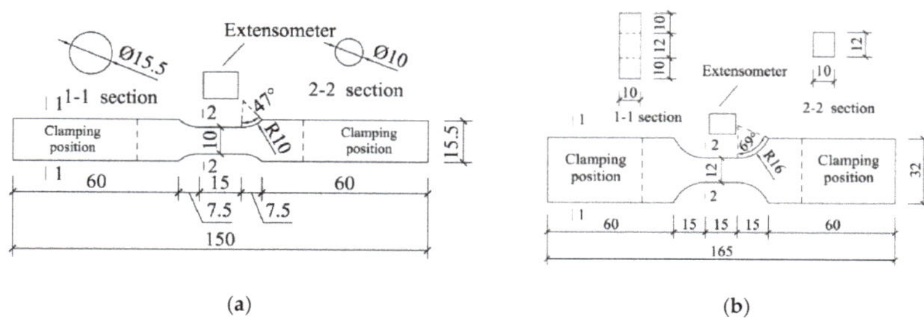

(a)

(b)

Figure 1. Dimensions of specimens. (**a**) Q235 and Q345; (**b**) S304 and S316.

2.2. Experiment Procedure

There were two stages during the whole experiment procedure. In the first stage, the specimens were heated to the preselected elevated temperatures and cooled down to room temperature according to three different heating–cooling treatments. In the second stage, the prepared specimens were tested under cyclic loading at room temperature.

2.2.1. Heating–Cooling Treatment

As shown in Figure 2, three heating–cooling treatment styles, with each including four stages per style, were employed in the preparation of specimens, which were marked as style 1, style 2, and style 3, respectively. In the three heat treatment styles, all the specimens were first heated at a rate of 20 °C/min to a temperature of 50 °C less than the target temperature with a temperature-controlled electric furnace SX-G36123 (Tianjin Zhonghuan Furnace Corporation, Tianjin, China). This temperature was maintained for 15 min before the next step. Then, the specimens were continually heated at a rate of 5 °C/min to a temperature of 10–15 °C less than the target temperature, and the temperature was maintained for 10 min at this time. Subsequently, the specimens were heated at a rate of 5 °C/min to the target temperature, and the heat soak times at the target temperature were 30 min, 180 min, and 30 min for heat treatment process styles 1–3, respectively. The heat processes described above were to ensure a uniform temperature distribution in the specimens. For heat treatment process style 1, ten elevated target temperatures were set, i.e., 100 °C, 200 °C, 300 °C, 400 °C, 500 °C, 600 °C, 700 °C, 800 °C, 900 °C, and 1000 °C. As a contrast, five elevated target temperatures were selected for heat treatment process styles 2 and 3, respectively, i.e., 100 °C, 300 °C, 500 °C, 700 °C, and 900 °C. After the specified heat soak times, the specimens were allowed to cool down to room temperature. A natural air cooling method was used in the heat treatment process styles 1 and 2, and a water cooling method was applied in heat treatment process style 3. Figure 3a,b show the prepared specimens.

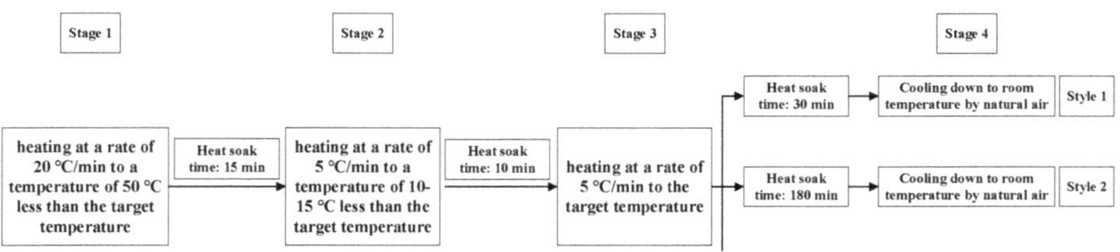

Figure 2. Heating–cooling treatment process.

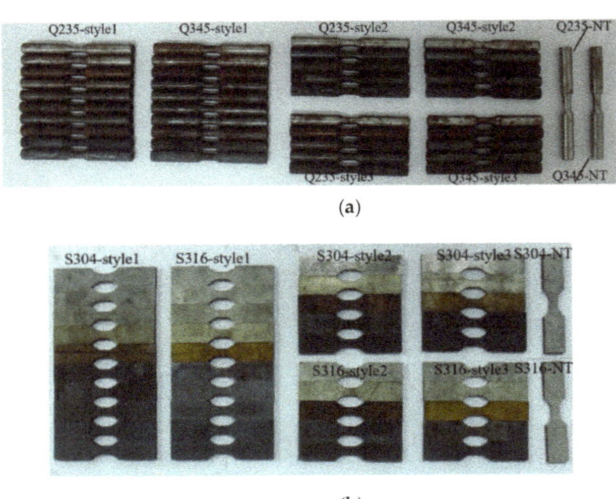

(a)

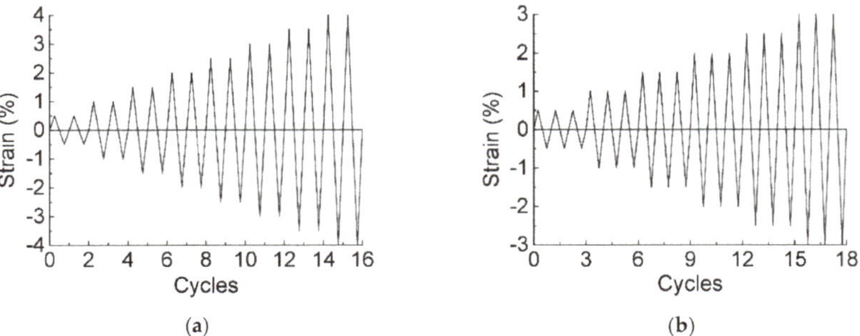

(b)

Figure 3. Prepared specimens. (**a**) Q235 and Q345; (**b**) S304 and S316.

2.2.2. Cyclic Test

The cyclic loads and monotonic loads were applied to the prepared specimens using the fatigue-testing machine Instron 8801 (Instron Corporation, Norwood, MA, USA). The applied loads were also recorded with the machine at the same time. A 5 mm range extensometer with a 12.5 mm gauge length was selected to measure the strain on the specimens.

The cyclic loading processes were controlled by strain at a rate of 0.0005/s, and the cyclic loading protocol was different for Q235/Q345 specimens and S304/S3016 specimens, as shown in Figure 4a,b. After the specified loading cycles, the specimens were monotonically loaded till failure at a strain rate of 2 mm/min.

Figure 4. Loading protocol. (**a**) Q235 and Q345; (**b**) S304 and S316.

3. Experimental Results and Discussion

3.1. Post-Fire Basic Features of Structural Steels

The first tensile stage in the first loading cycle of the hysteretic curve was extracted for each specimen to evaluate the initial elastic modulus E_s and the initial yielding strength f_y (nominal yielding strength for stainless steels) of the tested structural steels, as shown in Figure 5a,b. The ultimate strength f_u and fracture strength f_{ul} of each specimen after cyclic loading was analyzed using the monotonic loading process mentioned in Section 2.2.2, as shown in Figure 5c,d. To study the effect of heating–cooling treatments on the basic features of structural steels, the residual factors (R) were calculated and discussed based on the data from Figure 5 in Sections 3.1.1–3.1.3, which were defined as the ratio of the mechanical properties after cooling down from various elevated temperatures to that at room temperature without fire exposure.

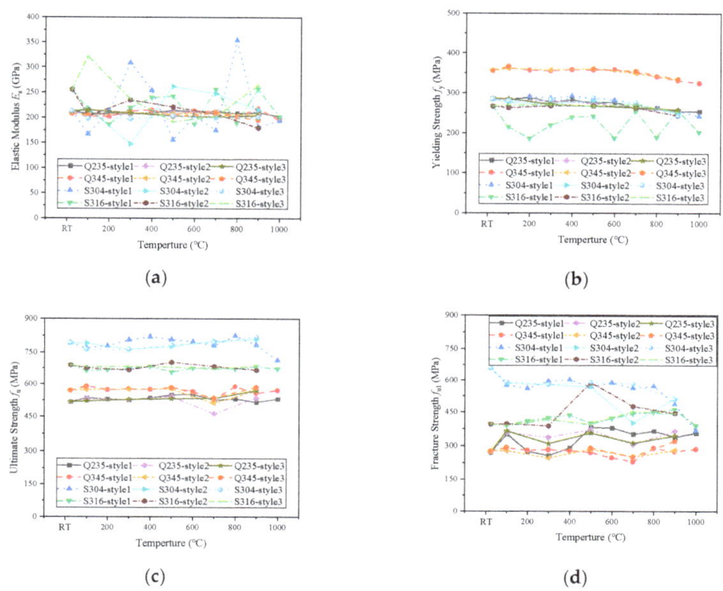

Figure 5. Basic features of structural steels. (**a**) Elastic modulus; (**b**) Yielding strength; (**c**) Ultimate strength; (**d**) Fracture strength.

3.1.1. Elastic Modulus

The post-fire elastic moduli (E_s) of the tested structural steels were calculated according to GB/T 22315-2008 [27]. Figure 6a,b show the elastic modulus residual factor (R_E) for all four types of structural steels. The Q235, Q345, and S304 steel specimens maintained their initial elastic moduli after exposure to temperatures up to 1000 °C and cooling down using different methods, with fluctuation being less than ±10%. By contrast, a 10–20% decrease in elastic moduli was found in the S316 steel specimens when treated with heating–cooling treatment styles 1 and 2. For the S316 steel specimens prepared with heating–cooling treatment style 3, the measured elastic modulus was initially increased by about 25% after exposure to 100 °C, and then it was dropped dramatically by about 10%, 25%, and 20% after exposure to 300 °C, 500 °C, and 700 °C, respectively. Moreover, the elastic modulus of the S316 steel specimen after exposure to 900 °C and cooling down by water regained its initial elastic modulus. The post-fire elastic moduli of the studied specimens indicated that the effect of target heat temperatures, heat soak times, and cooling-down methods on the elastic moduli of Q235, Q345, and S304 steels after exposure to elevated temperatures up to 1000 °C (900 °C for heating–cooling treatment styles 2 and 3) was negligible, while the influence of elevated temperatures and cooling-down methods on the elastic moduli of S316 steels was significant.

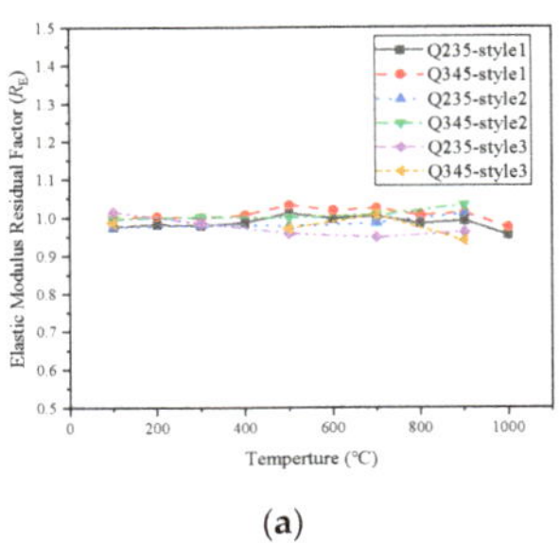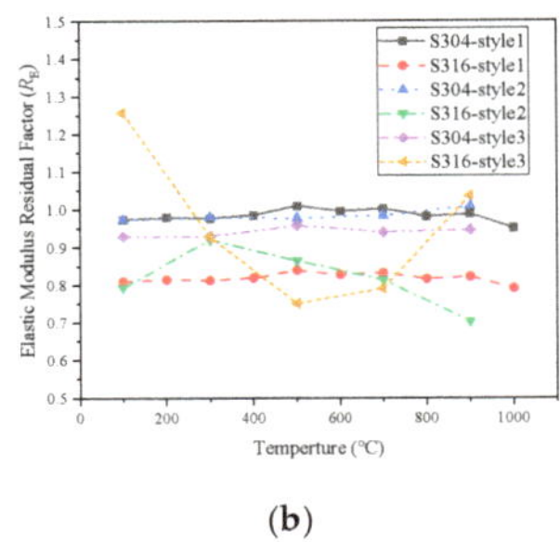

(a)

(b)

Figure 6. Elastic modulus residual factor. (**a**) Q235 and Q345; (**b**) S304 and S316.

3.1.2. Yielding Strength

The post-fire yielding strength (nominal yielding strength for stainless steels) residual factors (R_{fy}) are illustrated in Figure 7. No obvious change was found in the yielding strength of all four types of structural steels until the temperature reached 700 °C, after which the post-fire yielding strength of Q235 and Q345 steels was decreased slightly, and a reduction of about 10% was observed after exposure to 1000 °C (900 °C for heating–cooling treatment styles 2 and 3), as shown in Figure 7a. A decrease at a relatively rapid rate was found in the yielding strength of S304 and S316 steels after exposure to temperatures exceeding 700 °C, and reductions of approximately 15% were observed when the exposure temperature reached 1000 °C (900 °C for heating–cooling treatment styles 2 and 3), as shown in Figure 7b. The results showed that the critical temperature of yielding strength was about 700 °C for all four types of structural steels. The influences of different cooling methods on the residual yielding strength of the studied structural steels were insignificant in this experiment.

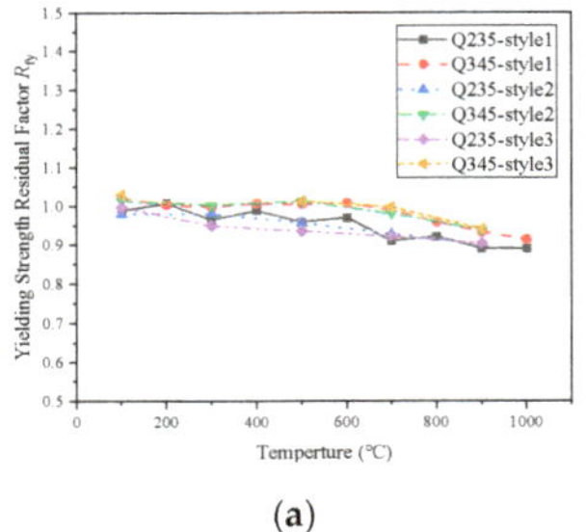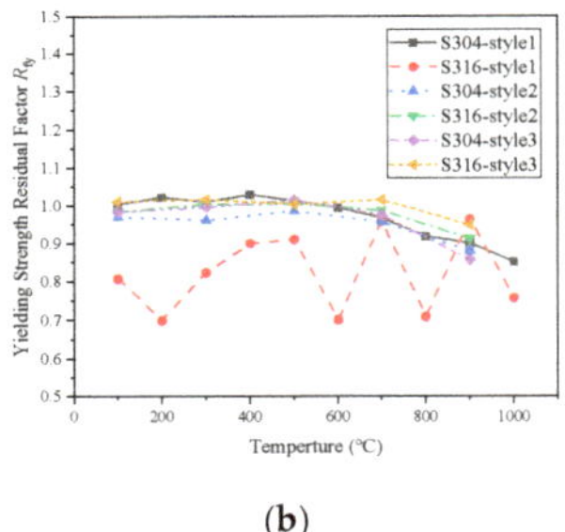

(a)

(b)

Figure 7. Yielding strength residual factor. (**a**) Q235 and Q345; (**b**) S304 and S316.

3.1.3. Ultimate Strength

The ultimate strength residual factors (R_{fu}) of the tested specimens are described in Figure 8a,b. The ultimate strength of all four types of structural steels experienced a fluctuation within a narrow range, and the maximum magnitude of the changes was within 10%. This phenomenon implied that the heating–cooling treatment styles discussed in this paper may have limited effects on the ultimate strength of all four types of structural steels after cyclic loading.

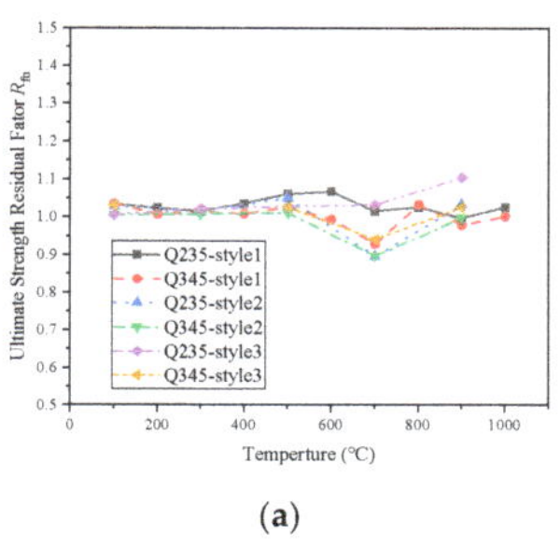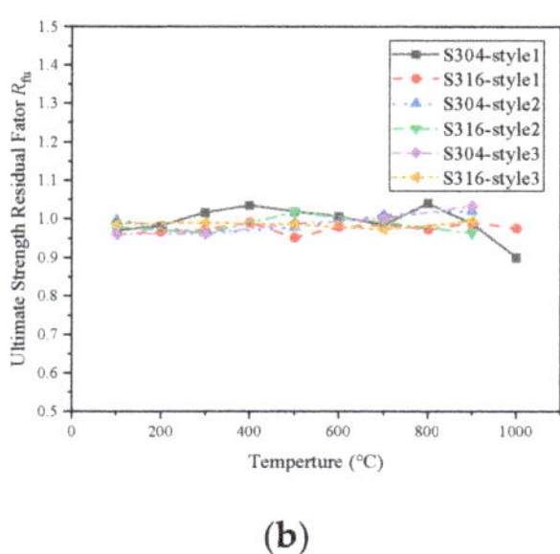

(a)   (b)

Figure 8. Ultimate strength residual factor. (**a**) Q235 and Q345; (**b**) S304 and S316.

3.1.4. Fracture Strength

The post-fire fracture strengths of all four types of structural steels were decreased compared to their corresponding ultimate strengths without fire influence, and the fracture strength drop ratios (D_{fu1}) were defined as $(f_u - f_{u1})/f_u$, as shown in Figure 9a,b. It can be found that all the fracture strength drop ratios were larger than 15%, which is the limit of the seismic design requirements [11]. This meant that the tested structural steels still exhibited good seismic ductility after exposure to heating–cooling processes. Moreover, the influence of steel grades on the fracture strength drop ratios was more significant than the target temperatures and cooling-down methods of heating–cooling treatments discussed in this paper.

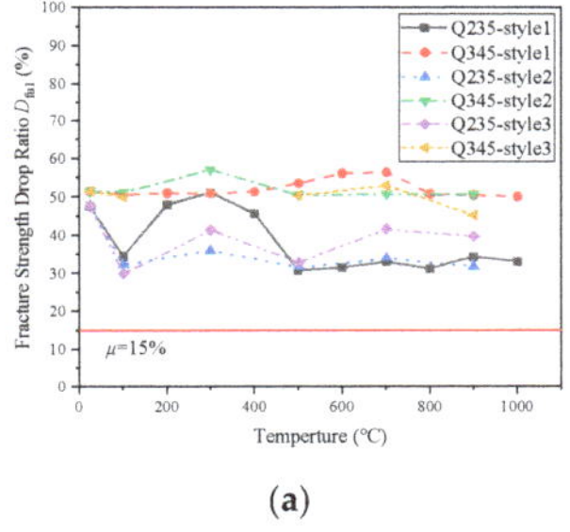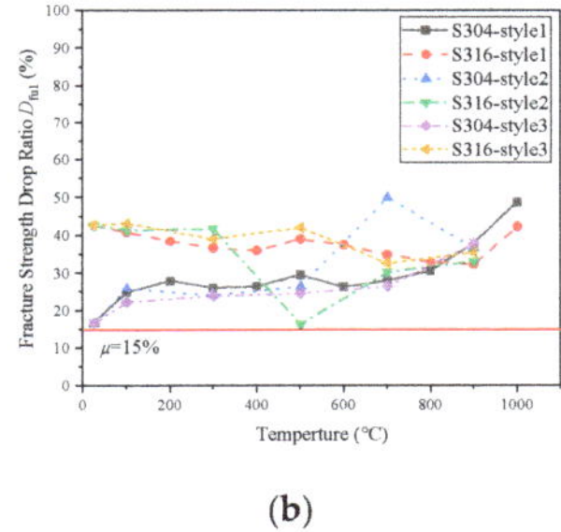

(a)   (b)

Figure 9. Fracture strength drop ratio. (**a**) Q235 and Q345; (**b**) S304 and S316.

3.2. Post-Fire hysteretic Performance of Structural Steels

3.2.1. Hysteretic Curve

In this section, the specimens exposure to room temperature are marked as S-NT, and other specimens are coded as shown in Figure 10. The hysteretic strain–stress curves of the tested specimens are shown in Figures 11–14, and subfigures (a–j) in each figure show the data at different elevated target temperatures.

All the tested specimens had plump hysteretic curves, which meant no buckling failure appeared in the specimens, which was in accordance with the failure modes observed during the experiment. Moreover, no obvious degradation of strength and stiffness was found in the hysteretic curves for the same strain levels. Therefore, all four types of structural steels had good ductility under cyclic loading despite the heating–cooling cycle they experienced.

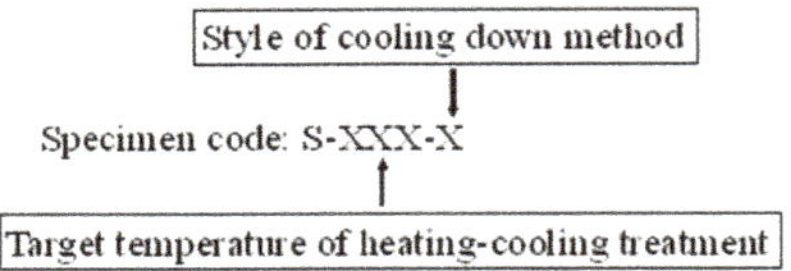

Figure 10. Specimen code.

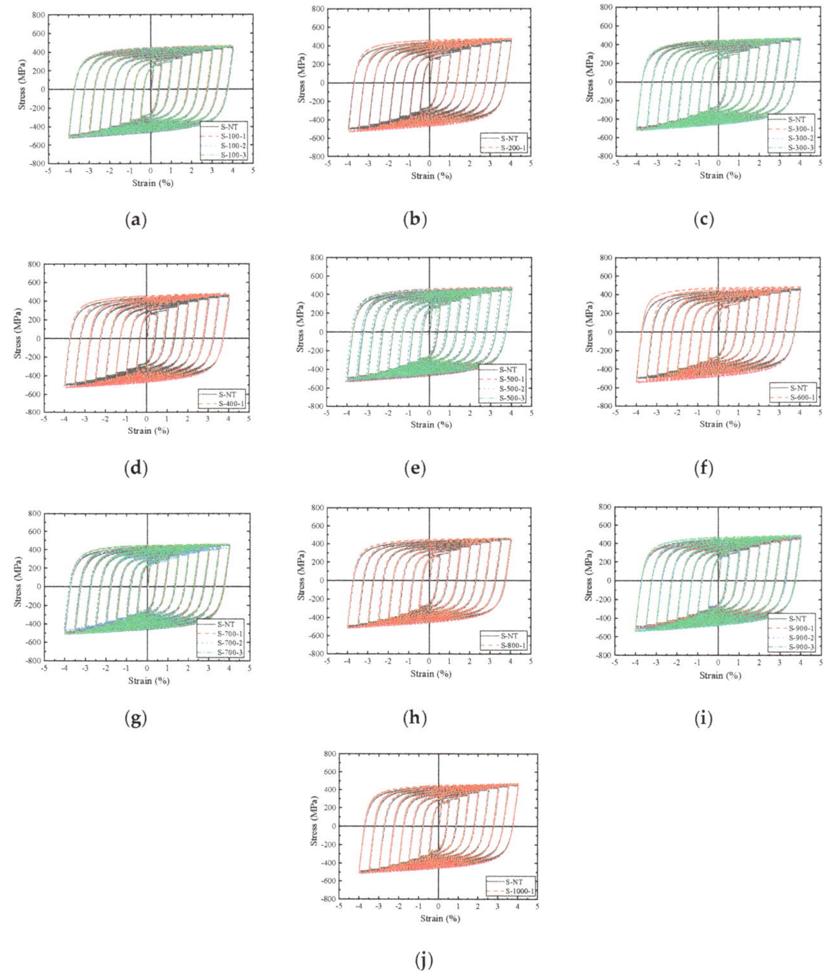

Figure 11. Hysteretic curves of Q235. (**a**) 100 °C; (**b**) 200 °C; (**c**) 300 °C; (**d**) 400 °C; (**e**) 500 °C; (**f**) 600 °C; (**g**) 700 °C; (**h**) 800 °C; (**i**) 900 °C; (**j**) 1000 °C.

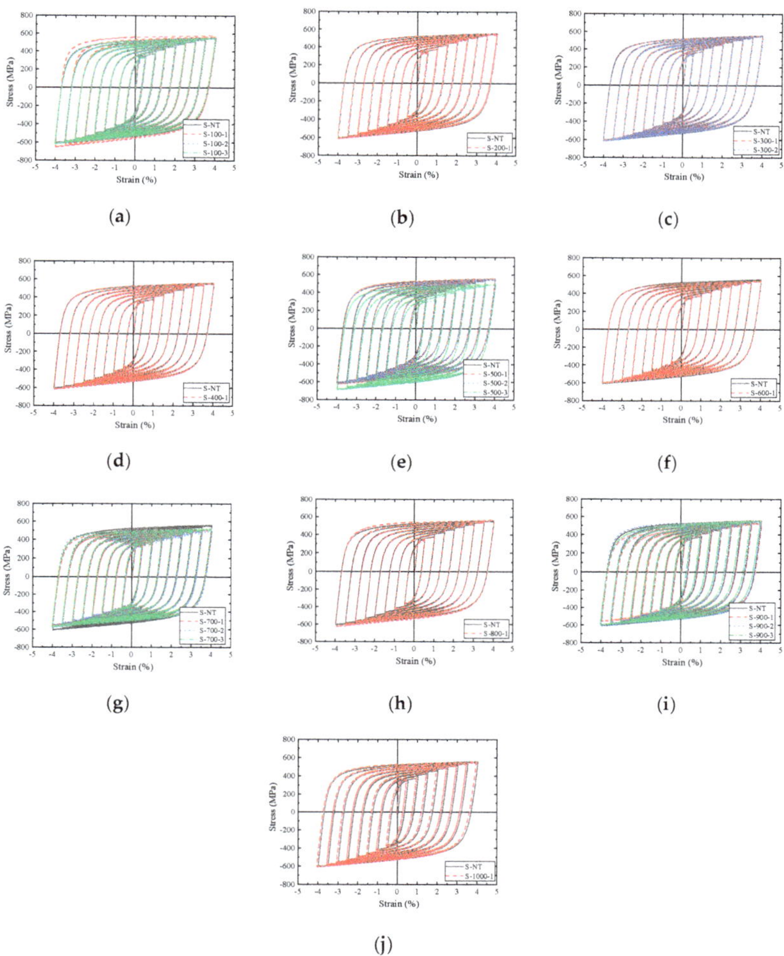

Figure 12. Hysteretic curves of Q345. (**a**) 100 °C; (**b**) 200 °C; (**c**) 300 °C; (**d**) 400 °C; (**e**) 500 °C; (**f**) 600 °C; (**g**) 700 °C; (**h**) 800 °C; (**i**) 900 °C; (**j**) 1000 °C.

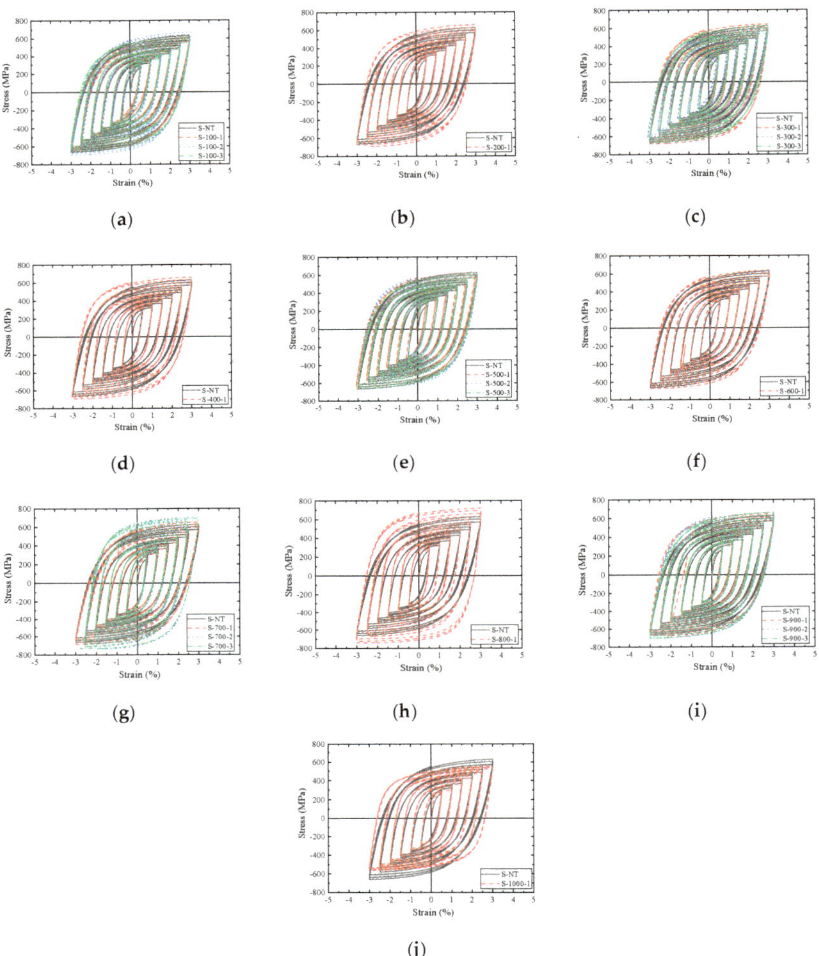

Figure 13. Hysteretic curves of S304. (**a**) 100 °C; (**b**) 200 °C; (**c**) 300 °C; (**d**) 400 °C; (**e**) 500 °C; (**f**) 600 °C; (**g**) 700 °C; (**h**) 800 °C; (**i**) 900 °C; (**j**) 1000 °C.

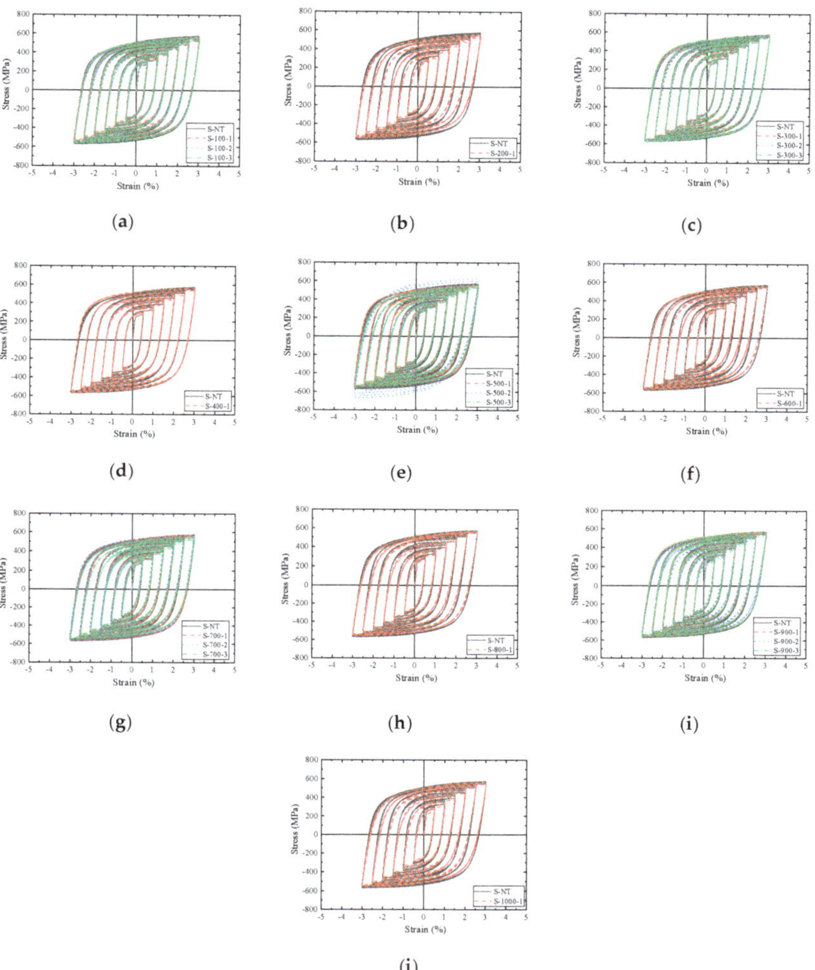

Figure 14. Hysteretic curves of S316. (**a**) 100 °C; (**b**) 200 °C; (**c**) 300 °C; (**d**) 400 °C; (**e**) 500 °C; (**f**) 600 °C; (**g**) 700 °C; (**h**) 800 °C; (**i**) 900 °C; (**j**) 1000 °C.

3.2.2. Masing Property

If the hysteresis loops of the metal obtained at different strain ranges coincide when they are plotted in relative coordinates, it means that the metal has a "Masing" property and the yield characteristics are the same [28]. To study the "Masing" properties of the tested structural steels in this paper, the hysteresis curves obtained in the last Section were reproduced as relative hysteretic curves in the relative coordinate system in this Section, and the relative hysteretic curves of typical specimens (room temperature, 900 °C) are shown in Figures 15–18 [29,30], and subfigures (a–d) show the relative hysteretic curves of the specimens prepared by different heating-cooling treatments.

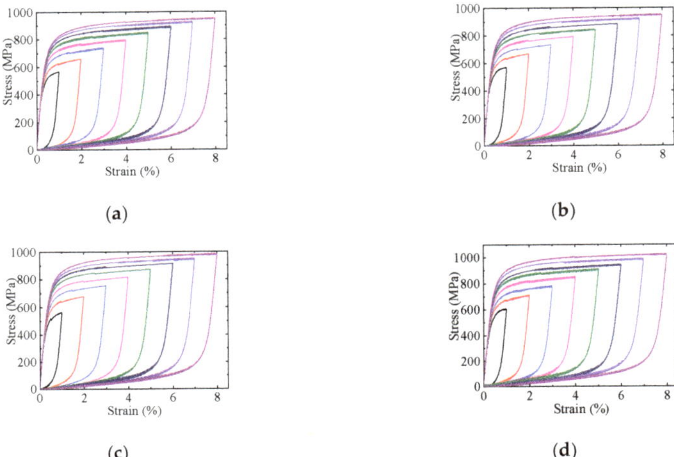

Figure 15. Relative hysteretic curves of Q235. (**a**) S-NT; (**b**) S-900-1; (**c**) S-900-2; (**d**) S-900-3.

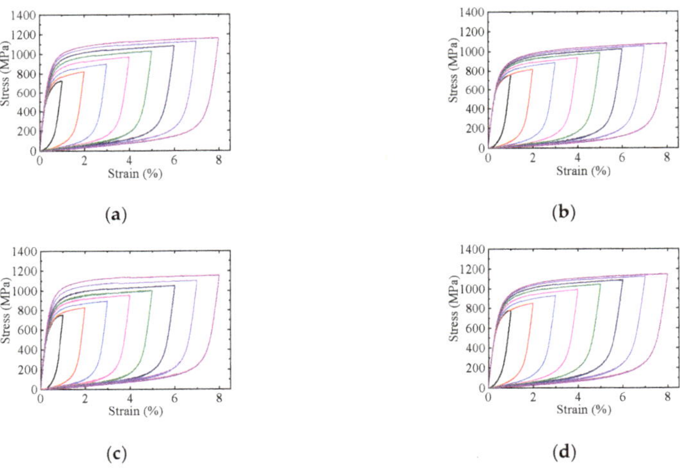

Figure 16. Relative hysteretic curves of Q345. (**a**) S-NT; (**b**) S-900-1; (**c**) S-900-2; (**d**) S-900-3.

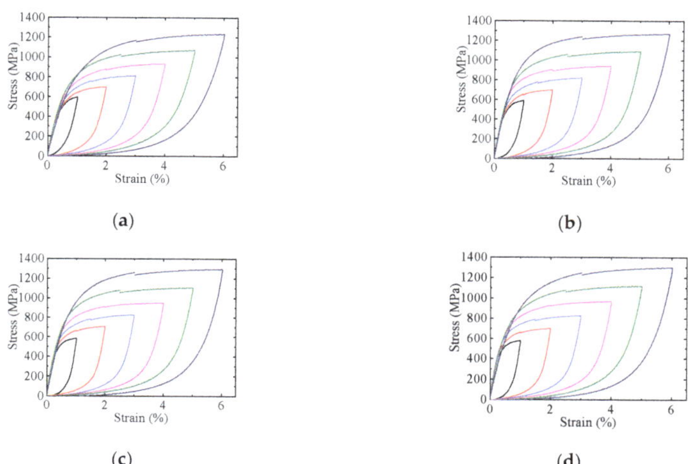

Figure 17. Relative hysteretic curves of S304. (**a**) S-NT; (**b**) S-900-1; (**c**) S-900-2; (**d**) S-900-3.

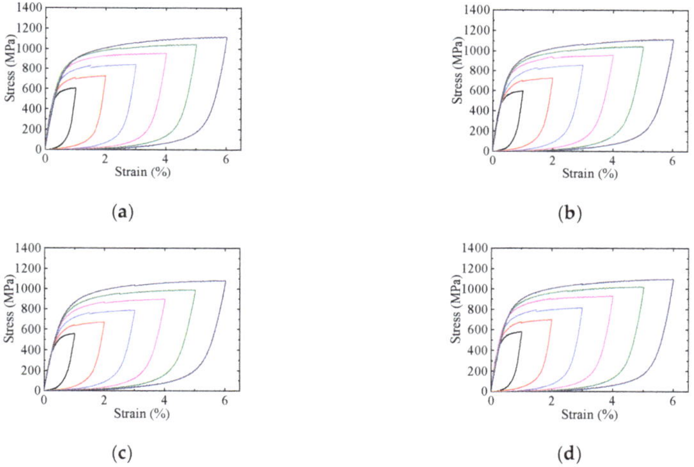

Figure 18. Relative hysteretic curves of S316. (**a**) S-NT; (**b**) S-900-1; (**c**) S-900-2; (**d**) S-900-3.

As indicated in Figures 15–18, the relative hysteretic curves obtained at different strain levels did not coincide with each other, which revealed that the four types of structural steels tested in this paper had non-Masing properties, which was unrelated to target heat temperatures, heat soak times, and cooling-down methods.

3.2.3. Energy Dissipation Capacity

Structural steels can dissipate energy when they experience plastic deformations, which can be measured as the area of hysteretic loops. The second loop at each strain level was extracted from the hysteretic curves of all four types of structural steel specimens and used to calculate the area of the hysteretic loops and evaluate their energy dissipation capacity. To save space, typical heat target temperatures were selected during the energy dissipation capacity evaluation, as shown in Figure 19.

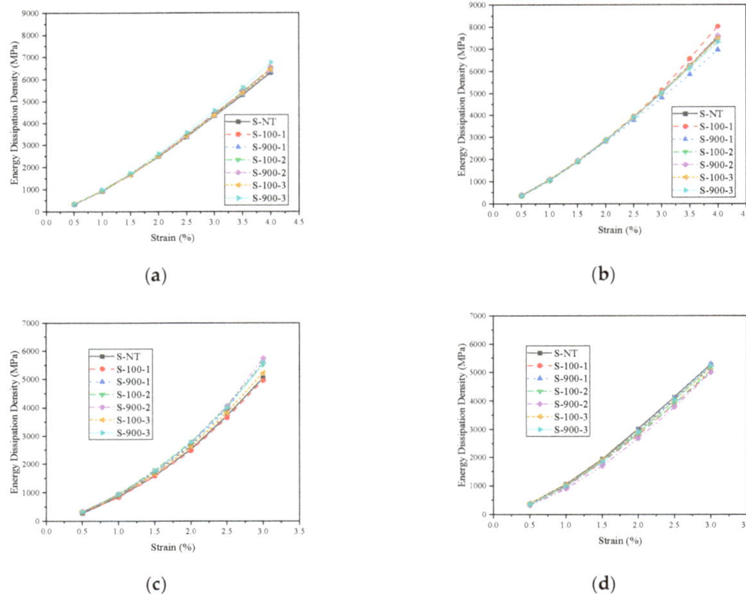

Figure 19. Energy dissipation density. (**a**) Q235; (**b**) Q345; (**c**) S304; (**d**) S316.

As shown in Figure 19a,b, for the Q235 and Q345 structural steel specimens prepared with heating–cooling treatment style 1, the energy dissipation capacity of the specimens with higher target heat temperatures was less than that of the specimens with lower target heat temperatures at the same strain level, and this difference became more obvious with an increase in strain amplitudes. This indicated that short-term heat treatment may weaken the energy dissipation capacity of Q235 and Q345 structural steels. On the contrary, for the Q235 and Q345 structural steel specimens prepared with heating–cooling treatment style 2, no obvious difference was found between the specimens with different target heat temperatures at the same strain amplitude, which meant long-term heat treatment may remove the effect of target heat temperatures on the energy dissipation capacity of Q235 and Q345 structural steels. For the Q235 and Q345 structural steel specimens prepared with heating–cooling treatment style 3, the energy dissipation capacity of the specimens with higher target heat temperatures was larger than that of the specimens with lower target heat temperatures at the same strain level owing to the hardening effect of the water cooling method. This hardening effect became more significant when the strain was beyond 1% in Q235 steel specimens, while the energy dissipation capacity of the specimens with higher target heat temperatures was once again less than that of the specimens with lower target heat temperatures at the same strain level, as seen in Q345 steel specimens when the strain was beyond 3%. This meant that the hardening effect of the water cooling method on energy dissipation was weaker than the negative effect of short-term heat treatment discussed above for the Q345 steel specimens.

As shown in Figure 19c,d, for the stainless steel specimens prepared with heating–cooling treatment style 1, the energy dissipation capacity of the specimens with higher target heat temperatures was larger than that of the specimens with lower target heat temperatures at the same strain level in both the S304 and S316 steel specimens. This difference was more obvious with an increase in strain amplitude. This meant that the short-term heat treatment effect may strengthen the energy dissipation capacity of S304 and S316 structural steels. For the specimens prepared with heating–cooling treatment style 2, there was no obvious difference in energy dissipation capacity between S304 steel specimens with different target heat temperatures, while the energy dissipation capacity of the specimens with higher target heat temperatures was less than that of the specimens with lower target heat temperatures at the same strain level in S316 steel specimens. The reason for this phenomenon was that long-term heat treatment may eliminate the strengthening effects that elevated temperatures have on the energy dissipation capacity of S304 steel specimens and may further weaken the energy dissipation capacity of the S316 steel specimens with higher target heat temperatures. For the S304 steel specimens prepared with heating–cooling treatment style 3, the energy dissipation capacity of the specimens with higher target heat temperatures was larger than that of the specimens with lower target heat temperatures at the same strain level, and the difference was increased along with the strain amplitude because of the hardening effect of the water cooling method. In contrast, this hardening effect was insignificant in the S316 steel specimens prepared with heating–cooling treatment style 3, and no obvious difference was found in the energy dissipation capacity between the specimens with different target heat temperatures.

3.2.4. Skeleton Curve

Skeleton curves of hysteretic loops can directly reflect the mechanical properties of structural steels under cyclic loading. The Ramberg and Osgood model [31] is a commonly used constitutive relation for cyclic models, as shown in Equation (1). Therefore, the Ramberg and Osgood model was employed to fit the skeleton curves in this paper.

$$\frac{\Delta\varepsilon}{2} = \frac{\Delta\varepsilon_e}{2} + \frac{\Delta\varepsilon_p}{2} = \frac{\Delta\sigma}{2E} + \left(\frac{\Delta\sigma}{2K'}\right)^{\frac{1}{n'}} \tag{1}$$

where $\Delta\varepsilon$ is total strain amplitude, $\Delta\varepsilon_e$ is elastic strain amplitude, $\Delta\varepsilon_p$ is plastic strain amplitude, $\Delta\sigma$ is stress amplitude, E is elastic modulus, K' is cyclic enhancement coefficient, and n' is cyclic enhancement exponent.

The elastic modulus E in Equation (1) can be obtained from Figure 5a. The cyclic enhancement coefficient K' and the cyclic enhancement exponent n' can be calculated by fitting the experimental skeleton curves via the Ramberg and Osgood model, as shown in Figure 20a,b. To check the accuracy of the Ramberg and Osgood model and the proposed parameters in Figure 20, a comparison of experimental skeleton curve points and fitting curves was conducted for all the specimens in this paper. The comparison results of typical specimens are shown in Figure 21 and subfigures (a–d) show the results for the four types of structural steels, respectively (E—experimental data, F—fit curves), which showed that the imitation had good results.

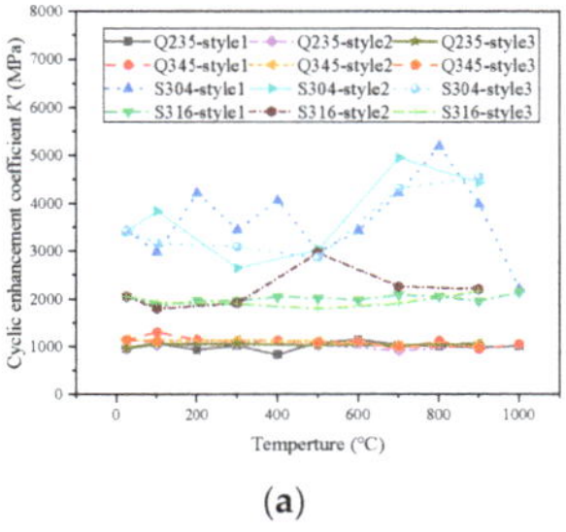
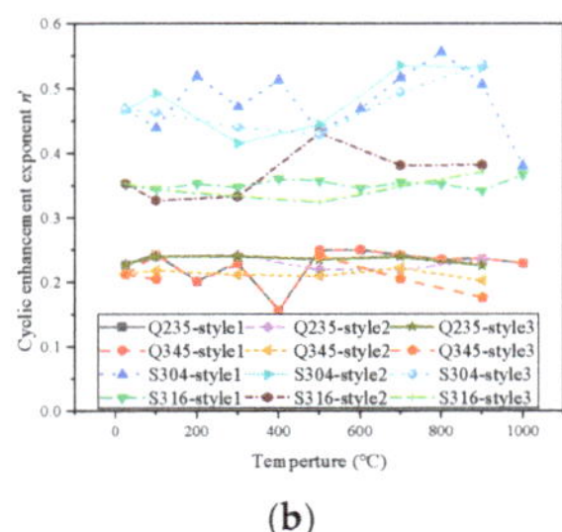

(a)   (b)

Figure 20. Parameters for Ramberg–Osgood relation. (**a**) Cyclic enhancement coefficient (K'); (**b**) Cyclic enhancement exponent (n').

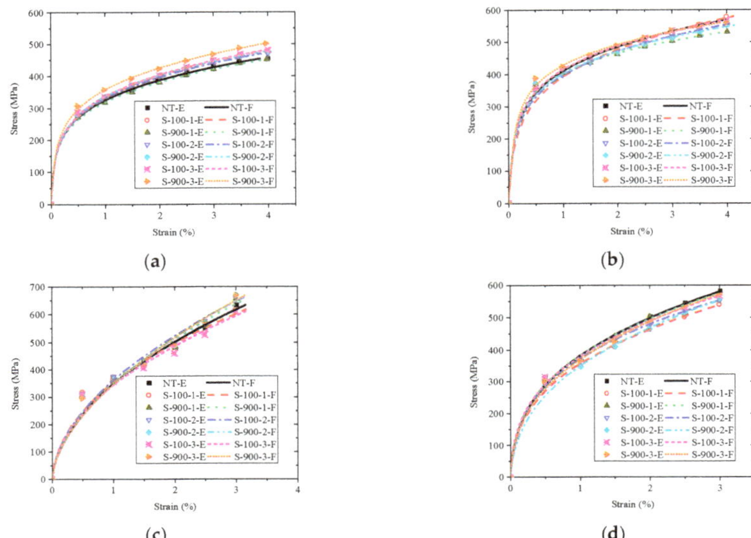

Figure 21. Comparison of experimental and fitting skeleton curves. (**a**) Q235; (**b**) Q345; (**c**) S304; (**d**) S316.

For the convenience of engineering application, a set of material parameters was proposed for Q235 and Q345 structural steels based on the data in Figures 5 and 20, as shown in Table 1. Due to the limit of the experimental data in this research, the results in Figures 5 and 20 show an obvious fluctuation, especially for S304 and S316 specimens. Therefore, the proposed values in Table 1 are rough estimates of material parameters, and it was hard to propose appropriate material parameters for S304 and S316 structural steels. More experimental data need to be collected to establish more accurate and convenient material models for Q235, Q345, S304, and S316 structural steels.

Table 1. Proposed material parameters for Q235 and Q345 structural steels.

Structural Steel	R_E	R_{fy}	K'	n'
Q235/Q345	0.95	0.9	1000	0.2

4. Conclusions

This paper experimentally investigated the cyclic performance of Chinese mild steel Grade Q235, Chinese high-strength steel Grade Q345, and Chinese stainless steel Grade S304 and S316 after exposure to various heating–cooling treatments. The main conclusions can be drawn as follows:

1. The effect of target heat temperatures, heat soak times, and cooling-down methods was negligible on the post-fire elastic moduli of Q235, Q345, and S304 structural steels, while the influence of elevated temperatures and cooling-down methods on the post-fire elastic moduli of S316 structural steels was significant.

2. The critical temperature of post-fire yielding strength was about 700 °C for all four types of structural steels, while the influences of different cooling-down methods on the residual yielding strength of the studied structural steels were insignificant in this experiment. The heating–cooling treatment styles discussed in this paper had limited effects on the ultimate strength of all four types of structural steels after cyclic loading.

3. All four types of structural steel exhibited good seismic capacity, including plump hysteretic curves and sufficient ductility, after exposure to heating–cooling treatments.

The energy dissipation capacity varied according to steel grades, target heat temperatures, heat soak times, and cooling-down methods, with the influence of steel grade on the ductility being more significant than the heating–cooling treatment styles discussed in this paper.

4. All four types of structural steels had non-Masing properties, which was unrelated to target heat temperatures, heat soak times, and cooling-down methods.

5. The Ramberg and Osgood model and corresponding model parameters can properly simulate the skeleton curves of all four types of structural steels after exposure to the heating–cooling treatments involved in this paper. Therefore, the proposed model can be applied as the constitutive relation of fire-affected steels in the post-fire seismic capacity evaluation of steel structures which are made of the four types of structural steels. However, more experimental data need to be collected to establish a more accurate and convenient material model for engineering applications.

Author Contributions: Formal analysis and writing—original draft preparation, P.D.; supervision and funding acquisition, H.L.; investigation and resources, X.X. All authors have read and agreed to the published version of the manuscript.

Funding: This research was funded by Hebei Province Science Fund for Distinguished Young Scholars, grant number E2021402006.

Institutional Review Board Statement: Not applicable.

Informed Consent Statement: Not applicable.

Data Availability Statement: Not applicable.

Conflicts of Interest: The authors declare no conflict of interest.

References

1. Lu, J.; Liu, H.; Chen, Z.; Liao, X. Experimental investigation into the post-fire mechanical properties of hot-rolled and cold-formed steels. *J. Constr. Steel Res.* **2016**, *121*, 291–310. [CrossRef]
2. Chen, Z.; Lu, J.; Liu, H.; Liao, X. Experimental study on the post-fire mechanical properties of high-strength steel tie rods. *J. Constr. Steel Res.* **2016**, *121*, 311–329. [CrossRef]
3. Qiang, X.; Bijlaard, F.S.K.; Kolstein, H. Post-fire mechanical properties of high strength structural steels S460 and S690. *Eng. Struct.* **2012**, *35*, 1–10. [CrossRef]
4. Kang, L.; Wu, B.; Liu, X.; Ge, H. Experimental study on post-fire mechanical performances of high strength steel Q460. *Adv. Struct. Eng.* **2021**, *24*, 2791–2808. [CrossRef]
5. Sajid, H.U.; Kiran, R. Post-fire mechanical behavior of ASTM A572 steels subjected to high stress triaxialities. *Eng. Struct.* **2019**, *191*, 323–342. [CrossRef]
6. Qiang, X.; Bijlaard, F.S.K.; Kolstein, H. Post-fire performance of very high strength steel S960. *J. Constr. Steel Res.* **2013**, *80*, 235–242. [CrossRef]
7. Chiew, S.P.; Zhao, M.S.; Lee, C.K. Mechanical properties of heat-treated high strength steel under fire/post-fire conditions. *J. Constr. Steel Res.* **2014**, *98*, 12–19. [CrossRef]
8. Wang, W.; Liu, T.; Liu, J. Experimental study on post-fire mechanical properties of high strength Q460 steel. *J. Constr. Steel Res.* **2015**, *114*, 100–109. [CrossRef]
9. Wang, F.; Lui, E.M. Experimental study of the post-fire mechanical properties of Q690 high strength steel. *J. Constr. Steel Res.* **2020**, *167*, 105966. [CrossRef]
10. Li, G.; Lyu, H.; Zhang, C. Post-fire mechanical properties of high strength Q690 structural steel. *J. Constr. Steel Res.* **2017**, *132*, 108–116. [CrossRef]
11. Zhou, H.; Wang, W.; Wang, K.; Xu, L. Mechanical properties deterioration of high strength steels after high temperature exposure. *Constr. Build. Mater.* **2018**, *199*, 664–675. [CrossRef]
12. Zhang, C.; Wang, R.; Song, G. Post-fire mechanical properties of Q460 and Q690 high strength steels after fire-fighting foam cooling. *Thin-Walled Struct.* **2020**, *156*, 106983. [CrossRef]
13. Gunalan, S.; Mahendran, M. Experimental investigation of post-fire mechanical properties of cold-formed steels. *Thin Wall Struct.* **2014**, *842*, 41–54. [CrossRef]
14. Wang, X.; Tao, Z.; Song, T.; Han, L. Stress-strain model of austenitic stainless steel after exposure to elevated temperatures. *J. Constr. Steel Res.* **2014**, *99*, 129–139. [CrossRef]
15. Li, X.; Lo, K.H.; Kwok, C.T.; Sun, Y.F.; Lai, K.K. Post-fire mechanical and corrosion properties of duplex stainless steel: Comparison with ordinary reinforcing-bar steel. *Constr. Build. Mater.* **2018**, *174*, 150–158. [CrossRef]

16. Huang, Y.; Young, B. Mechanical properties of lean duplex stainless steel at post-fire condition. *Thin-Walled Struct.* **2018**, *130*, 564–576. [CrossRef]
17. Smith, C.I.; Kirby, B.R.; Lapwood, D.G.; Cole, K.J.; Cunningham, A.P.; Preston, R.R. The reinstatement of fire damaged steel framed structures. *Fire Saf. J.* **1981**, *4*, 21–62. [CrossRef]
18. Wang, X.; Tao, Z.; Song, T.; Han, L. Mechanical properties of austenitic stainless steel after exposure to fire. Research and Applications in Structural Engineering, Mechanics and Computation. In Proceedings of the Fifth International Conference on Structural Engineering, Mechanics and Computation, Cape Town, South Africa, 2–4 September 2013.
19. Wang, X.; Tao, Z.; Hassan, M.K. Post-fire behaviour of high-strength quenched and tempered steel under various heating conditions. *J. Constr. Steel Res.* **2020**, *164*, 105785. [CrossRef]
20. Hai, L.; Sun, F.; Zhao, C.; Li, G.; Wang, Y. Experimental cyclic behavior and constitutive modeling of high strength structural steels. *Constr. Build. Mater.* **2018**, *189*, 1264–1285. [CrossRef]
21. Shi, Y.; Wang, M.; Wang, Y. Experimental and constitutive model study of structural steel under cyclic loading. *J. Constr. Steel Res.* **2011**, *67*, 1185–1197. [CrossRef]
22. SAC. *GB/T700-2006*; Carbon Structural Steels. China Standard Press: Beijing, China, 2006.
23. SAC. *GB/T1591-2008*; High Strength Low Alloy Structural Steels. China Standard Press: Beijing, China, 2008.
24. SAC. *GB/T1220-2007*; Stainless Steel Bars. China Standard Press: Beijing, China, 2007.
25. SAC. *GB/T3280-2015*; Cold Rolled Stainless Steel Plate, Sheet and Strip. China Standard Press: Beijing, China, 2015.
26. SAC. *GB/T15248-2008*; The Test Method for Axial Loading Constant-Amplitude Low-Cycle Fatigue of Metallic Materials. China Standard Press: Beijing, China, 2008.
27. SAC. *GB/T 22315-2008*; Metallic Material-Determination of Modulus of Elasticity and Poisson's Ratio. China Standard Press: Beijing, China, 2008.
28. Skelton, R.P.; Maier, H.J.; Christ, H.J. The Bauschinger effect, Masing model and the Ramberg–Osgood relation for cyclic deformation in metals. *Mater. Sci. Eng. A* **1997**, *238*, 377–390. [CrossRef]
29. Chen, Y.; Sun, W.; Chan, T. Effect of Loading Protocols on the Hysteresis Behaviour of Hot-Rolled Structural Steel with Yield Strength up to 420 MPa. *Adv. Struct. Eng.* **2013**, *16*, 707–719. [CrossRef]
30. Zhou, F.; Li, L. Experimental study on hysteretic behavior of structural stainless steels under cyclic loading. *J. Constr. Steel Res.* **2016**, *122*, 94–109. [CrossRef]
31. Ramberg, W.; Osgood, W.R. Description of Stress-Strain Curves by Three Parameters. *NACA Tech. Note No. 902*; 1943. Available online: https://ntrs.nasa.gov/citations/19930081614 (accessed on 8 May 2022).

Article

Lateral Buckling of Pipe-in-Pipe Systems under Sleeper-Distributed Buoyancy—A Numerical Investigation

Zechao Zhang [1], Zhihua Chen [2,*] and Hongbo Liu [2,3]

[1] China Three Gorges Corporation, Beijing 100089, China; zhang_zechao@ctg.com.cn
[2] School of Civil Engineering, Tianjin University, Tianjin 300072, China; hbliu@tju.edu.cn
[3] School of Civil Engineering, Hebei University of Engineering, Handan 056006, China
* Correspondence: zhchen@tju.edu.cn

Abstract: The crude oil in pipelines should remain at high temperature and pressure to satisfy the fluidity requirement of deep-sea oil transportation and consequently lead to the global buckling of pipelines. Uncontrolled global buckling is accompanied by pipeline damage and oil leakage; therefore, active buckling control of pipelines is needed. Pipe-in-pipe (PIP) systems have been widely used in deep-sea oil pipelines because of the protection and insulation characteristics of the outer pipe to the inner pipe. In this study, sleeper-distributed buoyancy is used as an active buckling control method for the global buckling of PIP systems with initial imperfections. The accuracy of this technique is verified by comparing the finite element model of a 3D pipeline with experimental data. The effects of buoyancy density, pipe–soil friction coefficient, initial imperfection, stiffness ratio of inner and outer pipes, and buoyancy unit interval on the global buckling performance are also analyzed. The critical buckling force and lateral displacement of this method are studied using an analytic solution, and the relevant calculation formulas are obtained and verified to provide a basis for its engineering application.

Keywords: pipe-in-pipe system; global buckling; the sleeper-distributed buoyancy method; active control; pipeline design

Citation: Zhang, Z.; Chen, Z.; Liu, H. Lateral Buckling of Pipe-in-Pipe Systems under Sleeper-Distributed Buoyancy—A Numerical Investigation. *Metals* **2022**, *12*, 1094. https://doi.org/10.3390/met12071094

Academic Editor: Giovanni Meneghetti

Received: 27 April 2022
Accepted: 22 June 2022
Published: 26 June 2022

Publisher's Note: MDPI stays neutral with regard to jurisdictional claims in published maps and institutional affiliations.

1. Introduction

Under the action of high temperature and high pressure, due to the thermal expansion and Poisson effect of pipeline materials, the pipeline inevitably has an elongation trend. However, the pipeline cannot expand freely due to the restriction of surrounding soil and rockfill. With the continuous increase in temperature and pressure, the pressure in the pipe gradually increases until a certain critical pressure and global buckling will happen. If the global buckling is not controlled, it is likely to cause excessive bending deformation and damage to the pipeline or induce local buckling propagation. Three main methods are applied to control the global buckling of pipelines [1,2]. The first one is to limit any possible buckling modes by using external loads [3]. Over burying or rockfill is usually used to limit the global buckling of pipelines [4]; however, its cost increases with the depth of pipeline laying, and the objective cannot be achieved when the depth is extremely large. Therefore, this method is only applicable to shallow water pipelines. The second one is to reduce the axial force of pipelines by decreasing the temperature of crude oil transportation or installing an expansion bend in pipelines, thus avoiding or slowing down the global buckling [5,6]. However, the installation of expansion bending is usually expensive and inconvenient. The third one is to stimulate controlled lateral buckling (vertical buckling is usually catastrophic) at the appropriate location [7]. Even with the increase in pipeline laying depth, most deep-sea pipelines can be directly laid on the rugged seabed, and the lateral buckling of high-temperature and -pressure pipelines is almost inevitable [8,9].

Controllable lateral buckling devices are used to control the buckling of pipelines at multiple planned locations rather than allowing uncontrollable and severe buckling to occur at only one location [10]. Therefore, thermal expansion can be evenly distributed into a series of buckling to ensure that the pipelines will not have a large axial movement. Various methods, including snake laying [11], vertical offset (sleeper method) [12], and local load reduction (distributed buoyancy method) [13], have been used to induce the buckling of pipelines at a certain location. The snake-shaped buckling distance is a half wavelength, which is defined as the distance among continuous snake-shaped vertices. Reasonably selecting the shape parameters ensures that the snake-shaped pipe-laying method can successfully induce pipeline buckling [14,15].

As shown in Figure 1, the sleeper method introduces large-diameter vertical support under a pipeline [9]. This support is perpendicular to the pipeline path to ensure that the pipeline can form a suspension in the local area, thus reducing the lateral pipe–soil resistance in the suspension section and facilitating the pipeline lateral buckling under low axial force [16]. The sleeper is usually composed of large-diameter pipes with antisinking plates to prevent the sinking of the pipeline. The number of sleeper layouts depends on the lateral buckling amplitude of the pipeline after buckling. The pipeline forms a suspension span near the sleepers because of the supporting effect of the latter under the former [17]. Pipe–soil interaction remains important even in areas far away from sleepers where unpredictable buckling may occur. Therefore, sleepers must be installed at equal intervals before pipe laying to induce pipeline buckling at multiple locations.

Figure 1. Sketch map of the sleeper method.

Distributed buoyancy method is applied in installing distributed buoyancy units on pipelines to reduce the underwater weight and, consequently, the lateral soil resistance of local submarine pipelines during lateral buckling.

The distributed buoyancy method can be divided into two categories: one is that the buoyancy is large and part of the pipeline leaves the seabed; the other is that the buoyancy is small and the pipeline is still in full contact with the seabed. For the first method, Peek [18] only provided the buoyancy value required for the lateral buckling of the rigid seabed by equivalent the straight pipeline to a fixed beam at both ends, and the initial imperfects were not considered. Sun [13] conducted a model test to simulate the lateral buckling of pipes. The model test compares the sleeper method with the first type of distributed buoyancy method. It was found that the distributed buoyancy method produces a larger buckling section, and the bending is smaller than the pipe cushion method; in this study, the pipe–soil interaction was not considered. Li Gang [19] used the finite element method to optimize the design parameters and layout scheme of the buoyancy module in the distributed buoyancy control technology, effectively reducing the critical buckling force of the global buckling, but the high temperature and high pressure in the pipeline were ignored.

In distributed buoyancy method, the pipeline length with an additional buoyancy unit is generally 60–200 m [20]. Pipe-in-pipe (PIP) systems are heavy; thus, additional materials are needed in building a buoyancy section to satisfy the design requirement. Lateral soil resistance is high under PIP systems; hence, sleepers become an attractive solution to initiate a lateral buckling easier than the on-bottom features. A previous project presented that the material cost for a 200 ft. (61 m)-long buoyancy section with a buoyant force equal to 90% of the PIP system submerged weight is approximately five times that for an 80 ft. (24 m)-long sleeper [13]. The outer diameter of the buoyancy section is significantly large (three times of PIP outer diameter), making the shoulder span also massive. A high contact pressure at the shoulders can limit the mobilization.

In the paper, sleeper and buoyancy methods are combined to control the global buckling of PIP systems. The SDB method combines the advantages of the sleeper method and distributed buoyancy method. A suspended section will be formed between pipelines and the seabed because of the existence of sleepers. At the part where the pipelines come in contact with the seabed, the lateral resistance of the seabed is decreased by the buoyancy unit, thus controlling the global lateral buckling of PIP systems.

2. Analysis Method

2.1. Geometric Model Detail

Figure 2 shows the basic composition of a PIP system. The pipes mainly consist of three parts, namely, inner pipe, outer pipe, and centralizer. The centralizer and the inner pipe are fixed, and the latter can slide along the inner wall of the outer pipe [19]. A certain gap exists between the centralizer and the outer pipe. According to engineering practice, the distance of this gap is generally from 1 mm to 10 mm. The distance among the central rings is generally 1 m to 2 m [20]. The gap between inner and outer pipes is usually filled with light, low-strength, and high-performance insulation materials. The outer pipe is wrapped with buoyancy blocks in the seabed contact section.

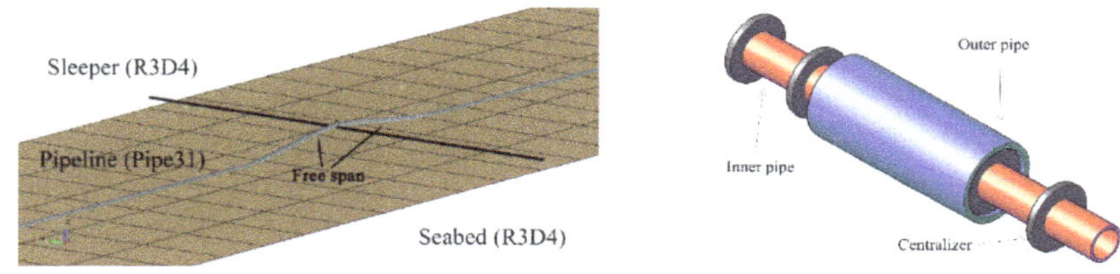

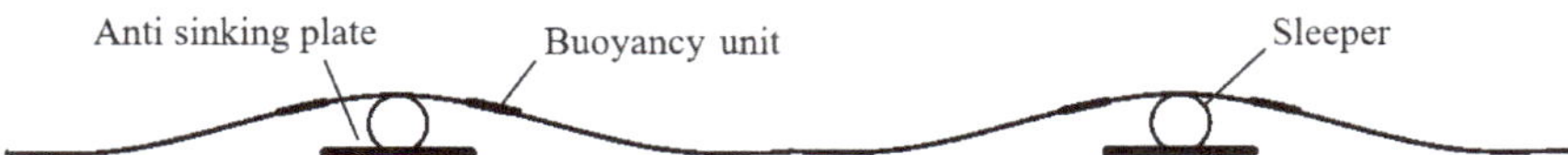

Figure 2. Sketch map of the SDB method.

2.2. FE Model Analysis

Figure 3 shows the FE models established in the software of ABAQUS 6.14. In the model, boundary constraints are end hinges, and the buoyancy is added as vertical upward force shows yellow. The sleeper is in the middle of the seabed under the pipeline.

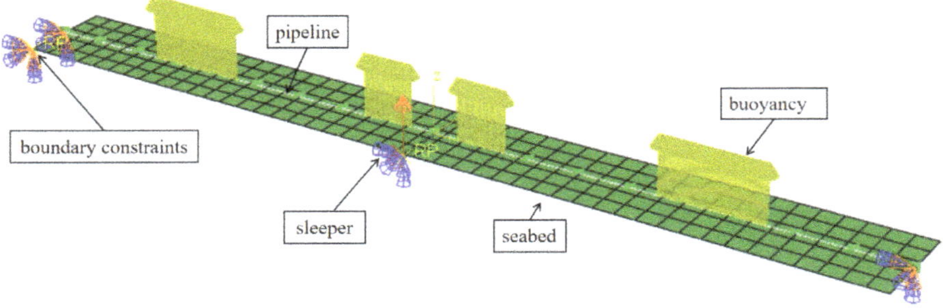

Figure 3. FE model of SDB.

First, the unit types of each part are determined. Pipe31 beam element is used to simulate the internal and external pipelines, R3D4 element is used to simulate the seabed, and a circular arc rigid surface is used to simulate the sleepers. In the model, a single node ITT element is used to simulate the centering ring to simulate the contact between the inner and outer pipes. The ITT element is attached to the node of the inner pipe beam element, and a corresponding virtual slip line is attached to the outer pipe. The slip line is composed of the outer pipe node.

Second, model constraints and loads are determined. Most parts of the pipeline in the model are directly placed on the seabed to provide vertical and horizontal constraints. The middle part of the pipes is supported by sleepers to form two end suspension sections on both sides of the sleepers. The distance between the two ends of the pipeline and the suspended section of the pipeline is up to 900 m, and the effect of end restraint on the global buckling section of the pipeline (suspended section) can be neglected. A simple support restraint is adopted at both ends of the pipeline. The Coulomb friction model is used, where the friction between the pipeline and seabed is set as μ_1, and the friction between the pipeline and sleeper is set as μ_2. The load of the pipeline model is uniform temperature field, internal pressure field, and load along the negative direction of Z direction (simulated gravity).

Third, the snake-shaped laying section of the pipeline is established. The serpentine paving section is represented as initial imperfection in the model. In this chapter, the shape of the initial imperfection is described by a sinusoidal function. The imperfection morphology equation can be expressed as follows:

$$h = h_0 \sin\left(\frac{\pi(l + 0.5l_0)}{l_0}\right) \tag{1}$$

where h_0 is the length of initial imperfection, and l_0 is the wavelength.

Fourth, analysis is implemented via four steps. The first step is that the pipeline comes in contact with the seabed under gravity. The second step is that the sleepers hold up the pipeline to the expected height. The third step is to simulate the buoyancy unit action by applying linear vertical load to the buoyancy unit of the pipeline. The fourth step is to apply uniform temperature and pressure to the inner pipe.

Finally, the calculation method of the model is determined. In this chapter, the dynamic solution considering artificial damping is adopted. Setting * Dynamic, Application = QUASI-STATIC in the Abaqus solution file. Although this method is dynamic, the quasistatic results can be obtained due to the influence of damping. Geometric stiffness is considered in the calculation, and the stiffness matrix of the system is updated continuously.

2.3. Model Validation

The pipelines in the literature [21] are selected for comparison to verify the FE model in this chapter, and the corresponding FE model is established for calculation. The pipeline in the literature does not contain a buoyancy unit; hence, the buoyancy value is set to 0 in the model, and the other parameters remain consistent. The parameters of the FE model are shown in Table 1.

The resulting force of the inner and outer axial forces represents the stability of the PIP system. This force increases gradually with the increase in temperature but not after reaching the critical buckling axial force, indicating that the pipeline has buckling. The critical buckling axial force obtained by numerical simulation is -118 N, and the experimental values are -123.4, -110.8, and -104.4 N. The average relative error is 4.6%. Therefore, the proposed model can capture the global buckling characteristics of the PIP system.

Table 1. Parameters in the literature.

	Outer Pipe	Inner Pipe
Diameter (mm)	20	10
Wall thickness (mm)	2	1.5
Materials	PMMA	Stainless steel
Modulus of elasticity (MPa)	2500	19,000
Coefficient of expansion (/°C)	1.20×10^{-4}	1.08×10^{-5}
Length (mm)	2000	
Axial friction coefficient between outer pipe and sand	0.5	-
Lateral friction coefficient between outer pipe and sand	0.3	-

3. Parametric Analysis

The main factors affecting the global buckling of the pipeline in the sleeper-distributed buoyancy (SDB) laying method are presented in this section. The following six parameters are selected for analysis: buoyancy density, pipe–soil friction coefficient, initial imperfection, rigidity ratio of inner and outer pipes, height of sleeper, and buoyancy unit interval. The range of the parameters is shown is Table 2.

Table 2. Range of parameters.

Inner Pipe (d) (mm)	250, 300, 350, 400, 450
Outer pipe (D) (mm)	500
Wall thickness (t) (mm)	10
Modulus of elasticity (E) (MPa)	206,000
Poisson ratio (δ)	0.3
Coefficient of linear expansion (α) (/°C)	1.01×10^{-5}
Initial defect (h_0/l_0)	0.001–0.020
Underwater weight (q) (N/m)	1000–3000
Friction coefficient of seabed (μ_1)	0.3–0.9
Friction coefficient of sleeper (μ_1)	0.1–0.6
Sleeper height (h_2) (m)	0.3–0.9
Buoyancy block distribution density	0–0.78

In the initial stage of heating (14 °C), lateral displacement occurs for the suspended section of the pipeline but not for the midpoint. This phenomenon is unique in the buckling of pipelines with sleepers. With increasing temperature, the lateral displacement in the middle of the pipeline also increases rapidly. When the temperature reaches 39 °C, wave peaks appear on both sides, and the lateral displacement occurs along the opposite direction of the midwave peaks. The global buckling length of the final pipeline is 200 m long, and the buckling displacement is symmetrical. Figure 4a illustrates the distribution of combined stresses along the length direction of the pipeline section, where each data point represents the average combined stresses of the four integral points of the section. As shown in the figure, peak stress is observed at the top of each buckling peak. Figure 4b displays the combined Mises stress of the pipeline at a 100 °C temperature increase.

In the follow-up analysis, the global buckling development model of the pipeline presents similar patterns. Hence, only the axial force–displacement curve is used to represent the global buckling of the pipeline. The following basic model parameters are discussed: $d = 254$ mm, $h_0/l_0 = 0.012$, $q = 1500$ N/m, $\mu_1 = 0.5$, $\mu_2 = 0.3$, and $h_2 = 0.5$ m. In the next section, the effects of various parameters on the global response of the SDB pipeline are discussed.

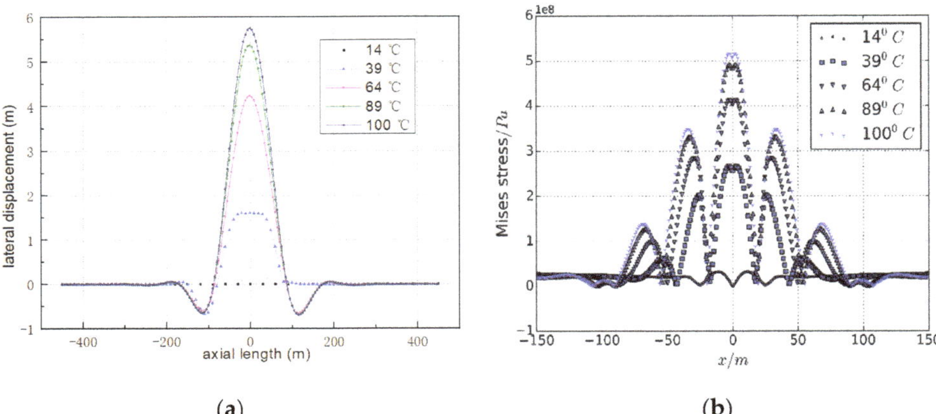

(a)   **(b)**

Figure 4. Result of a PIP system on global buckling. (**a**) Displacement development; (**b**) stress development.

3.1. Analysis of Buoyancy Density

The ratio of buoyancy to pipeline gravity is defined as α with values of 0.25, 0.27, 0.29, 0.31, and 0.33. The axial force and lateral displacement of the midpoint of the pipeline under the global buckling of the pipeline are shown in Figure 5 and Table 3. By comparison, the critical buckling axial force of the pipeline decreases with the increase in α. Comparison analysis shows that when buoyancy increases by 32% from 0.25 to 0.33, the axial force decreases by 10.2% from -1950 KN to -1750 KN, indicating that the critical buckling force can be significantly controlled by changing the number of buoyancy units.

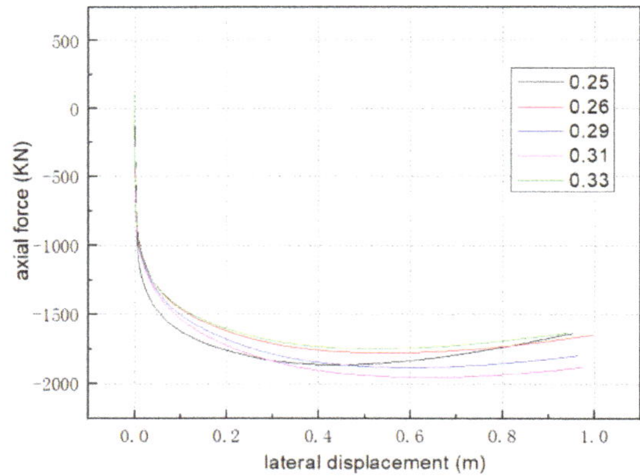

Figure 5. Influence of the buoyancy density in the axial force–lateral displacement graph.

Table 3. Critical axial force of the buoyancy density.

The Buoyancy Density	Critical Axial Force (KN)
0.25	-1950
0.27	-1910
0.29	-1870
0.31	-1810
0.33	-1750

3.2. Analysis of Pipe–Soil Friction Coefficient

Figure 6 and Table 4 depict that the effects of changing the friction between pipe and soil on the critical axial force of global buckling of the pipeline are mainly observed during the post-buckling stage but not before buckling. This phenomenon occurs because the influence of friction before buckling is relatively less due to the suspension section formed by the sleeper at the middle point. When buckling occurs, the pipeline is mainly subjected to sliding friction, at which time the friction coefficient plays a role. Consequently, the axial force of the pipeline increases with friction during the post-buckling stage.

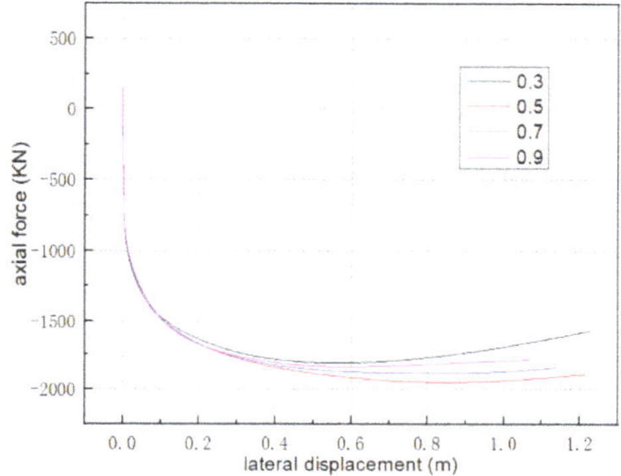

Figure 6. Influence of the pipe–soil friction coefficient in the axial force–lateral displacement graph.

Table 4. Critical axial force of pipe–soil friction coefficient.

Pipe–Soil Friction Coefficient	Critical Axial Force (KN)
0.3	−1800
0.5	−1830
0.7	−1870
0.9	−1900

3.3. Analysis of Initial Imperfection

The influence of initial imperfection wavelength on the critical buckling force–lateral displacement of the pipeline is shown in Figure 7 and Table 5. The graph illustrates that the axial force of the pipeline can be greatly affected by changing the initial imperfection wavelength. When the defect length of the pipeline increases by four times from 2 m to 10 m, the axial force also decreases by four times.

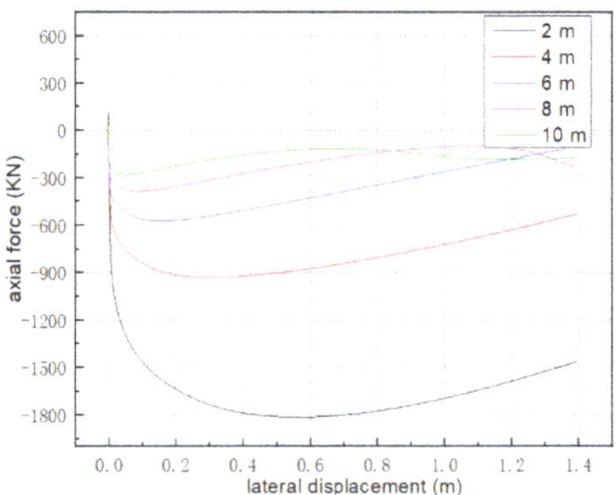

Figure 7. Influence of initial imperfection in the axial force–lateral displacement graph.

Table 5. Critical axial force of initial imperfection.

Initial Imperfection (m)	Critical Axial Force (KN)
2	−298
4	−430
6	−590
8	−920
10	−1800

3.4. Analysis of the Rigidity Ratio of Inner and Outer Pipes

Comparisons in Figure 8 and Table 6 show that the critical axial force of pipeline buckling decreases by 15.3% to −1630 KN from −1950 KN, whereas the rigidity ratio increases by 132% to 0.33 from 0.25. The change in rigidity ratio of inner and outer pipes is linear.

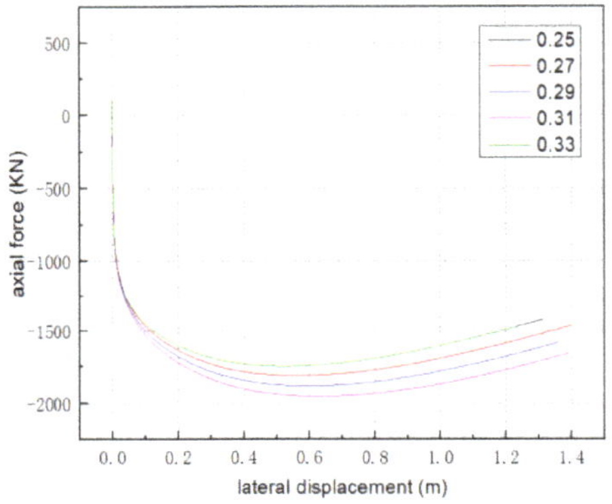

Figure 8. Influence of the rigidity ratio of inner and outer pipes in the axial force–lateral displacement graph.

Table 6. Critical axial force of the rigidity ratio of inner and outer pipes.

Rigidity Ratio of Inner and Outer Pipes	Critical Axial Force (KN)
0.25	−1950
0.27	−1880
0.29	−1810
0.31	−1740
0.33	−1630

3.5. Analysis of Sleeper Height

Comparison and analysis of the parameters of sleeper height in Figure 9 and Table 7 indicate that the change in sleeper height minimally influences the critical buckling axial force of pipelines. The sleeper height increases from 0.1 m to 0.9 m, and the change in the axial force of pipelines decreases from −1850 KN to −1750 KN. However, with the increase in sleeper height, the corresponding value of axial force decreases gradually when lateral displacement occurs. The change in sleeper height mainly affects the pre-buckling stage but has a low effect on the critical value of buckling.

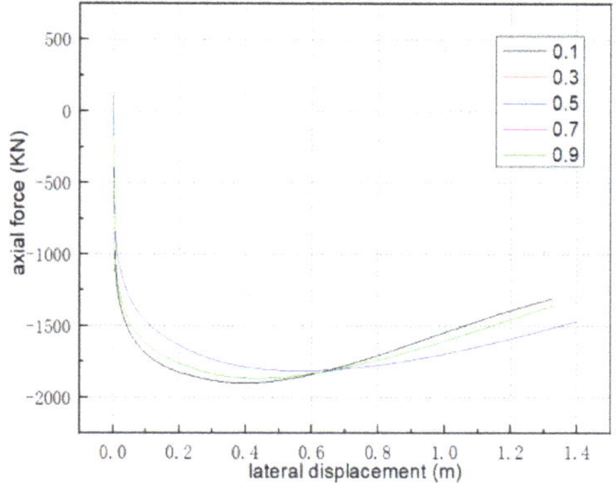

Figure 9. Influence of sleeper height in the axial force–lateral displacement graph.

Table 7. Critical axial force of sleeper height.

Sleeper Height (m)	Critical Axial Force (KN)
0.1	−1850
0.3	−1830
0.5	−1800
0.7	−1770
0.9	−1750

3.6. Analysis of Buoyancy Unit Interval

The distribution interval of the buoyancy unit is also studied. The total buoyancy under water is set to a fixed value. The buckling of pipelines with buoyancy units under different densities is analyzed by changing the distribution interval of the buoyancy unit. Figure 10 and Table 8 present that the change curve of the axial force–displacement of pipelines is inconsistent with other parameters only when the interval is 10 m. The axial force of other parameters is basically the same when the interval is 30–90 m, and the fluctuation is only approximately 1.5%. The lateral displacement of each pipeline is similar under each parameter. Under a certain total buoyancy force, the influence of changing the distribution interval of buoyancy units on the global buckling of pipelines is minimal.

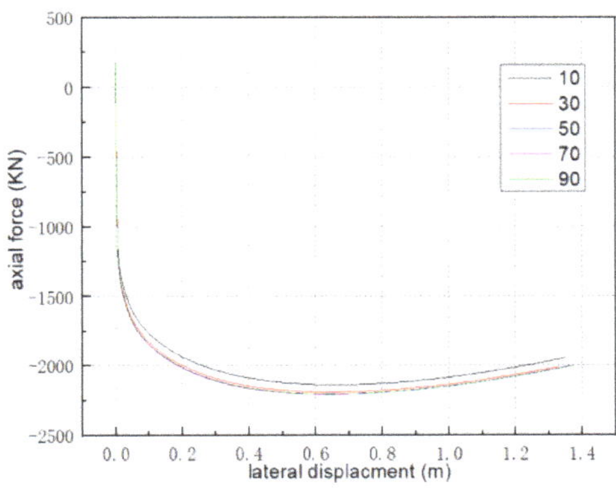

Figure 10. Influence of buoyancy unit interval in the axial force–lateral displacement graph.

Table 8. Critical axial force of buoyancy unit interval.

Buoyancy Unit Interval (m)	Critical Axial Force (KN)
10	−2100
30	−2250
50	−2250
70	−2250
90	−2250

4. Formula for Calculating the Critical Buckling Axial Force

4.1. Critical Buckling Force

The main factors affecting the global buckling of a PIP system under the distributed buoyancy method include bending stiffness, buoyancy specific gravity, initial imperfection, pipe–sleeper friction coefficient, and stiffness ratio. Therefore, for an initial defect of a particular shape, the critical buckling axial force of the pipeline can be expressed as follows:

On the basis of symmetry, the lateral buckling mode of the PIP system under the distributed buoyancy method is selected and shown in Figure 11. The X-0-Y coordinate is the projection surface of the pipeline on the seabed, in which the X direction is the axial direction of the pipeline, and the Y direction is the lateral direction of the pipeline. After buckling, the pipeline can be divided into three sections according to its deformation morphology and structural characteristics. The $0 \leq x \leq L_B$ section is the buoyancy unit installation section, where the pipeline buoyancy is defined as γW_s and the ratio of buoyancy module section length to buckling section pipeline length is β, that is, $\beta = L_B/L_1$. The pipeline displacement and strain of the buoyancy unit section are marked as "B" below. The main buckling section of the pipeline is section $0 \leq x \leq L_1$, in which $L_B < x \leq L_1$ does not contain a buoyancy unit. The pipeline $0 \leq x \leq L_1$ is separated from the seabed surface due to the sleeper, thereby resulting in a suspension span. The displacement and strain of the pipeline are expressed by the following standard "1." Section $L_B \leq x \leq L_2$ is the secondary buckling section, and the displacement and strain of the pipeline are expressed as "2" below. $L_2 \leq x \leq L_0$ is the axial slip section, and the displacement and strain of the pipeline are marked as "3." The virtual anchorage point is at $x = L_0$, and the floating weight of pipeline in $L_B \leq x \leq L_0$ section is W_S.

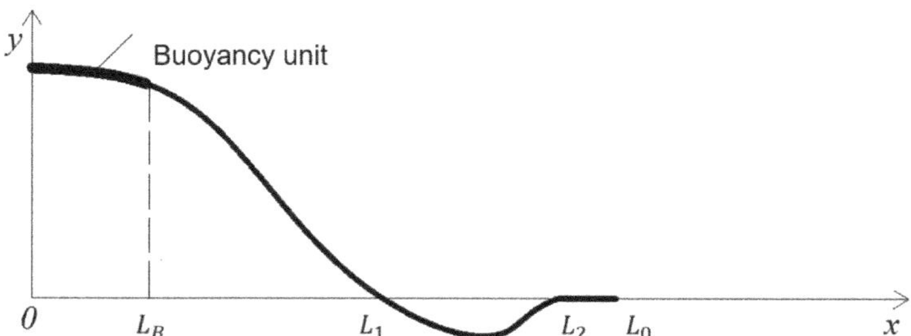

Figure 11. Force diagram of SDB.

According to the principle of virtual displacement, equilibrium equations can be established in different sections of the pipeline.

$$EIv_B'''' + Pv_B'' = -\gamma W_s \mu_s \tag{2}$$

$$EIv_1'''' + Pv_1'' = -(\gamma W_s + W_s)\mu_s \tag{3}$$

$$EIv_2'''' + Pv_2'' = -W_s \mu_L \tag{4}$$

$$P_0 - EA_t \varepsilon_B = P \tag{5}$$

$$P_0 - EA_t\varepsilon_1 = P \tag{6}$$

$$P_0 - EA_t\varepsilon_2 = P \tag{7}$$

$$EA_t u_0'' = -\mu_A W_S \tag{8}$$

$$\varepsilon_i = u_i' + \frac{1}{2}v_i'^2 \quad i = B, 1, 2 \tag{9}$$

where v_i is the lateral displacement of the pipeline in the corresponding section, u_i is the axial displacement of the pipeline in the corresponding section, μ_A is the axial friction coefficient of seabed and pipeline, μ_L is the lateral friction coefficient of seabed and pipeline, and μ_s is the friction coefficient of pipeline and sleeper. On the basis of the hypothesis of small deformation, the second derivative contribution of axial displacement is neglected. The satisfiable formula (9), $A_t = A_{in} + A_{out}$ is fulfilled.

According to the mathematical forms of Equations (2)–(4), the general solution can be obtained as a lateral displacement function that satisfies the following forms:

$$v_B(x) = A_1 \cos(\lambda x) + A_2 \sin(\lambda x) + A_3 x + A_4 - \frac{\gamma\omega_1}{2\lambda^2}x^2 \tag{10}$$

$$v_1(x) = A_5 \cos(\lambda x) + A_6 \sin(\lambda x) + A_7 x + A_8 - \frac{\gamma\omega_2}{2\lambda^2}x^2 \tag{11}$$

$$v_2(x) = A_9 \cos(\lambda x) + A_{10} \sin(\lambda x) + A_{11} x + A_{12} - \frac{\gamma\omega_3}{2\lambda^2}x^2 \tag{12}$$

where

$$\lambda^2 = \frac{P}{EI} \tag{13}$$

$$\omega_3 = \frac{\mu_L W_s}{EI}, \ \omega_2 = \frac{(\gamma W_s + W_s)\mu_s}{EI}, \ \omega_1 = \frac{\gamma W_s \mu_s}{EI}. \tag{14}$$

Equation (8) is introduced into Equations (4)–(6) to obtain the axial and lateral displacements of buckling pipelines.

$$u_B'(x) = \frac{(P_0 - P)}{EA_t} - \frac{1}{2}v_B'^2(x) \tag{15}$$

$$u_1'(x) = \frac{(P_0 - P)}{EA_t} - \frac{1}{2}v_1'^2(x) \tag{16}$$

$$u_2'(x) = \frac{(P_0 - P)}{EA_t} - \frac{1}{2}v_2'^2(x) \tag{17}$$

The axial displacement function of the pipeline in the slip section can be obtained by integrating Equation (7).

$$u_3(x) = -\frac{u_A W_s}{EA_t}\frac{x^2}{2} + B_1 x + B_2 \tag{18}$$

In the above formula, B_1, B_2, and $A_i (i = 1\text{–}12)$ are undetermined displacement coefficients. The axial and lateral displacements of the pipeline should satisfy the following boundary conditions and connection conditions at the boundary of the pipeline and at the junction of each section:

Lateral displacement v_B should be satisfied at $x = 0$ as follows:

$$v_B' = 0, \tag{19}$$

$$v_B''' = 0. \tag{20}$$

Lateral displacements v_B and v_1 satisfy the continuous boundary conditions at $x = L_B$ as follows:

$$v_B(L_B) = v_1(L_B), \tag{21}$$

$$v'_B(L_B) = v'_1(L_B), \tag{22}$$

$$v''_B(L_B) = v''_1(L_B), \tag{23}$$

$$v'''_B(L_B) = v'''_1(L_B). \tag{24}$$

Lateral displacements v_1 and v_2 satisfy the following continuous boundary conditions at $x = L_1$:

$$v_1(L_1) = 0, \tag{25}$$

$$v_2(L_1) = 0, \tag{26}$$

$$v'_1(L_1) = v'_2(L_1), \tag{27}$$

$$v''_1(L_1) = v''_2(L_1), \tag{28}$$

$$v'''_1(L_1) = v'''_2(L_1). \tag{29}$$

Lateral displacement v_1 satisfies the following continuous boundary conditions at $x = L_2$:

$$v_2(L_2) = 0, \tag{30}$$

$$v'_2(L_2) = 0, \tag{31}$$

$$v''_2(L_2) = 0. \tag{32}$$

Axial displacement u_B should satisfy at $x = 0$ as follows:

$$u_B(0) = 0 \tag{33}$$

Axial displacements u_B and u_1 satisfy continuous boundary conditions at $x = L_B$ as follows:

$$u_B(L_B) = u_1(L_B), \tag{34}$$

$$u'_B(L_B) = u'_1(L_B). \tag{35}$$

Axial displacements u_1 and u_2 satisfy the following continuous boundary conditions at $x = L_1$:

$$u_1(L_1) = u_2(L_1), \tag{36}$$

$$u'_1(L_1) = u'_2(L_1). \tag{37}$$

Axial displacements u_2 and u_3 satisfy the following continuous boundary conditions at $x = L_2$:

$$u_2(L_2) = u_3(L_2), \tag{38}$$

$$u'_2(L_2) = v'_3(L_2). \tag{39}$$

Axial displacement u_3 satisfies the following continuous boundary conditions at $x = L_0$:

$$v_3(L_0) = 0, \tag{40}$$

$$v'_3(L_0) = 0. \tag{41}$$

The lateral displacement functions (10)–(12) are introduced into the lateral displacement boundary conditions (18)–(24) and (25)–(30) to obtain the undetermined coefficients of the lateral displacement function.

The lateral displacement function $v_B(x)$ coefficient of buoyancy unit in lateral displacement buckling is

$$A_1 = -\frac{\omega_1\{[2\cos(\lambda L_1)+(\gamma-1)C\cos(\beta\lambda L_2)]\sin(\lambda L_2)\}}{\lambda^4\sin(\lambda L_2)} + \frac{\omega_1\{[2\cos(\lambda L_1)+(\gamma-1)\sin(\beta\lambda L_2)]\cos(\lambda L_2)-[(\gamma-1)\beta+2]\lambda L_1+\lambda L_2\}}{\lambda^4\sin(\lambda L_2)}, \tag{42}$$

$$A_2 = A_3 = 0, \tag{43}$$

$$
\begin{aligned}
A_4 = &- \frac{\omega_1\left\{[(\gamma-1)\beta(\beta-2)-1](\lambda L_1)^2 \sin(\lambda L_2)\right\}}{\lambda^4 \sin(\lambda L_2)} + \\
&\frac{\omega_1\left\{[4\sin(\lambda L_2-\lambda L_1)\cos(\lambda L_1)-2(\gamma-1)\sin(\beta\lambda L_1)]\sin(\lambda L_1)\sin(\lambda L_2)\right\}}{2\lambda^4 \sin(\lambda L_2)} + \\
&\frac{\omega_1\left\{2(\gamma-1)\cos(\beta\lambda L_1)\cos(\lambda L_2)\sin(\beta\lambda L_1)2(\gamma-1)\beta\lambda L_1\cos(\lambda L_1)\right\}}{2\lambda^4 \sin(\lambda L_2)} + \frac{\omega_1\left\{2(2\lambda L_1-L_2)\cos(\lambda L_1)+2(\gamma-1)\sin(\lambda L_2)\right\}}{2\lambda^4 \sin(\lambda L_2)}.
\end{aligned}
\tag{44}
$$

The displacement function $v_1(x)$ coefficient of the main buckling section without buoyancy is

$$
\begin{aligned}
A_5 = &\frac{\omega_2\left\{[2\sin(\lambda L_1)+(\gamma-1)\sin(\beta\lambda L_2)]\cos(\lambda L_2)\right\}}{\lambda^4 \sin(\lambda L_2)} \\
&- \frac{\omega_2\left\{[2\cos(\lambda L_1)\sin(\lambda L_2)]+(\gamma-1)\beta\lambda L_1+2\lambda L_1-\lambda L_2\right\}}{\lambda^4 \sin(\lambda L_2)},
\end{aligned}
\tag{45}
$$

$$A_6 = \frac{\omega_2(\gamma-1)\sin(\beta\lambda L_1)}{\lambda^4}. \tag{46}$$

$$A_7 = -\frac{\omega_2\beta L_1(\gamma-1))}{\lambda^4}. \tag{47}$$

$$
\begin{aligned}
A_8 = &- \frac{\omega_2\left\{[-2(\gamma-1)\beta-1](\lambda L_1)^2 \sin(\lambda L_2)+2(\gamma-1)\sin(\beta\lambda L_1)\cos(\lambda L_1-\lambda L_2)\right\}}{2\lambda^4 \sin(\lambda L_2)} - \\
&\frac{\omega_2\left\{4\sin(\lambda L_2-\lambda L_1)\cos(\lambda L_1)-2[(\gamma-1)\beta\lambda L_1+2\lambda L_1-\lambda L_2](\gamma-1)\cos(\lambda L_1)\right\}}{2\lambda^4 \sin(\lambda L_2)}.
\end{aligned}
\tag{48}
$$

The lateral displacement function $v_2(x)$ coefficient of the secondary buckling section is

$$A_9 = \frac{\omega_3\left\{[2\sin(\lambda L_1) + (\gamma-1)\sin(\beta\lambda L_2)]\cos(\lambda L_2) - [(\gamma-1)\beta+2]\lambda L_1+\lambda L_2\right\}}{\lambda^4 \sin(\lambda L_2)}, \tag{49}$$

$$A_{10} = \frac{\omega_3[(\gamma-1)\sin(\beta\lambda L_1) + 2sin(\lambda L_1)]}{\lambda^4}, \tag{50}$$

$$A_{11} = -\frac{\omega_3(\gamma\beta-\beta+2)L_1)}{\lambda^4}, \tag{51}$$

$$
\begin{aligned}
A_{12} = &\frac{\omega_3\left\{[2(\gamma-1)\beta+2]\lambda L_1\lambda L_2\sin(\lambda L_2)-(\lambda L_2)^2\sin(\lambda L_2)\right\}}{2\lambda^4 \sin(\lambda L_2)} + \\
&\frac{\omega_3\left\{2[(\gamma\beta-\beta+2)\lambda L_1-\lambda L_2]\cos(\lambda L_2)-4\sin(\lambda L_1)-2(\gamma-1)\sin(\beta\lambda L_1)\right\}}{2\lambda^4 \sin(\lambda L_2)}.
\end{aligned}
\tag{52}
$$

Lateral displacement functions (10)–(12) are introduced into boundary conditions (24) and (38) to obtain the dimensionless parameters lambda λL_1 and lambda λL_2, which must satisfy the following characteristic equations:

$$
\begin{aligned}
&2[(\gamma-1)\cos(\lambda L_1 - \lambda L_2) - \gamma + 1]\sin(\beta\lambda L_1) + 44\sin(\lambda L_2 - \lambda L_1)\cos(\lambda L_1) + \\
&4[\sin(\lambda L_2) - \sin(\lambda L_1)] + 2\{[(\gamma-1)\beta + 2][\cos(\lambda L_2) - \cos(\lambda L_1)]\}(\lambda L_1) - \\
&2[\cos(\lambda L_2) - \cos(\lambda L_1)]\lambda L_2 - (2\gamma\beta - 2\beta + 3)(\lambda L_1)^2 \sin(\lambda L_2) + 2(\gamma\beta - \beta + 2) \\
&\lambda L_1\lambda L_2\sin(\lambda L_2) - (\lambda L_2)^2 \sin(\lambda L_2) = 0,
\end{aligned}
\tag{53}
$$

$$(\gamma-1)\sin(\beta\lambda L_1) - \sin(\lambda L_2) + 2\sin(\lambda L_1) - [(\gamma\beta - \beta + 2)\lambda L_1 - \lambda L_2]\cos(\lambda L_2) = 0. \tag{54}$$

Formula (15) is integrated and introduced into boundary condition (34) to obtain the axial displacement function of the buoyancy section as follows:

$$u_B(x) = \frac{(P_0 - p)}{EA_t}x - \frac{1}{2}\int_0^x v_B'(x)dx. \tag{55}$$

Formula (16) is integrated from L_B to X, and boundary condition (34) is introduced to obtain the axial displacement function of the main buckling section of the pipeline without buoyancy module as follows:

$$u_1(x) = \frac{(P_0 - P)}{EA_t}(x - L_B) - \frac{1}{2}\int_{L_B}^{x} v_B'(x)dx + u_B(L_B) \tag{56}$$

Formula (17) is integrated from L_1 to X and introduced into boundary condition (37) to obtain the axial displacement function of the secondary buckling section of the pipeline as follows:

$$u_2(x) = \frac{(P_0 - P)}{EA_t}(x - L_1) - \frac{1}{2}\int_{L_1}^{x} v_2'(x)dx + u_1(L_1). \tag{57}$$

The expressions of the axial displacement function can be obtained by introducing the slip section axial displacement function (18) into boundary conditions (40), (41) as follows:

$$u_3(x) = \left[\frac{\mu_A W_S}{2EA_t}(x + L_0 - 2L_2) - \frac{(P_0 - P)}{EA}\right](L_0 - x). \tag{58}$$

The relationship between the axial load of the buckling section of the pipeline and the axial load of the virtual anchorage point can be obtained by introducing the upper formula to the boundary condition (41) as follows:

$$P - P_0 = \mu_A W_S(L_0 - L_1). \tag{59}$$

The lateral buckling equation can then be obtained as

$$P_0 = P + \mu_A W_S L_2\left[-1 + \sqrt{1 + \frac{EA_t}{\mu_A W_S L_2^2}\left(\int_{L_B}^{x} v_B'^2(x)dx + \int_{L_1}^{x} v_2'^2(x)dx + \int_{0}^{x} v_B'^2(x)\right)}\right], \tag{60}$$

$$P = \left(\frac{\lambda L_i}{L}\right)^2 EI, \tag{61}$$

where $EI = E_1 I_1 + E_2 I_2$.

4.2. Formula Validation

The following parameters of the PIP system are selected to verify the accuracy of the formulas, as shown in Table 9.

Table 9. Parameters of the PIP system.

Parameter of the PIP System	Value
Inner pipe diameter (mm)	40
Outer pipe diameter (mm)	46
Inner pipe wall thickness (mm)	5
Outer pipe wall thickness (mm)	5
Modulus of elasticity (Gpa)	207
Poisson ratio	0.3
Pipe floating weight (KN/m)	3300
Design temperature (°C)	50
Coefficient of thermal expansion (1/°C)	1.17×10^{-5}
Axial friction coefficient	0.5
Lateral friction coefficient	0.5

A comparison of the FE model result and analytic solutions in Table 10 shows that the error of the analytic solutions is consistent with that of the FE model, indicating that the formulas have high accuracy in forecasting the global buckling of the PIP system.

Table 10. Comparison of the results.

	Buckling Length	Axial Force in Buckling Segment	Maximum Lateral Displacement
FE model	110.5	9.53×10^3	6.87
Analytic solutions	114.8	9.68×10^3	6.92
Error%	3.8	1.5	0.7

5. Conclusions

In this study, a FE model for the active control buckling of PIP systems under the SDB method is established, and the effects of different parameters on the global buckling are analyzed. The accuracy of the model is verified by existing projects. On the basis of the analysis results for the FE model, the formulas for calculating the critical buckling axial force and lateral displacement of this type of pipeline are obtained by analytical solution. The main conclusions are as follows:

1. The proposed SDB method to control the global buckling of PIP systems can solve the instability caused by the high center of gravity of the pipeline under the sleeper control method and the extremely large outer diameter of pipelines wrapped by buoyancy unit under the buoyancy method.
2. Geometric model and FE models of PIP systems under combined control by the SDB method are proposed, and the accuracy of the models is verified. Parametric analysis shows that the bending stiffness, buoyancy ratio, initial defects, and stiffness ratio of inner and outer pipes are the main factors affecting the critical buckling force of pipelines.
3. The critical buckling axial force and lateral displacement formulas of pipelines are deduced and validated by analytical solution. A comparison shows that the formulas have high accuracy and can meet the requirements for engineering design.

Author Contributions: Conceptualization, Z.C.; methodology, H.L.; software, Z.Z. and H.L.; validation, Z.Z.; formal analysis, Z.Z. and H.L.; investigation, Z.Z. and H.L.; resources, Z.C.; data curation, Z.Z.; writing—original draft preparation, Z.Z.; writing—review and editing, Z.Z. and H.L.; supervision, Z.C. All authors have read and agreed to the published version of the manuscript.

Funding: This research received no external funding.

Institutional Review Board Statement: The study did not require ethical approval.

Informed Consent Statement: Not applicable.

Data Availability Statement: Not applicable.

Acknowledgments: The authors are grateful for the support provided by the National Basic Research Program of China under grant no. 2014CB046801. Scientific Research grant project funded by China Three Gorges Corporation (WWKY-2020-0741).

Conflicts of Interest: The authors declare no conflict of interest. The funders had no role in the design of the study, in the collection, analyses, or interpretation of data, in the writing of the manuscript, or in the decision to publish the results.

References

1. Wang, Z.; Chen, Z.; Liu, H. On lateral buckling of subsea pipe-in-pipe systems. *Int. J. Steel Struct.* **2015**, *15*, 881–892. [CrossRef]
2. Liu, R.; Li, C. Determinate dimension of numerical simulation model in submarine pipeline global buckling analysis. *Ocean Eng.* **2018**, *152*, 26–35. [CrossRef]
3. Zeng, X.G.; Duan, M.L.; Che, X.Y. Analysis on upheaval buckling of buried subsea PIP pipeline. *The Ocean Eng.* **2014**, *32*, 72–77.
4. Zeng, X.; Duan, M.; Che, X. Critical upheaval buckling forces of imperfect pipelines. *Appl. Ocean. Res.* **2014**, *45*, 33–39. [CrossRef]
5. Wang, Z.; Chen, Z.; Liu, H. Numerical study on upheaval buckling of pipe-in-pipe systems with full contact imperfections. *Eng. Struct.* **2015**, *99*, 264–271. [CrossRef]
6. Liu, R.; Hao, X.; Wu, X.; Yan, S. Numerical studies on global buckling of subsea pipelines. *Ocean Eng.* **2014**, *78*, 62–72. [CrossRef]

7. Karampour, H.; Albermani, F.; Gross, J. On lateral and upheaval buckling of subsea pipelines. *Eng. Struct.* **2013**, *52*, 317–330. [CrossRef]

8. Zhang, Z.; Liu, H.; Chen, Z. Lateral Buckling Theory and Experimental Study on Pipe-in-Pipe Structure. *Metals* **2019**, *9*, 185. [CrossRef]

9. Wang, Z.; Tang, Y.; Feng, H.; Zhao, Z.; Liu, H. Model Test for Lateral Soil Resistance of Partially Embedded Subsea Pipelines on Sand during Large-Amplitude Lateral Movement. *J. Coast. Res.* **2017**, *33*, 607–618.

10. Zechao, Z.; Zhihua, C.; Hongbo, L.; Zhe, W.; Kaiyue, L. Study of Dynamic Effect on Lateral Buckling of Pipe-in-Pipe System with Initial Imperfections. *J. Tianjin Univ.* **2019**, *52*, 404–412.

11. Reddy, N.R. Lateral Buckling Behaviour of Snake-Lay Pipeline with Vertical Support at Crown. In *International Conference on Offshore Mechanics and Arctic Engineering*; American Society of Mechanical Engineers: New York, NY, USA, 2013; Volume 55362, p. V04AT04A010.

12. Proc, K.; Chaney, L. Thermal Load Reduction of Truck Tractor Sleeper Cabins. *Sae Int. J. Commer. Veh.* **2008**, *1*, 268–274. [CrossRef]

13. Sun, J.; Paul, J.; Han, S. Thermal Expansion/Global Buckling Mitigation of HPHT Deep-Water Pipelines, Sleeper or Buoyancy. In Proceedings of the 22nd International Offshore and Polar Engineering Conference, Rhodes, Greece, 17–22 June 2012.

14. Rundsag, J.O.; Tørnes, K.; Cumming, G.; Rathbone, A.D.; Roberts, C. Optimised Snaked Lay. In Proceedings of the Eighteenth International Offshore and Polar Engineering Conference, Vancouver, BC, Canada, 6–11 July 2008.

15. Che, X.Y.; Duan, M.L.; Zeng, X.G.; Gao, P.; Pang, Y.Q. Experimental Study and Numerical Simulation of Global Buckling of Pipe-in-Pipe Systems. *Appl. Math. Mech.* **2014**, *35*, 188–201.

16. Alrsai, M.; Karampour, H.; Albermani, F. Numerical study and parametric analysis of the propagation buckling behaviour of subsea pipe-in-pipe systems. *Thin Walled Struct.* **2018**, *125*, 119–128. [CrossRef]

17. Westgate, Z.J.; Randolph, M.F.; White, D.J.; Brunning, P. Theoretical, numerical and field studies of offshore pipeline sleeper crossings. In Proceedings of the 2nd International Symposium on Frontiers in Offshore Geotechnics, Perth, Australia, 7–12 November 2011; pp. 845–850.

18. Peek, R.; Yun, H. Flotation to trigger lateral buckles in pipelines on a flat seabed. *J. Eng. Mech.* **2007**, *133*, 442–451. [CrossRef]

19. Gang, L. Multi-Objective Optimization Design for Lateral Buckling Control of Subsea Pipelines by Distributed Buoyancy Sections. *Appl. Math. Mech.* **2016**, *9*, 945–955.

20. Denniel, S.; Ross, R.A.L. Method and Apparatus for Mounting Distributed Buoyancy Modules on a Rigid Pipeline. U.S. Patent 8,573,888, 5 November 2013.

21. Cumming, G.; Rathbone, A. Euler Buckling of Idealised Horizontal Pipeline Imperfections. In *International Conference on Offshore Mechanics and Arctic Engineering*; American Society of Mechanical Engineers: New York, NY, USA, 2010; Volume 49132, pp. 293–300.

Article

Probabilistic Seismic Assessment of CoSPSW Structures Using Fragility Functions

Zhilun Tan [1], Qiuhong Zhao [1,2,*], Yu Zhao [3] and Cheng Yu [4]

[1]   School of Civil Engineering, Tianjin University, Tianjin 300350, China; zltan@tju.edu.cn
[2]   Key Laboratory of Coast Civil Structure and Safety of Ministry of Education, Tianjin University, Tianjin 300350, China
[3]   China Railway Design Corporation Co., Ltd., Tianjin 300308, China; yuzhao@tju.edu.cn
[4]   Department of Mechanical Engineering, University of North Texas, Denton, TX 76207, USA; cheng.yu@unt.edu
*   Correspondence: qzhao@tju.edu.cn

Abstract: The corrugated steel plate shear wall (CoSPSW) is a new type of steel plate shear wall, in which corrugated wall plates instead of flat wall plates are adopted. The lateral stiffness and shear buckling capacity of the shear wall system could be significantly enhanced, and then, wall plate buckling under gravity loads would be mitigated. This paper presents a study on the probabilistic assessment of the seismic performance and vulnerability of CoSPSWs using fragility functions. The damage states and corresponding repair states of CoSPSWs were first established from experimental data. Then, incremental dynamic analyses were conducted on the CoSPSW structures. The structural and nonstructural fragility functions were developed, based on which the seismic performance and vulnerability of the CoSPSWs were obtained and compared with the conventional steel plate shear walls (SPSWs). It was shown that for various repair states, the 25th percentile PGA of the CoSPSW was always higher than the SPSWs with the same wall thickness and boundary frame, which indicated that the CoSPSW has a lower damage potential and better seismic performance than the SPSW.

Keywords: corrugated steel plate shear wall; damage state; repair state; fragility function; probabilistic assessment

Citation: Tan, Z.; Zhao, Q.; Zhao, Y.; Yu, C. Probabilistic Seismic Assessment of CoSPSW Structures Using Fragility Functions. *Metals* **2022**, *12*, 1045. https://doi.org/10.3390/met12061045

Academic Editors: Carlos Capdevila-Montes and Giovanni Meneghetti

Received: 1 May 2022
Accepted: 17 June 2022
Published: 18 June 2022

Publisher's Note: MDPI stays neutral with regard to jurisdictional claims in published maps and institutional affiliations.

1. Introduction

Steel plate shear walls (SPSWs) were proposed in the 1970s as new lateral load-resisting systems for mid- to high-rise buildings, with advantages such as high lateral strength, ductility, and energy dissipation capacity [1–3]. Originally, the SPSW was composed of flat infill wall plates, boundary beams, and boundary columns, while the wall plates were generally slender and tended to buckle under low lateral loads or gravity loads in a serviceability limit state, which was related to wall out-of-plane deformation and an unpleasant buckling noise. The ultimate lateral strength of the wall plate would then be achieved through the yielding of the diagonal tension field formed due to wall buckling under lateral loads [4], which also resulted in high anchoring forces on the boundary frame [5]. On the other hand, wall buckling under gravity loads or during construction is not permitted by the Chinese code [6], which prohibits a synchronized installation of the wall plate with the rest of the story, and this hinders the construction schedule, especially for high-rise buildings.

Corrugated steel plate shear walls (CoSPSWs) could be a viable solution in that sense; they consist of horizontally or vertically corrugated wall plates and a boundary frame [7], as shown in Figure 1. The corrugation significantly increases the out-of-plane stiffness and buckling strength of the wall plate, which would then resist the lateral loads through in-plane shear instead of tension field action. In addition, due to the "accordion effects", axial stiffness will be at its minimum when perpendicular to the rib and greatly enhanced when parallel to the rib. In addition, horizontal or vertical corrugation could either minimize or resist the vertical stresses transferred to the wall plates from gravity loads and thus avoid buckling during construction. Therefore, corrugated wall plates could be erected simultaneously with the rest of the story and considerably increase construction speed, especially for high-rises.

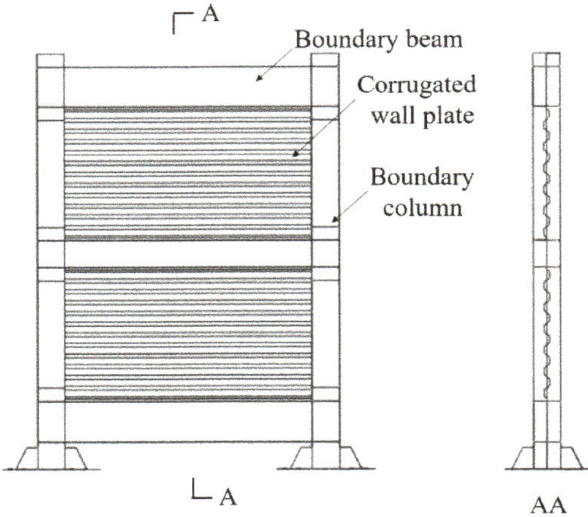

Figure 1. Corrugated steel plate shear wall.

Experimental [8–17] and numerical analysis [18–23] have been conducted on the cyclic and lateral behavior of CoSPSWs and have revealed that the hysteresis curve pinching of flat wall plates could be obviously improved and energy dissipation could be increased accordingly [20,21], and the corrugated wall plates could even achieve shear yielding sometimes, which significantly lowered the anchoring forces on the boundary frame [15]. Furthermore, extended research has been conducted on corrugated steel shear walls with perforations or openings [24–27], a reduced beam section [28,29], a semi-rigid boundary frame [30], and a low-yield-point steel wall panel [31], as well as on the application of corrugated steel shear walls in a modular structural design [32]. A performance-based seismic design (PBSD) procedure has also been proposed [33], and time-history analyses have revealed that the inter-story drift of the CoSPSW with PBSD distributed more smoothly than the CoSPSW with a traditional design, which helped to avoid weak stories.

In order to acquire a rational estimation of potential seismic losses, the seismic performance of structures could be quantitatively evaluated through a probabilistic assessment method, which is realized by developing analytical fragility functions. Fragility functions are statistical models that could predict the possibility of structures reaching or exceeding different damage states under different earthquake intensities; therefore, they could characterize the seismic performance of structures quantitatively and describe the relationship between the earthquake intensity and the damage state of structures.

Researchers have conducted probabilistic seismic assessment on various wall structures and evaluated their seismic performance by developing fragility functions for different damage states or repair states. Baldvins et al. [34] established twelve damage states and five repair states of SPSWs based on experimental results and observations and developed the fragility functions for each repair state. Negar et al. [35] established five damage states and the corresponding repair states of steel-plate concrete composite shear walls based on experimental results and observations and derived the fragility functions for the damage states associated with concrete crushing and faceplate fracture. Wang et al. [36] conducted incremental dynamic analyses on coupled low-yield-point steel plate shear walls, based on which six damage states were established and fragility functions were developed for different damage states.

In addition, probabilistic seismic assessment was conducted on structural frames retrofitted with different types of steel walls or panels using fragility functions for different damage states or repair states. Zhang et al. [37] conducted time-history analyses on retrofitted SPSW systems in order to develop the fragility functions and found that the SPSW with low-yield-point steel wall plates had lower seismic vulnerability than the SPSW with ordinary steel wall plates. Jiang et al. [38,39] conducted incremental dynamic analyses on steel frames retrofitted with steel panels and developed the fragility functions and validated the effectiveness of adding steel panels to reduce the seismic vulnerability of existing steel frame buildings. Bu et al. [40] conducted incremental dynamic analyses on steel frames equipped with four different types of steel slit shear walls and developed the fragility functions, and the type with best performance was identified for all damage states. Currently, research on the probabilistic seismic assessment of CoSPSWs is still very limited.

In this paper, the seismic performance of CoSPSWs will be quantitatively evaluated through the probabilistic assessment method using fragility functions, which has not been reported in current literature. The damage states and corresponding repair states of CoSPSWs will first be established through experimental results and test observations. Then, incremental dynamic analyses will be conducted on a 10-story CoSPSW using PBSD, based on which analytical fragility functions will be developed for the different damage and repair states of structural and nonstructural components. Finally, the seismic performance and vulnerability of the CoSPSW and the SPSW are evaluated and compared probabilistically using the fragility functions.

2. Damage States and Repair States

2.1. Establishment of Damage States and Repair States

The seismic performance and the vulnerability of the structures can be assessed through the probabilistic assessment, which estimates the probabilistic structural response, i.e., the engineering demand parameter (EDP) as a function of ground motion intensity, i.e., the intensity measure (IM), and uses quantitative measures to evaluate the structural and nonstructural performance under seismic loads [37]. Therefore, first it is necessary to determine the engineering demand parameter, which could be peak inter-story drift, peak floor acceleration or ductility, etc. As the peak inter-story drift and the peak floor acceleration are used for seismic loss estimation in HAZUS-MH MR5 [41], these two structural response factors were chosen as the engineering demand parameter (EDP) in this study. The geometry and peak inter-story drift of 19 corrugated steel plate shear wall specimens from nine cyclic tests are summarized and shown in Table 1 below.

From the test observations on the 19 specimens listed in Table 1 and the relevant damage states of SPSW [34], 12 damage states (DS1–DS12) and 5 repair states (RS1–RS5) were established to describe the damage degree of the CoSPSW specimens and determine the corresponding repair methods, as shown in Tables 2 and 3. For damage states DS1,2,4,5, and 6, structural repair is not necessary as there is no permanent residual deformation. For damage states DS 3 and 9, the corrugated web plate suffers relatively severe damage and cannot recover the initial state. Thus, the infilled corrugated web plate needs to be replaced. For DS7,8,10, and 11, FEMA 352 (FEMA 2000) [42] recommends that the frame members should be repaired when the buckling degree exceeds tolerances and offers details for various repair methods depending on the extent of the damage. For DS12, the structure suffers extreme damage and the VBEs and HBEs, or the entire CoSPSW, need to be replaced.

Table 1. Test data of 19 corrugated steel plate shear wall specimens.

Specimen No.		Thickness (mm)	Aspect Ratio (L/H)	Height-to-Thickness Ratio (H/t)	Peak Inter-Story Drift (%)
Emami [9]	S-2	1.25	1.33	1184	6.10
	S-3	1.25	1.33	1184	6.10
Hos [11]	C-30	1.25	1.20	1261	6.35
	C-45	1.25	1.20	1261	6.35
	C-60	1.25	1.20	1261	6.35
Ding [12]	CSPSW-1	1.60	1.22	1687	4.00
	CSPSW-2	2.00	1.22	1687	2.20
	CSPSW-3	1.60	0.83	1687	4.00
	CSPSW-4	1.66	1.22	1687	2.10
Sudeok Shon [13]	FR-TR-V	3.20	1.69	555	8.00
	FR-TR-H	3.20	1.69	555	8.00
Cao [14]	S-1	2.93	0.96	425	2.53
	S-2	2.47	0.71	628	1.48
Zhao [15]	S-2	2.00	1.00	550	5.00
	S-3	2.00	1.00	550	4.00
	S-4	2.00	1.00	550	4.50
Wang [16]	SPSW-2	3.00	0.70	660	1.90
	SPSW-3	3.00	0.70	660	2.30
Jin [17]	CoSPSW	5.00	1.00	180	2.50

Table 2. Damage states.

Damage State	Description	Damage State	Description
1	Elastic Web Plate Buckling for Slender Corrugated Web Plate	7	HBE Local Buckling Requiring Repair
2	Initial Corrugated Web Plate Yielding	8	VBE Local Buckling Requiring Repair
3	Significant Plastic Deformation of Corrugated Web Plate	9	Corrugated Web Plate Cracking
4	Initial HBE and/or VBE Yielding	10	HBE and HBE-to-VBE Connection Cracking
5	Initial VBE Local Buckling	11	VBE Cracking
6	Initial HBE Local Buckling	12	Connection and/or Boundary Frame Failure

Table 3. Repair states and corresponding damage states.

Repair State	Description	Corresponding Damage States
1	Cosmetic repair	1,2,4,5,6
2	Replace Web Plate	3,9
3	HBE and Connection Repair	7,10
4	VBE Repair	8,11
5	Replace Boundary Elements or Frame	12

The inter-story drifts are also recorded from the test observations on the 19 specimens listed in Table 4. The median value, coefficient of variation, and the number of data points of the inter-story drift for each damage state and repair state are listed in Tables 4 and 5, respectively. Figure 2 shows the distribution of the repair states and the corresponding inter-story drift. As shown in Table 4, the median drift increases with the damage state, which indicates that the ordering of the states is proper and consistent with the damage sequence. From Table 5 and Figure 2, it can be known that for RS1, RS2, and RS4 the results are more accurate because the number of data points is more abundant, with a smaller dispersion. For RS3, there are fewer data points and related descriptions; so, the determination of this repair state is more subjective. For RS5, as the ultimate performance of the CoSPSW is obviously affected by the wall corrugation configuration that varies with each group of test specimens, the dispersion of the data points is more obvious. Figure 2 also shows that the inter-story drift gradually increases when the repair state varies from RS1 to RS5, which further shows that the established repair state sequence is reasonable and consistent with the ideal failure sequence of the CoSPSW. It is worth noting that the height-to-thickness ratio of the specimens collected in this paper ranges from 425 to 1261, and the aspect ratio ranges from 1 to 1.33.

Table 4. The inter-story drift data of 19 tested specimens of each damage state.

DS	Median (%)	COV	Data Points
1	0.19	0.15	5
2	0.25	0.32	4
3	0.42	0.39	10
4	0.26	0.45	5
5	-	-	0
6	-	-	0
7	1.07	0.20	2
8	1.25	0.17	5
9	1.37	0.35	10
10	-	-	0
11	3.00	0.10	3
12	4.10	0.39	10

Table 5. The inter-story drift data of 19 tested specimens of each repair state.

RS	Median (%)	COV	Data Points
1	0.19	0.38	14
2	0.80	0.63	20
3	1.07	0.20	2
4	1.51	0.40	8
5	4.10	0.39	10

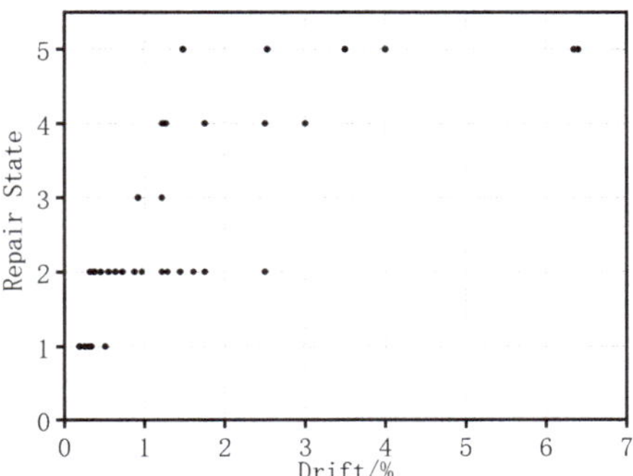

Figure 2. Distribution diagram of repair state data and inter-story drift.

2.2. Calibration of Recommended Values of Inter-Story Drift

The damage and repair probability are estimated based on the threshold values of the structural response parameters, e.g., inter-story drift. These threshold values of different repair states are generally obtained from test observations and adjusted through the maximum likelihood method. The values of inter-story drift determined by the maximum likelihood method would be more reasonable as they can avoid the random errors caused by using the median and standard deviation of the sample data directly.

Table 6 lists the median value and standard deviation value of inter-story drift, calculated through the maximum likelihood method using Matlab. It can be seen from Table 6 that the calculated value θ_x is close to the median value of the inter-story drift in Table 5. In addition, the calculated value β_x is further adjusted for the convenience of practical application, according to the recommendation of ATC-58 [43], as in Equation (1) below. Finally, the adjusted θ_x and β values are appropriately rounded up to θ_{rec} and β_{rec} for the convenience of the design and application, as shown in Table 6.

$$\beta = \sqrt{\beta_x^2 + \beta_u^2} \tag{1}$$

where β_x is the standard deviation value calculated by the maximum likelihood method, and β_u is taken as 0.25 in order to consider uncertainty in the statistics and the limitation of the number of experimental data.

Table 6. Repair state statistics and recommended values (%).

Repair State	θ_x	β_x	β	θ_{rec}	β_{rec}
RS1	0.23	0.31	0.39	0.20	0.40
RS2	0.80	0.63	0.67	0.80	0.50
RS3	1.06	0.14	0.29	1.00	0.40
RS4	1.76	0.38	0.46	1.80	0.40
RS5	4.31	0.47	0.54	4.30	0.50

Table 7 summarizes the recommended inter-story drift θ_{rec} of each repair state and the corresponding description of the CoSPSWs and SPSWs, in which the recommended inter-story drift and repair states for the SPSWs were established by Baldvins et al. [34], based on experimental observations. As seen in Table 7, no inter-story drift was recommended for RS4 because there were not enough experimental data. The recommended inter-story drift of "Repair boundary column" for the SPSW is lower than that of the CoSPSW because the anchoring forces from the yielding of the diagonal tension field in the flat wall plates in the SPSW would increase the damage degree of the boundary column.

Table 7. Repair states and recommended inter-story drift of CoSPSWs and SPSWs.

Repair States	Recommended Inter-Story Drift		Description	
	CoSPSW	SPSW	CoSPSW	SPSW
RS1	0.002	0.004	Repair infill wall surface	
RS2	0.008	0.006	Replace infill wall	
RS3	0.01	0.015	Repair boundary beam and beam-column connection	Repair boundary column
RS4	0.018	-	Repair boundary column	Repair boundary beam and beam-column connection
RS5	0.043	0.0275	Replace boundary beam, column, or frame	

Tables 8 and 9 show the damage degrees of drift-sensitive and acceleration-sensitive nonstructural members, according to the recommendation of HAZUS-MH MR5 [41], which are divided into four grades, namely "slight", "moderate", "severe", and "complete". The seismic performance and vulnerability assessment of the structure will be conducted in later sections using these repair states/damage states and the corresponding inter-story drift or peak floor acceleration.

Table 8. Damage states for nonstructural drift-sensitive components.

Damage States	Slight	Moderate	Extensive	Complete
Inter-story drift	0.004	0.008	0.025	0.050

Table 9. Damage states for nonstructural acceleration-sensitive components.

Damage States	Slight	Moderate	Extensive	Complete
Peak floor acceleration	0.3 g	0.6 g	1.2 g	2.4 g

3. Incremental Dynamic Analyses

3.1. Structural Models

In order to obtain the relationship between the EDP and the IM, incremental dynamic analyses were conducted on 10-story CoSPSW structures designed with PBSD [33], based on which the fragility function was developed, and the probabilistic vulnerability was assessed. For comparison, two models were studied: (i) a 10-story CoSPSW with a span of 4.5 m and a story height of 3.2 m and (ii) a 10-story SPSW using a flat steel plate with the same thickness and boundary frame as that of the 10-story CoSPSW. The floor plan is shown in Figure 3. The dead (live) load was 4.0 (2.0) kN/m^2 on the floors and 4.5 (2.0) kN/m^2 on the roof. The construction site is Class II (i.e., medium-stiff soil) within a region of Seismic Intensity 8 and Design Earthquake Group I [44]. The floor was 120 mm thick cast-in-place concrete, and the beams and columns were steel welded H-sections with rigid beam-column connections. Q235B grade steel (235 MPa design yield strength) and Q345B grade steel (345 MPa design yield strength) were used for the wall plates and frame members, respectively. The corrugation depth and wavelength of the corrugated wall plate were 100 mm and 400 mm, respectively, and the width of the subpanels was uniform. The design parameters and member sections are shown in Tables 10 and 11.

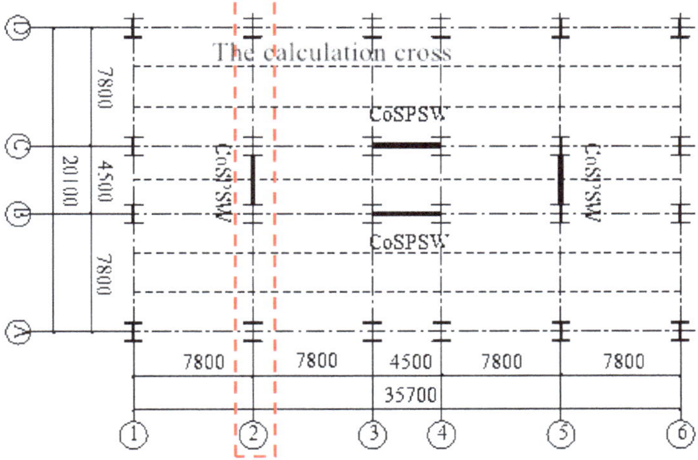

Figure 3. Floor Plan of the Structure.

Table 10. Design Parameters for the structural model.

Design Parameter	Value
Horizontal seismic influence factor	0.463
Spectral acceleration Sa/g	1.106
Fundamental period T/s	1.152
Yield drift θ_y	0.005
Target drift θ_u	0.025
Ductility coefficient μ_s	5
Ductility reduction coefficient R_μ	5
Energy correction factor γ	0.360
ζ	2.745
V_y/W	0.152

Table 11. Member Sections from of the structural model.

Floor	Columns in the Wall Plate Span *	Columns Outside the Wall Plate Span *	Wall Plate Thickness (mm)
10	H450 × 450 × 12 × 20	H350 × 350 × 16 × 18	1.8
9	H500 × 500 × 14 × 22	H400 × 400 × 16 × 18	3.0
8	H550 × 550 × 18 × 26	H400 × 400 × 16 × 20	4.0
7	H600 × 600 × 24 × 32	H450 × 450 × 16 × 22	4.9
6	H600 × 600 × 32 × 42	H450 × 450 × 20 × 26	5.6
5	H650 × 650 × 36 × 48	H500 × 500 × 20 × 26	6.3
4	H650 × 650 × 48 × 56	H500 × 500 × 24 × 28	6.9
3	H700 × 700 × 48 × 56	H550 × 550 × 24 × 28	7.3
2	H750 × 750 × 56 × 64	H550 × 550 × 28 × 32	7.6
1	H750 × 750 × 68 × 72	H600 × 600 × 28 × 36	7.8

* Note: H sections: overall depth (mm) × flange width (mm) × web thickness (mm) × flange thickness (mm).

3.2. Finite Element Models

Two 3-span structural models with wall plates in the interior span and moment frames in the exterior spans were modeled using finite element software ABAQUS (SIMULIA, 2014), as shown in Figure 4. Both the frame beams and the columns were modeled using beam element B31, while the corrugated wall plates were simplified into an equivalent braces model [45] using truss element T3D2, according to the principle of equivalent lateral stiffness and strength. The area A_{eb} and the yield strength f_{eb} of the equivalent brace model were calculated from Equation (2) and Equation (4), respectively [45]. The flat wall plates were simplified into a strip model, in which 10 equally spaced discrete tension-only strips oriented at 45° relative to the vertical were represented. The truss element T3D2 was used to model the strip, and the area of strip A_s was calculated from Equation (5). The out-of-plane displacements were restrained at all the beam-column joints, and the columns were fixed at the base. A bilinear elastoplastic constitutive model that considered the strain hardening was adopted for the steel material of the infill wall plate and the boundary members, with an elastic modulus E = 206 GPa, a strain hardening modulus E_h = 0.01E, a Poisson's ratio v = 0.3, and yield strengths of 235 MPa and 345 MPa, respectively.

$$A_{eb} = \frac{K_P l_{eb}}{2E \cos^2 \beta} \tag{2}$$

$$K_P = \frac{0.95 E q_{co} l_n t}{2 h s_{co}(1 + v)} \tag{3}$$

$$f_{eb} = \frac{\tau_{cr} l_n t}{2 A_{eb} \cos \beta} \tag{4}$$

$$A_s = \frac{t_w \sqrt{L_e^2 + H_e^2}}{n_2} \cos[45° - \tan^{-1}(H_e/L_e)] \tag{5}$$

where l_{eb} is the length of the equivalent brace; q_{co} and s_{co} are the wavelength and corrugation developed length, respectively; t_w is the thickness of the wall plate; L_e and H_e are the net width and net height of the wall plate; and n_2 is the number of tension-only strips, which was taken as 10 in this study.

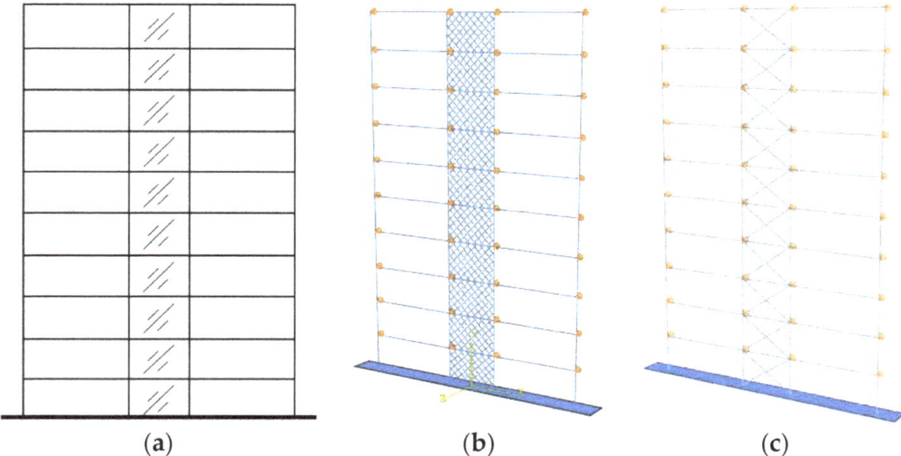

Figure 4. Finite Element Model: (**a**) the 3-span structure, (**b**) the SPSW model, and (**c**) the CoSPSW model.

The accuracy of the finite element model was validated by comparing the numerical analysis results with the experimental results. In order to accomplish the validation, the cyclic experimental specimen of a single-story CoSPSW specimen [9] and a three-story SPSW specimen [46] were considered. As shown in Figure 5, the hysteretic curve of the finite element analysis correlates well with the experiment results, which means that the equivalent braces model and the tension-only strip model are appropriate and precise in reflecting the responses of the CoSPSW and SPSW, respectively.

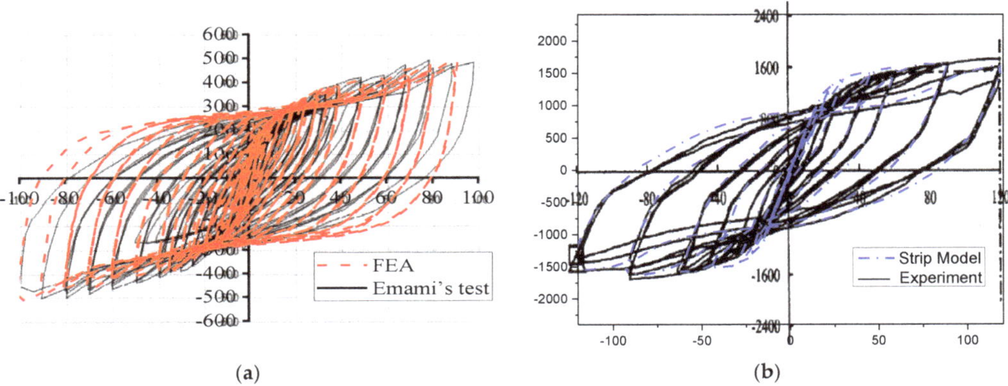

Figure 5. Validation of the finite element analysis: (**a**) validation of CoSPSW model, (**b**) validation of SPSW model.

3.3. Incremental Dynamic Analyses

Incremental dynamic analyses (IDA) were carried out on both the CoSPSW and the SPSW structures, in which additional masses were applied at the beam-column joints, and the earthquake excitation was input through the base. As shown in Table 12, ten earthquake excitation records with a magnitude of higher than 6.5, an effective period larger than 4 s, and different spectrum characteristics were selected, according to the Chinese seismic design specification [44]. The peak ground acceleration (PGA) of the seismic excitations was adjusted from 0.1 g to 1.2 g, with an interval of 0.1 g for the incremental dynamic analyses.

Table 12. Selected Ground Motions.

Record No.	Record	Minimum Frequency (Hz)	PGA (g)	PGV * (cm/s)
RSN0169	IMPVALL.H_H-DLT352	0.09	0.35	33
RSN0174	IMPVALL.H_H-E11230	0.10	0.38	45
RSN0752	LOMAP_CAP090	0.25	0.51	38
RSN0767	LOMAP_G03090	0.13	0.56	45
RSN0953	NORTHR_MUL279	0.15	0.49	67
RSN0960	NORTHR_LOS270	0.13	0.47	41
RSN1111	KOBE_NIS090	0.13	0.48	47
RSN1485	CHICHI_TCU045N	0.05	0.51	46
RSN1602	DUZCE_BOL090	0.06	0.81	66
RSN1787	HECTOR_HEC090	0.04	0.33	45

Note: * PGV: peak ground velocity.

In this study, the inter-story drift and the peak floor acceleration were chosen as the engineering demand parameter (EDP), and the peak ground acceleration (PGA) was chosen as the intensity measure (IM), with reference to previous studies on the probabilistic assessment of various structures, including SPSWs [37]. The probabilistic seismic assessment was then employed to relate the EDPs to the IMs and derive the fragility functions. The mean and standard deviation of the EDP for a given IM were estimated by regression analysis, and the relationship between the mean EDP and IM is as follows:

$$EDP = a(IM)^b \text{ or } \ln(EDP) = b\ln(IM) + \ln(a) \tag{6}$$

where the constants a and b are the regression coefficients obtained from the IDA analysis. It is assumed that the remaining variability in ln (EDP) for a given IM has a constant variance for all IMs, and the standard deviation can be obtained as follows [47]:

$$\zeta_{EDP|IM} = \sqrt{\frac{\sum\limits_{i=1}^{N}\left[\ln(EDP_\text{i}) - \ln\left(a(IM_\text{i})^b\right)\right]^2}{N-2}} \tag{7}$$

where N is the number of EDP-IM pairs, and EDP_i and IM_i are the values of the i-th pair.

From the incremental dynamic analyses, a large number of peak inter-story drift, peak floor acceleration, and corresponding PGA, i.e., EDP-IM, data pairs were obtained, and ln[IM = PGA] − ln[EDP = Drift] and ln[IM = PGA] − ln[EDP = PFA] are plotted in Figures 6 and 7, respectively. The values of parameters a and b can be obtained by performing regression analysis on the ln [IM = PGA] and ln [EDP = Drift] from the incremental dynamic analysis results. The coefficient R^2, which describes the quality of the fitting effect of the two parameters, was obtained through linear regression analyses, and R^2 was near 0.8 for the two EDP-IM pairs of the SPSW and CoSPSW structures, indicating a strong correlation. Therefore, the Drift-PGA and PFA-PGA data pairs were appropriate for developing the fragility functions. The lognormal distribution parameters a and b, the standard deviation $\zeta_{EDP|IM}$, and the functions were then determined and are shown in Table 13.

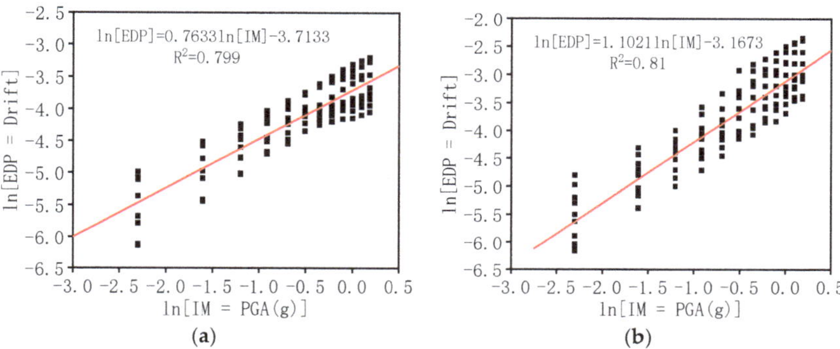

Figure 6. Relationship between ln[IM = PGA] and ln[EDP = Drift]: (**a**) CoSPSW, (**b**) SPSW.

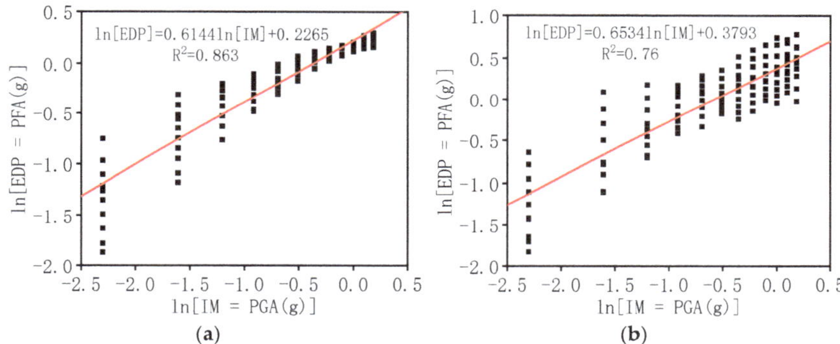

Figure 7. Relationship between ln[IM = PGA] and ln[EDP = PFA]: (**a**) CoSPSW, (**b**) SPSW.

Table 13. Regression coefficients, standard deviation, and function for EDP–IM data pairs.

Model	Selected		Regression Coefficients		$\zeta_{EDP|IM}$	Function
	EDP	IM	a	b		
CoSPSW	Drift	PGA	0.0244	0.7633	0.2792	Drift = 0.0244(PGA)$^{0.7633}$
	PFA	PGA	1.3054	0.6144	0.1882	PFA = 1.3054(PGA)$^{0.6144}$
SPSW	Drift	PGA	0.0421	1.1021	0.2919	Drift = 0.0421(PGA)$^{1.1021}$
	PFA	PGA	1.4613	0.6534	0.2810	PFA = 1.4613(PGA)$^{0.6534}$

4. Probabilistic Assessment

4.1. Fragility Functions

Assuming that the engineering demand parameter (EDP), i.e., the peak inter-story drift and the peak floor acceleration, has a lognormal distribution for a given intensity measure (IM), then the fragility function defining the probability of the EDP reaching or exceeding a certain limit state (LS) under a given IM is as follows

$$P(EDP \geq LS|IM) = 1 - \int_0^{LS} \frac{1}{\sqrt{2\pi} \cdot \zeta_{EDP|IM} \cdot EDP} \cdot e^{\left[-\frac{1}{2} \left(\frac{\ln(EDP) - \ln(a(IM)^b)}{\zeta_{EDP|IM}} \right)^2 \right]} \cdot d(EDP) \tag{8}$$

Accordingly, ln (*EDP*) can be considered to have a standard normal distribution, and Equation (8) could be simplified into the following equation, using the standard normal cumulative distribution function $\Phi\,(\cdot)$:

$$P(EDP \geq LS|IM) = 1 - \Phi\left(\frac{\ln(LS) - \ln\left(a(IM)^b\right)}{\zeta_{EDP|IM}}\right) \tag{9}$$

The damage and repair probability were determined based on the threshold values of the EDP, e.g., the inter-story drift and peak-floor acceleration, which initiate different damage states and repair states. These threshold values are LS in Equations (8) and (9) and are shown in Tables 7–9. According to the considered threshold values of the repair/damage states and the regression coefficients obtained through IDA analysis, the fragility curves for the structural and nonstructural components were obtained for the 10-story CoSPSW and SPSW, respectively, and are shown in Figures 8–10.

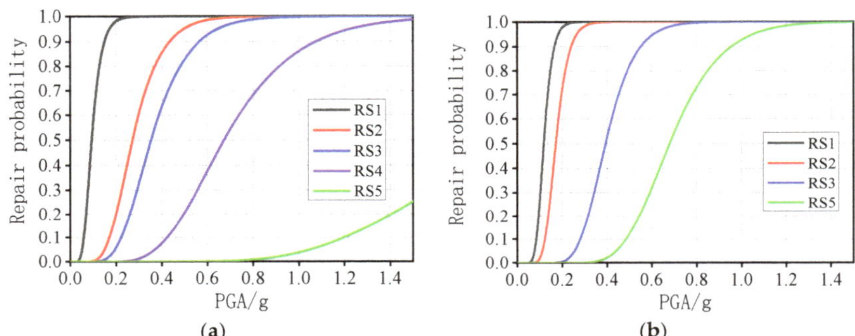

Figure 8. Structural Fragility Curves: (**a**) CoSPSW, (**b**) SPSW.

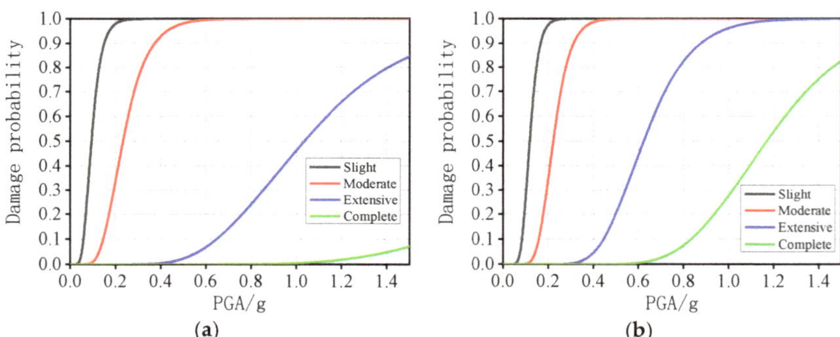

Figure 9. Drift-sensitive nonstructural components fragility curves: (**a**) CoSPSW, (**b**) SPSW.

4.2. Seismic Vulnerability

When the variation is similar, the seismic vulnerabilities of the different structures could be compared through the 25th percentile PGA values, i.e., the PGA values corresponding to exceeding the probability of 25% [48]. The larger the 25th percentile PGA values, the better the structural performance in earthquakes, as it indicates that the structure needs to undergo higher levels of seismicity to suffer the damage level. The seismic vulnerability of the CoSPSW and the SPSW structures were then assessed by comparing the 25th percentile PGA values of the structural and nonstructural fragility curves for the damage/repair states, as shown in Figures 11–13.

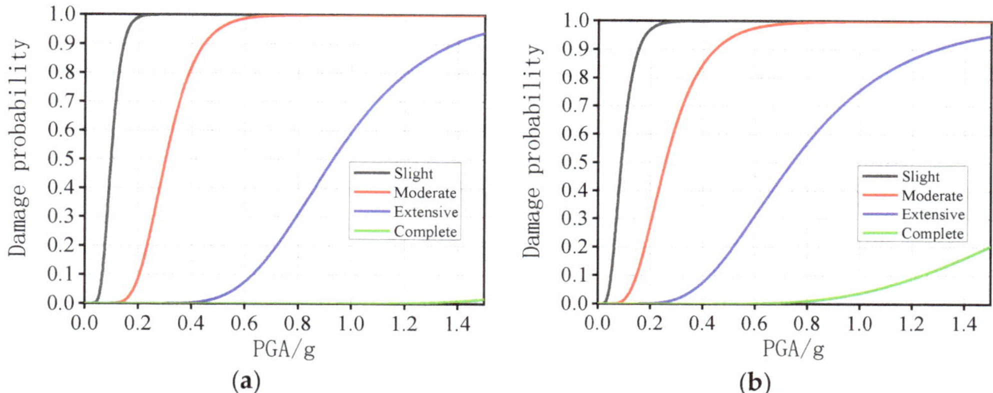

Figure 10. Acceleration-sensitive nonstructural components fragility curves: (**a**) CoSPSW, (**b**) SPSW.

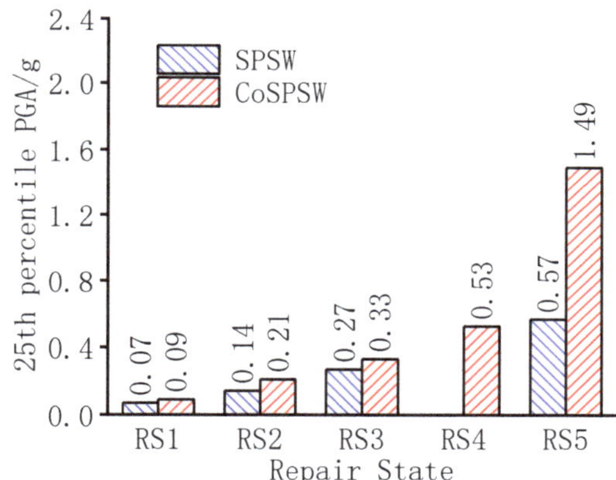

Figure 11. The 25th percentile PGA from structural fragility curves.

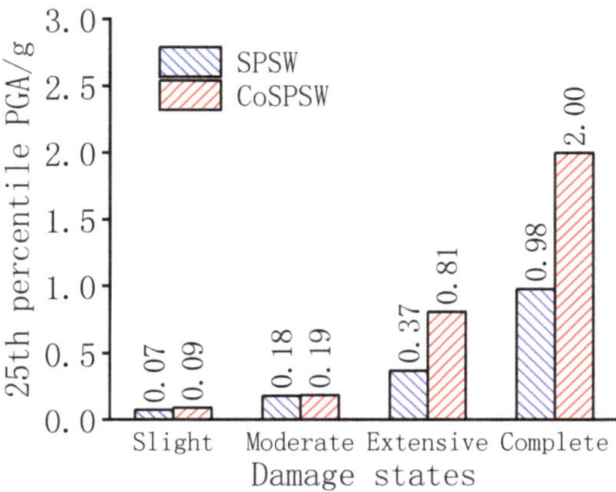

Figure 12. The 25th percentile PGA from drift-sensitive nonstructural fragility curves.

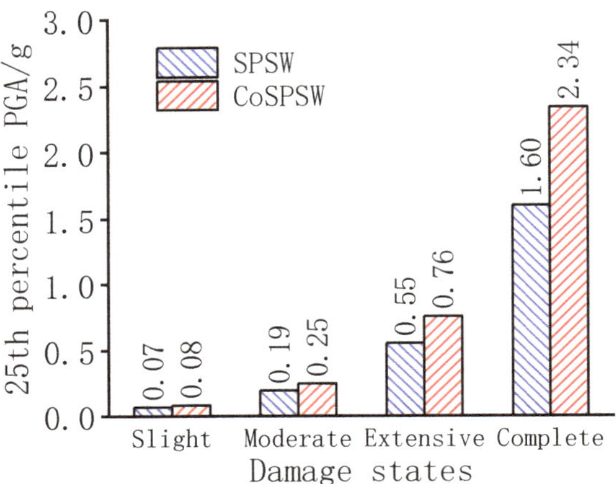

Figure 13. The 25th percentile PGA from acceleration-sensitive nonstructural fragility curves.

It is clear from Figures 11–13 that from the RS1 to the RS5 repair states and from the "Slight" to the "Complete" damage states, the 25th percentile PGA values for both the CoSPSW structure and the SPSW structure increase significantly, indicating that the structures would experience more serious damages with the increase in seismic intensity. Furthermore, from the RS1 to the RS5 repair states, the 25th percentile PGA values of the fragility curves of the CoSPSW structure are 22%, 33%, 18%, and 62% higher than those of the SPSW structure, respectively. For the same repair method concerning the repair of the boundary column (the RS4 of the CoSPSW), the 25th percentile PGA value is 0.53 g, which is 55% higher than the RS3 of the SPSW. This indicates that for a given damage state, the CoSPSW is required to undergo higher levels of seismicity to suffer the same damage degree as the SPSW. For the drift-sensitive nonstructural damage state of "Slight" to "Complete", the 25th percentile PGA values of the fragility curves of the CoSPSW structure are 22%, 5%, 54%, and 51% higher than those of the SPSW structure, respectively. For the acceleration-sensitive nonstructural damage state of "Slight" to "Complete", the 25th percentile PGA values of the fragility curves of the CoSPSW structure are 13%, 24%, 28%, and 32% higher than those of the SPSW structure, respectively.

Therefore, as the number of damage states/repair states increases, the 25th percentile PGA values of the CoSPSW become obviously higher than those of the SPSW, which indicates that the CoSPSW has a lower seismic vulnerability and probability of structural and nonstructural damage than the SPSW structure, especially for the repair state RS5 for the structural members and the "extensive" and "complete" damage states for the non-structural members.

5. Conclusions

(1) The inter-story drift data on the seismic performance of the corrugated steel plate shear walls (CoSPSWs) were collected and analyzed. The observations on the cyclic behavior of the 19 CoSPSW test specimens have resulted in the development of twelve damage states. Five repair states were also proposed to consider the difficulty of repairing the observed damage following the earthquake, ranging from cosmetic repair to member or boundary frame replacement.

(2) In order to avoid the random error caused by the direct use of the median and standard deviation of the test data, the maximum likelihood method was used to calculate the median value and the standard deviation value of the inter-story drift. According to ATC58, the recommended median value θ_{rec} and the standard deviation value β_{rec} of the inter-story drift corresponding to the different repair states of the CoSPSW were derived and were used as the threshold value of each repair state in the fragility function.

(3) Incremental dynamic analyses (IDA) were conducted on a 10-story CoSPSW structure and a conventional steel plate shear wall (SPSW) structure with the same wall plate thickness and boundary frame. From the IDA results, the relationship between the EDP (peak inter-story drift/peak floor acceleration) and the IM (PGA) were obtained. The coefficient R^2 was approximately 0.8, which meant that the data-fitting was relatively precise. Based on the relationship between the EDP and the IM, the regression coefficients in the fragility function were obtained.

(4) According to the recommended threshold values of the repair/damage states and the regression coefficients obtained through incremental dynamic analyses (IDA), fragility curves were obtained for the 10-story CoSPSW structure and the SPSW structure, respectively. It was shown that the 25th percentile PGA of the CoSPSW was higher than that of the SPSW for various repair states and damage states, especially for the repair state RS5 for the structural members and the "extensive" and "complete" damage states for the non-structural members. Therefore, the CoSPSW structure had a lower damage degree and better seismic behavior than the SPSW structure.

Author Contributions: Q.Z. provided the idea and gave professional seismic guidance on the corrugated steel plate shear wall; Z.T. analyzed the engineering data and wrote the paper draft; Y.Z. conducted the finite element analyses and prepared the paper draft; C.Y. conducted the validation of the finite element model and provided supervision of the writing process. All authors have read and agreed to the published version of the manuscript.

Funding: This research was funded by the National Natural Science Foundation of China (NSFC) under grant number 51878447, 51678406, 51378340.

Institutional Review Board Statement: Not applicable.

Informed Consent Statement: Not applicable.

Data Availability Statement: Not applicable.

Conflicts of Interest: The authors declare no conflict of interest.

References

1. Montgomery, C.J.; Medhekar, M.; Lubell, A.S. Unstiffened Steel Plate Shear Wall Performance under Cyclic Loading. *J. Struct. Eng.* **2001**, *127*, 973–975. [CrossRef]
2. Berman, J.W.; Bruneau, M. Plastic Analysis and Design of Steel Plate Shear Walls. *J. Struct. Eng.* **2003**, *129*, 1448–1456. [CrossRef]
3. Berman, J.W. Seismic behavior of code designed steel plate shear walls. *Eng. Struct.* **2011**, *33*, 230–244. [CrossRef]
4. Driver, R.G. Cyclic Test of Four-Story Steel Plate Shear Wall. *J. Struct. Eng.* **1998**, *124*, 112–120. [CrossRef]
5. Berman, J.W.; Bruneau, M. Capacity Design of Vertical Boundary Elements in Steel Plate Shear Walls. *Eng. J. Am. Inst. Steel Constr. Inc.* **2008**, *45*, 57–71.
6. *JGJ/T 380-2015*; Technical Specifications for Steel Plate Shear Walls. China Construction Industry Press: Beijing, China, 2016. (In Chinese)
7. Berman, J.W.; Celik, O.C.; Bruneau, M. Comparing hysteretic behavior of light-gauge steel plate shear walls and braced frames. *Eng. Struct.* **2005**, *27*, 475–485. [CrossRef]
8. Berman, J.W.; Bruneau, M. Experimental investigation of light-gauge steel plate shear walls. *J. Struct. Eng.* **2005**, *131*, 259–267. [CrossRef]
9. Emami, F.; Mofid, M.; Vafai, A. Experimental study on cyclic behavior of trapezoidally corrugated steel shear walls. *Eng. Struct.* **2013**, *48*, 750–762. [CrossRef]
10. Stojadinovic, B.; Tipping, S. Structural testing of corrugated sheet steel shear walls. In Proceedings of the19th International Specialty Conferences on Cold-Formed Steel Structures, Rolla, MO, USA, 14–15 October 2008.
11. Hosseinzadeh, L.; Emami, F.; Mofid, M. Experimental investigation on the behavior of corrugated steel shear wall subjected to the different angle of trapezoidal plate. *Struct. Des. Tall Spec. Build.* **2017**, *2*, e1390. [CrossRef]
12. Ding, Y.; Deng, E.F.; Zong, L.; Dai, X.; Lou, N. Cyclic tests on corrugated steel plate shear walls with openings in modularized-constructions. *J. Constr. Steel Res.* **2017**, *138*, 675–691. [CrossRef]
13. Sudeok, S.; Mina, Y.; Seungjae, L. An Experimental Study on the Shear Hysteresis and Energy Dissipation of the Steel Frame with a Trapezoidal-Corrugated Steel Plate. *Materials* **2017**, *10*, 261. [CrossRef]
14. Cao, Q.; Huang, J. Experimental study and numerical simulation of corrugated steel plate shear walls subjected to cyclic loads. *Thin-Walled Struct.* **2018**, *127*, 306–317. [CrossRef]

15. Qiu, J.; Zhao, Q.H.; Yu, C.; Li, Z.X. Experimental studies on cyclic behavior of corrugated steel plate shear walls. *J. Struct. Eng.* **2018**, *144*, 4018200. [CrossRef]
16. Wang, W.; Ren, Y.; Lu, Z. Experimental study of the hysteretic behaviour of corrugated steel plate shear walls and steel plate reinforced concrete composite shear walls. *J. Constr. Steel Res.* **2019**, *160*, 136–152. [CrossRef]
17. Jin, S.S.; Wang, Q.Y.; Zhou, J.; Bai, J.L. Numerical and experimental investigation of assembled multi-grid corrugated steel plate shear walls. *Eng. Struct.* **2022**, *251*, 113544. [CrossRef]
18. Emami, F.; Mofid, M. On the hysteretic behavior of trapezoidally corrugated steel shear walls. *Struct. Des. Tall Spec. Build.* **2014**, *23*, 94–104. [CrossRef]
19. Edalati, S.A.; Yadollahi, Y.; Pakar, I.; Emadi, A.; Bayat, M. Numerical study on the performance of corrugated steel shear walls. *Wind Struct.* **2014**, *19*, 405–420. [CrossRef]
20. Kalali, H.; Hajsadeghi, M. Hysteretic performance of SPSWs with trapezoidally horizontal corrugated web-plates. *Steel Compos. Struct.* **2015**, *19*, 277–292. [CrossRef]
21. Zhao, Q.H.; Sun, J.H.; Li, Y.; Li, Z.X. Cyclic analyses of corrugated steel plate shear walls. *Struct. Des. Tall Spec. Build.* **2017**, *26*, e1351. [CrossRef]
22. Dou, C.; Jiang, Z.Q.; Pi, Y.L.; Guo, Y.L. Elastic shear buckling of sinusoidally corrugated steel plate shear wall. *Eng. Struct.* **2016**, *121*, 136–146. [CrossRef]
23. Farzampour, A.; Mansouri, I.; Hu, J.W. Seismic behavior investigation of the corrugated steel shear walls considering variations of corrugation geometrical characteristics. *Int. J. Steel Struct.* **2018**, *18*, 1297–1305. [CrossRef]
24. Bahrebar, M.; Kabir, M.; Zirakian, M.; Hajsadeghi, M.; Lim, J. Structural performance assessment of trapezoidally-corrugated and centrally-perforated steel plate shear walls. *J. Constr. Steel Res.* **2016**, *122*, 584–594. [CrossRef]
25. Bahrebar, M.; Lim, J.; Clifton, G.; Zirakian, T.; Shahmohammadi, A.; Hajsadeghi, M. Perforated steel plate shear walls with curved corrugated webs under cyclic loading. *Structures* **2020**, *24*, 600–609. [CrossRef]
26. Farzampour, A.; Yekrangnia, M. On the behavior of corrugated steel shear walls with and without openings. In Proceedings of the Second European Conference on Earthquake Engineering and Seismology, Istanbul, Turkey, 25–29 August 2014.
27. Farzampour, A.; Laman, J.A. Behavior prediction of corrugated steel plate shear walls with openings. *J. Constr. Steel Res.* **2015**, *114*, 258–268. [CrossRef]
28. Farzampour, A.; Mansouri, I.; Lee, C.H.; Sim, H.B.; Hi, J.W. Analysis and design recommendations for corrugated steel plate shear walls with a reduced beam section. *Thin-Walled Struct.* **2018**, *132*, 658–666. [CrossRef]
29. Masoud, H.A.; Mahna, S. Seismic behavior of steel plate shear wall with reduced boundary beam section. *Thin-Walled Struct.* **2018**, *116*, 169–179. [CrossRef]
30. Fang, J.; Bao, W.; Ren, F.; Guan, T.; Xue, G.; Jiang, J. Experimental study of hysteretic behavior of semi-rigid frame with a corrugated plate. *J. Constr. Steel Res.* **2020**, *174*, 106–119. [CrossRef]
31. Shariati, M.; Faegh, S.S.; Mehrabi, P.; Bahavarnia, S.; Zandi, Y.; Masoom, D. Numerical study on the structural performance of corrugated low yield point steel plate shear walls with circular openings. *Steel Compos. Struct.* **2019**, *33*, 569–581. [CrossRef]
32. Yu, Y.J.; Chen, Z.H. Rigidity of corrugated plate sidewalls and its effect on the modular structural design. *Eng. Struct.* **2018**, *175*, 191–200. [CrossRef]
33. Zhao, Y. Lateral Behavior and Design Method of Corrugated Steel Plate Shear Walls. Master Thesis, Tianjin University, Tianjin, China, 2018. (In Chinese).
34. Baldvins, N.M.; Berman, J.W.; Lowes, L.N.; Janes, T.M. Fragility functions for steel plate shear walls. *Earth. Spectra.* **2012**, *28*, 405–426. [CrossRef]
35. Negar, M.; Epackachi, S. Fragility functions for steel-plate concrete composite shear walls. *J. Constr. Steel Res.* **2019**, *167*, 105776. [CrossRef]
36. Wang, M.; Guo, Y.; Yang, L. Damage indices and fragility assessment of coupled low-yield-point steel plate shear walls. *J. Build. Eng.* **2021**, *11*, 103010. [CrossRef]
37. Zhang, J.; Zirakian, T. Probabilistic assessment of structures with SPSW systems and LYP steel infill plates using fragility function method. *Eng. Struct.* **2015**, *85*, 195–205. [CrossRef]
38. Jiang, L.; Hong, Z.; Hu, Y. Effects of various uncertainties on seismic risk of steel frame equipped with steel panel wall. *Bul. Earth. Eng.* **2018**, *16*, 5995–6012. [CrossRef]
39. Hu, Y.; Zhao, J.H. Seismic risk assessment of steel frames equipped with steel panel wall. *Struct. Des. Tall Spec. Build.* **2017**, *26*, e1368. [CrossRef]
40. Bu, H.; He, L.; Jiang, H. Seismic fragility assessment of steel frame structures equipped with steel slit shear walls. *Eng. Struct.* **2021**, *249*, 113328. [CrossRef]
41. *HAZUS-MH MR5*; Earthquake Loss Estimation Methodology—Technical and User's Manual. Department of Homeland Security, Federal Emergency Management Agency, Mitigation Division: Washington, DC, USA, 2010.
42. *FEMA 352*; Recommended Post-Earthquake Evaluation and Repair Criteria for Welded Steel Moment-Frame Buildings. Safety Council for the Federal Emergency Management Agency: Washington, DC, USA, 2000.
43. *ATC-58*; Guidelines for Seismic Performance Assessment of Buildings. Safety Council for the Federal Emergency Management Agency, Applied Technology Council: Redwood, CA, USA, 2009.
44. *GB50011-2010*; Code for Seismic Design of Buildings. China Architecture and Building Press: Beijing, China, 2010. (In Chinese)

45. Jiang, W.W.; Jin, H.J.; Sun, F.F. Research on simplified analysis models of non-buckling corrugated steel shear walls. *Pro. Steel Build. Struct.* **2019**, *21*, 61–71. (In Chinese)
46. Park, H.G.; Kwack, J.H.; Jeon, S.W.; Kim, W.K.; Choi, I.R. Framed steel plate wall behavior under cyclic lateral loading. *J. Struct. Eng.* **2007**, *133*, 378–388. [CrossRef]
47. Baker, J.W.; Cornell, C.A. Vector-valued ground motion intensity measures for probabilistic seismic demand analysis. In *PEER Report*; College of Engineering, University of California: Berkeley, CA, USA, 2006.
48. Zhang, J.; Huo, Y.; Brandenberg, S.J.; Kashighandi, P. Effects of structural characterizations on fragility functions of bridges subject to seismic shaking and lateral spreading. *Earth. Eng. Eng. Vibr.* **2008**, *7*, 369–382. [CrossRef]

metals

MDPI

Article

Study on Shear Strength of Partially Connected Steel Plate Shear Wall

Yuqing Yang *, Zaigen Mu * and Boli Zhu

School of Civil and Resource Engineering, University of Science and Technology Beijing, Beijing 100083, China; zboly@ustb.edu.cn
* Correspondence: yqyang@ustb.edu.cn (Y.Y.); zgmu@ces.ustb.edu.cn (Z.M.)

Abstract: The paper proposes partially connected steel plate shear walls, in which the infill plates and frames are connected by discretely distributed fish plates at the corners and at the centers. The high lateral resistance of a steel plate shear wall has led to its widespread use in the design of structural shear resistance. In this paper, finite element models of the partially connected steel plate shear walls are established by the finite element method, and the effect of the different partial connections on the shear strength is firstly investigated. Moreover, the variation of the shear strength with the plate-to-frame connectivity ratio is analyzed numerically, and the effect of the connectivity ratio on the development of the tensile field is studied. Based on the numerical analysis results, the effect of the connectivity ratio on shear strength is evident at low levels. When the connectivity ratio is over 80%, the shear strength of the partially connected steel plate shear wall can reach 95% of that of the fully connected steel plate shear wall. When the connection ratio is at a low level, the advantages of the central connection on the shear strength of the structures are higher than those with corner connections. Furthermore, the fitting formula for the partially connected steel plate shear wall is obtained by changing the connectivity ratio and width-to-height ratio of the examples, which can predict the shear capacity of the partially connected steel plate shear wall with different partial connections.

Keywords: partially connected steel plate shear wall; shear strength; connectivity ratio; finite element method; design formula

Citation: Yang, Y.; Mu, Z.; Zhu, B. Study on Shear Strength of Partially Connected Steel Plate Shear Wall. *Metals* **2022**, *12*, 1060. https://doi.org/10.3390/met12071060

Academic Editor: Denis Benasciutti

Received: 27 April 2022
Accepted: 15 June 2022
Published: 21 June 2022

Publisher's Note: MDPI stays neutral with regard to jurisdictional claims in published maps and institutional affiliations.

1. Introduction

The infill plate of steel plate shear wall (SPSW) structures is usually fully connected to the boundary elements by fish plates or bolts (Figure 1). The steel plate shear wall has high lateral resistance and energy dissipation capability [1], hence, it is a better choice for lateral resistant members of structures and is widely used in multi-rise and super-tall structures. In 1931, Wanger [2] studied the post-buckling strength of thin aluminum alloys and proposed the "Theory of pure diagonal tension". Then, Basler [3] considered uniform tensile stresses in the tensile field and presented the shear force of a thin steel web. In 1983, Thorburn et al. [4] applied the theory of tensile field to thin steel plate shear walls. Since then, many types of thin steel plate shear walls were developed, including unstiffened, stiffened, buckling-restrained, perforated, composited, low-yield-point steel or aluminum [5–10], and metal shear panels show good performance [11]. Among them, unstiffened steel plate shear walls are the most basic and common. The development of the tension field of the infill plate affects the bearing capacity of the structure. Therefore, the performance of the infill plate after buckling and the lateral resistance of the infill plate with different boundary conditions are the focus of the study.

Yu et al. [12] proposed a new connection type, allowing for reducing the requirement of bolt hole accuracy for traditional bolted connections. The infill plate is connected to boundary frames through angles steel, which can facilitate the installation and maintenance of steel plate shear walls. Paslar et al. [13] studied six different connection types of plate-to-frame connections and evaluated their structural performance by comparing them with the fully connected plate, as shown in Figure 2. The infill plate connected to the columns reduces the bearing capacity more than the plate connected to the beams. When the ratio of plate-to-frame connection is above 80%, all lateral resistance properties can reach 95% of the full connection type, including lateral load capacity, ductility, and stiffness. Azandariani et al. [14] conducted a parametric analysis with a three-story steel plate shear wall and studied the base shear capacity of SPSWs with partial plate-to-column connections of different width-to-height (aspect) ratios and high-to-thickness ratios. The results show that partial plate-to-column connections can reduce the additional effect on the columns, but the stiffness and shear strength of the structure decreased by approximately 25% on average.

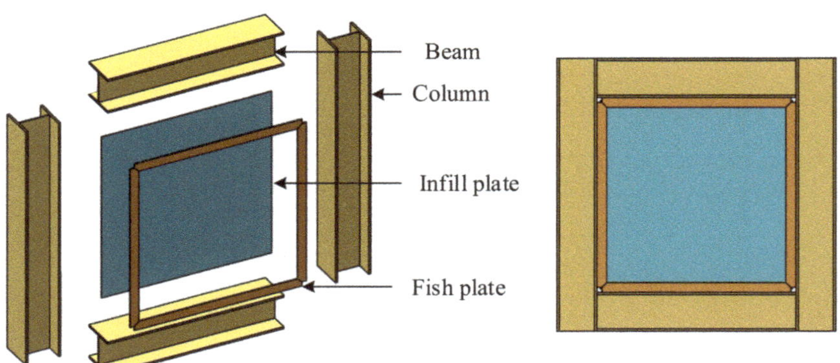

Figure 1. Steel plate shear wall.

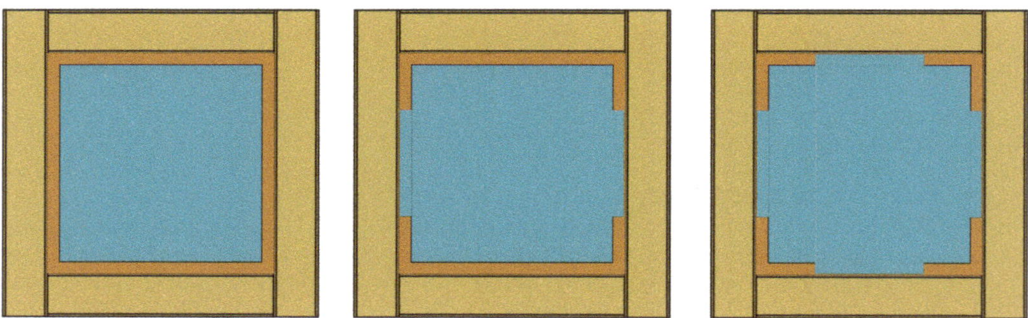

Figure 2. Steel plate shear wall with partial connection.

Wei et al. [15,16] proposed a buckling-restrained SPSW with four-corner connections, where the structure is only connected to the boundary frame at the four corners of the infill plate. Du et al. [17] proposed to partially connect the infill plate of the buckling-restrained SPSW to the concrete-filled steel tubular columns, including the four-corner connection and partial column connection. Mirsadeghi and Fanaie [18] investigated the effect of the plate-to-column connectivity ratio on structural performance. Through experiments and numerical calculations, when the ratio of the plate-to-frame connection exceeds 70%, the plate develops entire parallel tensile strips.

There are some papers that demonstrate that the finite element software ABAQUS [19] can simulate the shear behavior of the steel plate shear wall by using a reasonable set of material parameters and boundary conditions [20–22]. Moreover, numerical simulations are in good agreement with the experimental results and save time and the cost of tests [23]. Thus, finite element simulations can be used to predict the behavior of the structures.

Currently, the primary focus of existing research is a quantitative analysis of the overall behavior of the partially connected steel plate shear wall (PC-SPSW). However, the results of the investigation on the shear strength of the infill plate are less encouraging, and the bearing capacity calculation formulae that take the connectivity ratio into account are unavailable. This study proposes eight different types of PC-SPSW. The effect of connectivity ratios on the shear behavior of the infill plate is studied by the finite element (FE) method, and the calculation formula of structural shear strength is provided.

2. Theoretical Analysis and Finite Element Modeling

2.1. Shear Strength of Infill Plate

The shear strength results of the finite element model are compatible with the test results and the theoretical calculation formula Equation (1). The shear strength of a thin infill plate with width b, height h, and thickness t can be calculated by the following equations [24].

$$F_{\text{plate}} = bt\tau_{\text{cr}} + \frac{1}{2}bt\sigma_t \sin(2\theta) \tag{1}$$

$$\begin{cases} \tau_{\text{cr}} = \frac{K\pi^2 E}{12(1-v)^2}\left(\frac{t}{b}\right)^2 \\ \sigma_t = -\frac{3}{2}\tau_{\text{cr}}\sin 2\theta + \sqrt{\sigma_y^2 + \left(\frac{9}{4}\sin^2 2\theta - 3\right)\tau_{\text{cr}}^2} \end{cases} \tag{2}$$

where τ_{cr} is the elastic buckling shear stress; σ_t is field tensile stress; σ_y is yield stress of steel plate; θ is the tension field inclination with respect to the vertical axis, which generally can be simplified to take $45°$; K is the elastic buckling coefficient and can be calculated by the following formula when the four edges of the plate are simply supported [25]; v is the Poisson's ratio, which is taken as 0.3.

$$\begin{cases} K = 5.35 + 4\left(\frac{b}{h}\right)^2 & \frac{b}{h} \geq 1 \\ K = 4 + 5.35\left(\frac{b}{h}\right)^2 & \frac{b}{h} < 1 \end{cases} \tag{3}$$

The specification [26] ignores the elastic buckling load of the infill plate, thus field tensile stress σ_t can be taken as the yield stress σ_y. Thus, the theoretical value of the shear strength of the infill plate can be calculated by Equation (4).

$$F_{\text{plate}} = \frac{1}{2}bt\sigma_y \sin(2\theta) \tag{4}$$

The beam-only connected steel plate shear wall can be seen as a type of PC-SPSW with a connectivity ratio of 0 to the column, as shown in Figure 3a. Ozcelik and Clayton [27] propose a method for calculating the base shear of the steel plate shear wall with a beam-connected web plate, as shown in Equations (5)–(7).

$$F_{\text{plate}} = \frac{1}{2}\left(1 - \frac{\tan \theta}{b/h}\right)bt\sigma_y \sin(2\theta) \tag{5}$$

$$\theta = \eta \tan^{-1}\left(\frac{b}{h}\right) \tag{6}$$

$$\eta = 0.55 - 0.03\left(\frac{b}{h}\right) \geq 0.51 \tag{7}$$

The shear strength of the steel plate shear wall which is partially connected to columns from two corners can be calculated using Equation (8) [18]. Where h_{nc} is not connected to a length of the plate in Figure 3b.

$$F_{\text{plate}} = \frac{1}{2}(1 - h_{nc}\tan \theta)bt\sigma_y \sin(2\theta) \tag{8}$$

Wei et al. [15] assume the diagonal formation of tension and compression fields, illustrated in Figure 3c and equivalent tension and compression strips are used to calculate the shear strength of the structure. By comparing Equations (4), (5), and (8), it can be seen that the shear strength of PC-SPSW is related to the fully connected steel plate shear walls with the connectivity ratio γ, such as Equation (9).

$$F = [1 - f(\gamma)]F_{\text{plate}} = [1 - f(\gamma)]\frac{1}{2}\sigma_y bt \sin(2\theta) \tag{9}$$

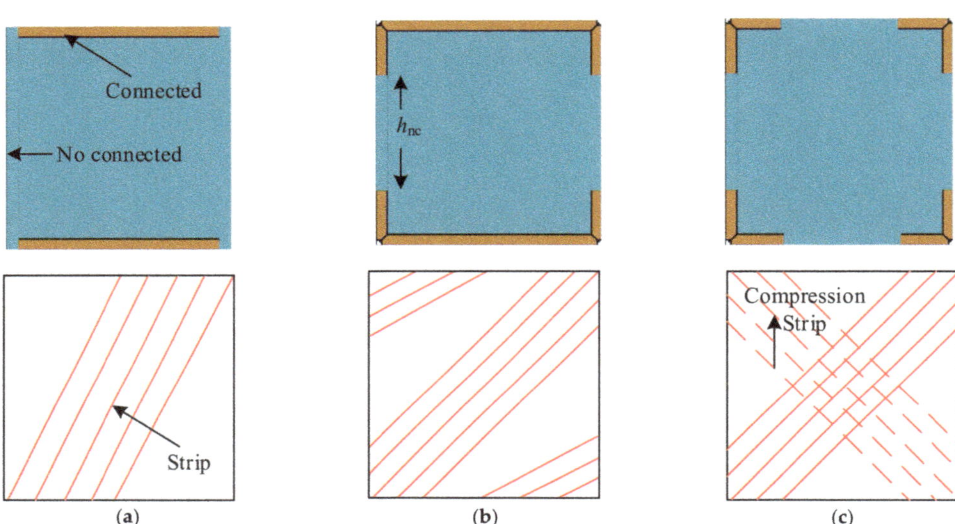

Figure 3. Simplified mechanical model for PC-SPSW: (**a**) Only beam connection; (**b**) Partially column connection; (**c**) Four-corner connection.

2.2. Basic Model

First, a base model of SPSW with the pin-connected frame was established to study the shear capacity of the infill plate, and the results of the finite element model were compared with the theoretical equation of shear strength to verify the feasibility of the finite element model practicality.

Table 1 lists the parameters of the steel plate shear wall model. The sections of the beams and columns are H750 mm × 600 mm × 18 mm × 36 mm and H600 mm × 600 mm × 18 mm × 36 mm, respectively, ensuring that the frame has enough rigidity to allow the steel plate material properties can be fully developed. The connection type of beam-to-column is pin linked. As a result, the influence of bearing capacity from the frame is eliminated.

Table 1. Parameters of SPSW model (Unit: mm).

Section of Column	Section of Beam	h of Plate	b of Plate	t of Plate
H600 mm × 600 mm × 18 mm × 36 mm	H750 mm × 600 mm × 18 mm × 36 mm	3000	3000	10

A base model of the steel plate shear wall is established in ABAQUS 6.13 developed by Dassault Systèmes, Waltham, MA, USA [19]. The beams and columns use beam element B31. The infill plate uses shell element S4R, which is suitable for thin panels with hourglass control [28]. The thickness direction of the plate uses the Simpson integral with a thickness integration point of 5. The beam-column connection is a PIN connection by MPC constraint, and the infill plate-to-frame connection is simulated by a TIE connection. The effect of fish plates and residual stresses of the infill plate is not considered. The yield stresses of the frame and the infill plate are 345 MPa and 240 MPa, respectively. The elastic modulus E of steel is 206,000 MPa and von Mises criterion is used as the yield criterion.

2.3. Nonlinear Analysis

Steel plate shear walls deform in buckling before yielding, forming a tension field to resist the shear forces. As the tension field develops, the out-of-plane deformation increases. The structure reaches its ultimate state when the plate yields the material. Therefore, considering the geometric and material nonlinearities can accurately determine the shear capacity of the structure [29].

Furthermore, steel plates are not ideal flat plates and have initial defects due to manufacturing, transportation and installation. Initial imperfections can be obtained in three ways: the first is to actually measure the initial deformation of the plate; the second is to use trigonometric stochastic theory to describe the deformation distribution; and the third is to use the first-order buckling mode as the distribution model, assuming that the initial imperfections are distributed according to the most unfavorable distribution. The thin plate is very susceptible to buckling, and the amplitude of the initial imperfection does not have a significant effect on the analysis results [30]. Thus, the first-order buckling mode of the infill plate is used to approximate the distribution mode of the initial imperfection with a maximum geometric imperfection amplitude of $b/1000$ [30].

The influence of different mesh sizes on the structure was analyzed, as shown in Figure 4. When the mesh size was taken to be 400 mm or even less, the finite element results were in error with the theoretical values by less than 5%. The ratio of FE results to the theoretical value is basically 1 when 25 mm is taken, but it takes much time. Therefore, we ended up with a mesh size of 100 mm to minimize computational time while maintaining computational accuracy. The S4R element has no thickness, therefore, the mesh size of the model is 100 mm (length) $\times$ 100 mm (width).

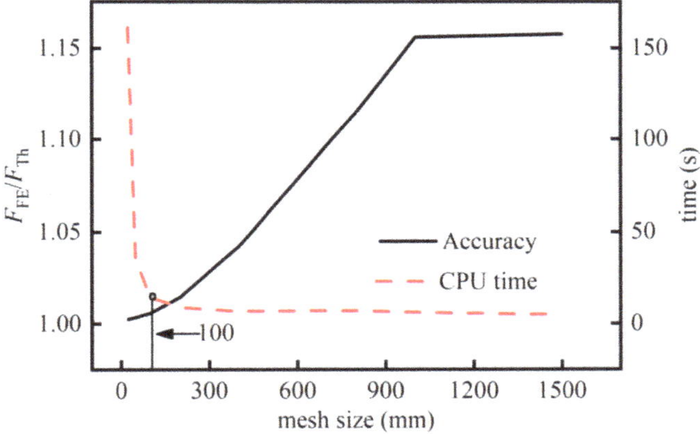

Figure 4. Curves of mesh size versus calculation accuracy and time.

The buckling analysis uses the subspace iterative method by ABAQUS/Buckle, the results are shown in Figure 5a. Then, the static analysis is solved by Newton's iterative algorithm in ABAQUS/Standard with considering geometric nonlinearity. In addition, material nonlinearity is considered and the stress–strain relationship of steel is an ideal elastic–plastic model without strain hardening, as shown in Figure 5b. The horizontal displacement is applied to the top beam to simulate the structure under shear load action and the maximum drift angle is 3%.

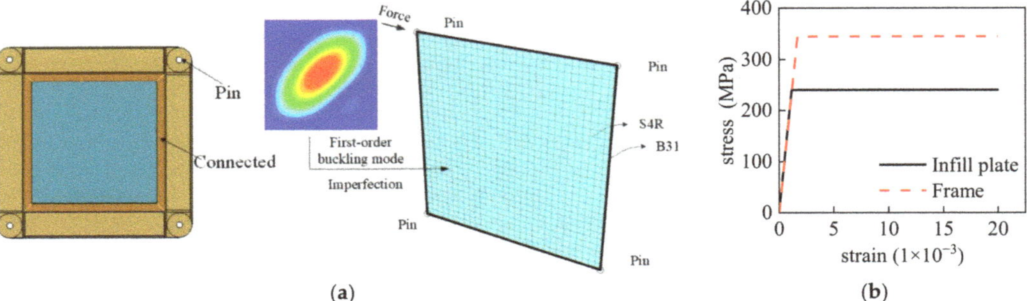

(**a**)　　　　　(**b**)

Figure 5. Finite element model of SPSW: (**a**) FE model of steel plate shear wall; (**b**) stress–strain relationship.

2.4. Verifying of FE Model

Figure 6 shows the load–displacement curve, von Mises stress, and plastic strain distribution of the SPSW model. At point A, the structure reaches the maximum bearing capacity and the infill plate has fully yielded. Then, with the horizontal displacement continuing to increase, the bearing capacity remains basically unchanged. The steel is assumed to be an ideal elastoplastic material without considering the hardening and failure of the material, so the ultimate bearing capacity of the structure at point B is equal to the theoretical shear strength of the infill plate. It should be noted that although the stresses in the plate all reach the yield stress of 240 MPa in Figure 6c, the plastic strain of the plate is not uniform and the strain changes along the diagonal direction are large, as can be seen in Figure 6d. The theoretical shear strength of the infill plate from Equation (4) is 3600 kN, and the ultimate bearing capacity obtained from the finite element model is 3623 kN. Varying the model parameters, the finite element results match the theoretical equation values, which can indicate the feasibility of the modeling and can be used for the subsequent analysis.

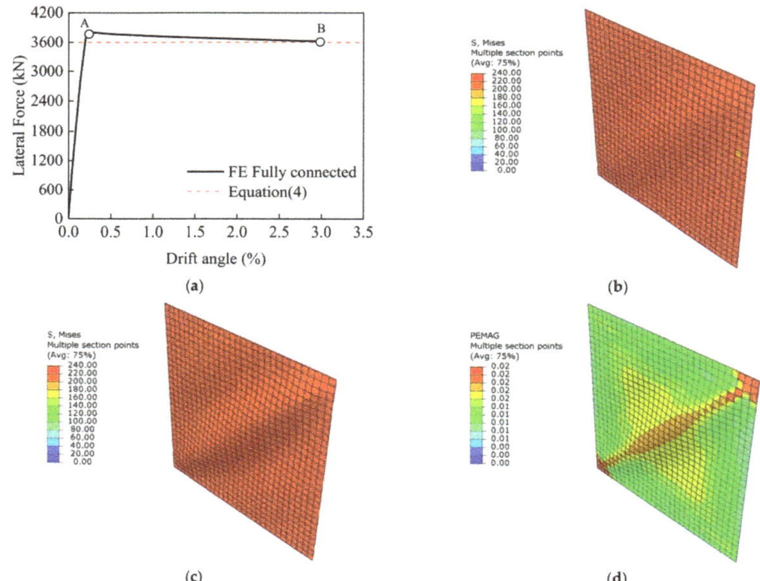

Figure 6. Load–displacement curve and stress of SPSW: (**a**) Load–displacement curve; (**b**) Stress distribution at point A; (**c**) Stress distribution at point B; (**d**) Plastic strain distribution at point B.

2.5. Description of Partial Connections

Based on the above model, eight different types of plate-to-frame connections are proposed, as shown in Figure 7. S1 is the plate that is completely attached to the beams, however, it is partially connected to columns from the center. S1R differs from the S1 in that it is partially connected to columns from two corners, with each connection segment being the same length. S2 is connected to a column in two segments, as opposed to S1, which has a one-segment connection. S3 and S4 plates are partially connected to both beams and columns in the same connection types.

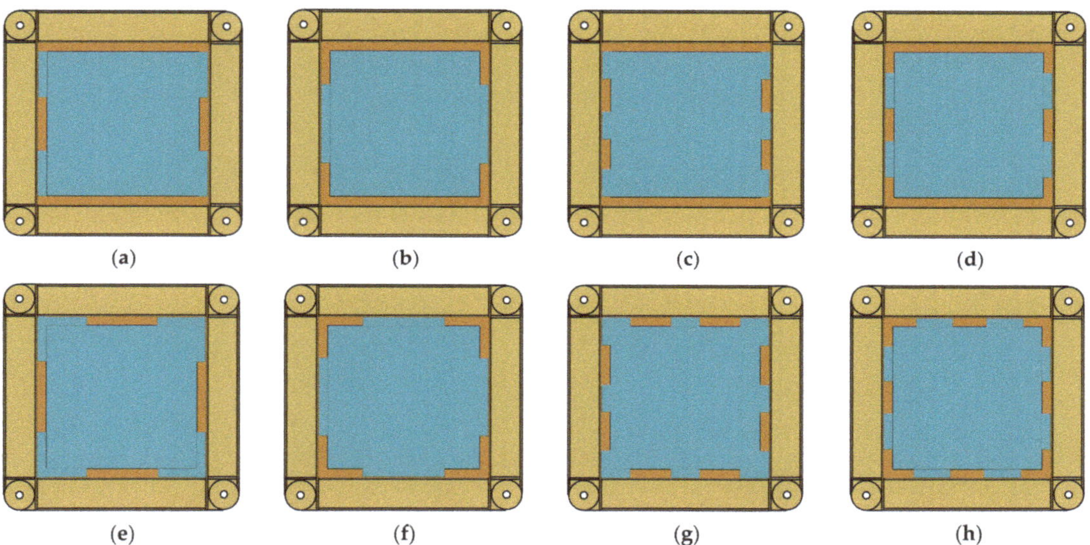

Figure 7. Different types of infill plate-to-frame connection: (**a**) S1; (**b**) S1R; (**c**) S2; (**d**) S2R; (**e**) S3; (**f**) S3R; (**g**) S4; (**h**) S4R.

The ratio of the attached length of the plate to the full length of the plate is defined as the plate-to-frame connectivity ratio, abbreviated as connectivity ratio γ. Table 2 list the models with different connectivity ratios. For instance, the S1-P13 model indicates that the S1 connection form with a connectivity ratio of about 13%; the S2-P100 means that the S2 connection form and the plate are fully connected to the columns.

Table 2. Models with different connectivity ratios.

Model	Connectivity Ratio γ of S1, S1R (%)	Connectivity Ratio γ of S3, S3R (%)	Model	Connectivity Ratio γ of S2, S2R (%)	Connectivity Ratio γ of S4, S4R (%)
P0	0 (Beamonly)	-	P0	0 (Beamonly)	-
P13	13	13	P20	20	20
P27	27	27	P40	40	40
P40	40	40	P60	60	60
P53	53	53	P80	80	80
P67	67	67	P100	100 (full)	100 (full)
P80	80	80	-	-	-
P93	93	93	-	-	-
P100	100 (full)	100 (full)	-	-	-

3. Numerical Results

3.1. Structural Performance of S1 and S1R Models

S1 and S1R are both fully connected to the beams, but different from the connection to the columns. The results of the shear strength of S1 and S1R are shown in Table 3, and the relationship between shear strength and the connectivity ratio is shown in Figure 8. The connectivity ratio can obviously affect the bearing capacity of steel plate shear wall structures. When the connectivity ratio is 0, S1 and S1R are beam-only connected, and the strengths of the models are approximately 53% reduced to that of the full connection model. Although the beam-only type can avoid the influence of the tension field on the columns, it also loses nearly half of the shear strength. In this way, the material properties of the infill plate cannot be fully developed. With an increase in the connectivity ratio, the strength of the structures gradually increases and approaches the shear strength of the fully connected plate. At the same connectivity ratio, the strength of S1 is higher than that of S1R.

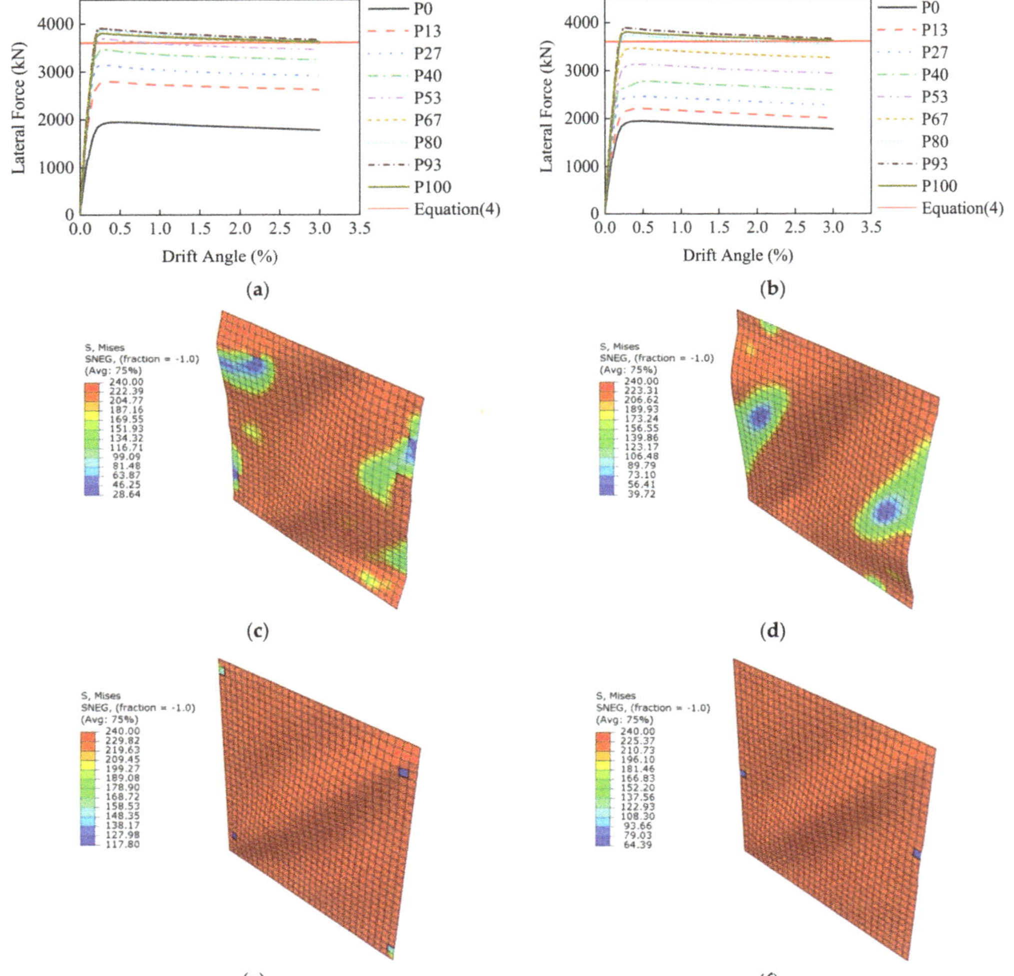

Figure 8. Load-displacement curves and stress distributions at 3% drift of S1 and S1R models: (**a**) S1; (**b**) S1R; (**c**) S1-P13; (**d**) S1R-P13; (**e**) S1-P80; (**f**) S1R-P80.

Table 3. Results of shear strength for S1 and S1R models.

Model	Connectivity Ratio γ (%)	S1		S1R	
		Value (kN)	Decrease Ratio (%)	Value (kN)	Decrease Ratio (%)
P0	0 (Beamonly)	1726	−52.91	1726	−52.91
P13	13	2615	−28.66	1997	−45.52
P27	27	2910	−20.62	2270	−38.07
P40	40	3251	−11.32	2589	−29.37
P53	53	3457	−5.70	2935	−19.94
P67	67	3585	−2.20	3259	−11.10
P80	80	3643	−0.62	3542	−3.38
P93	93	3665	−0.02	3660	−0.16
P100	100 (full)	3666	0	3666	0

From the stress diagram in Figure 8, when the connectivity ratio is small, the S1-P13 and S1R-P13 have a large area of the plate that does not reach the yield stress, of which S1R is larger. The steel cannot fully yield resulting in low bearing capacity. The development of the tension field was seriously affected in the unattached part due to the lack of anchorage from boundary elements. This effect decreases with an increase in connectivity ratio. The shear strength of the S1-P67, for which the connectivity ratio is 67%, almost reaches the full connection. For S1R, the shear strength of which almost reaches the fully connected type is the S1R-P80 model. When the beam is fully connected, increasing the connectivity ratio from the center can improve ultimate bearing capacity more than that from the corners.

3.2. Structural Performance of S2 and S2R Models

S2 is fully connected to the beams, but the center of the infill plate is partially connected to the columns in two segments. S2R is partially connected to the columns from the center and two corners. The ultimate bearing capacity results of S2 and S2R are shown in Table 4. The shear strength of S2 is higher than that of S2R with the same connectivity ratio. The shear strength of the S2 and S2R model gradually increases with an increase in the connectivity ratio. The shear strength of the S2 type with a connection ratio of 60% is close to the strength of the fully connected model. While the S2R-type requires a connection ratio of 80% for its shear strength to be approximately equal to the fully connected case. From Figure 9c,d at the connection of 20%, S2 and S2R extend a local low-stress region around the unconnected part and the area of the low-stress region of S2R is larger than that of S2. When the connection rate reaches 80%, in Figure 9e,f the steel plates of S2 and S2R are basically all in the plastic state.

Table 4. Results of shear strength for S2 and S2R models.

Model	Connectivity Ratio γ (%)	S2		S2R	
		Value (kN)	Decrease Ratio (%)	Value (kN)	Decrease Ratio (%)
P0	0 (Beamonly)	1776	−51.55	1776	−51.55
P20	20	2986	−18.54	2513	−31.45
P40	40	3356	−8.45	2985	−18.57
P60	60	3579	−2.37	3394	−7.41
P80	80	3652	−0.38	3626	−1.09
P100	100 (full)	3666	0	3666	0

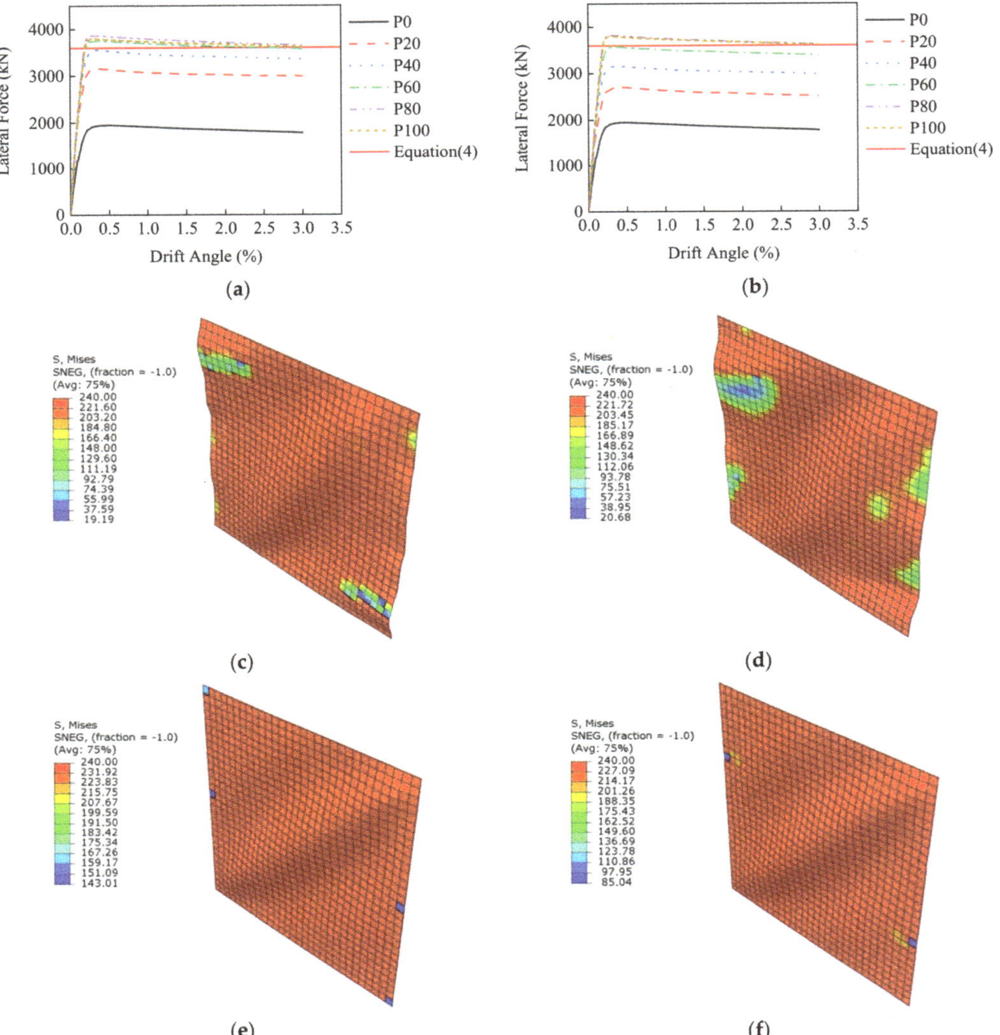

Figure 9. Load–displacement curves and stress distributions at 3% drift of S2 and S2R models: (**a**) S2; (**b**) S2R; (**c**) S2-P20; (**d**) S2R-P20; (**e**) S2-P80; (**f**) S2R-P80.

3.3. Structural Performance of S3 and S3R Models

The plate of S3 is connected to beams and columns from the centers, while S3R is connected from the corners. Table 5 displays the shear strengths of S3 and S3R. S3 and S3R have comparable shear capacities. The shear strength of the S3-P80 is almost as high as the full connection model, but the S3R-P80 has approximately 8% less strength. The stress distribution of the two types is obviously not the same in Figure 10c,d. S3-P13 has a visible tension field in the middle of the infill plate. The middle of the plate is yielded, whereas the unattached edge of the four corners has a large area in low stress. There is a large area of unyielding in the middle of the S3R-P13, where the yield area is applied only at the four corners without a tension field. When the connection ratio reaches 80, the plates of S3 and S3R almost yield, and both have an obvious tension field in Figure 10e,f.

Table 5. Results of shear strength for S3 and S3R models.

Model	Connection Ratio γ (%)	S3		S3R	
		Value (kN)	Decrease Ratio (%)	Value (kN)	Decrease Ratio (%)
P0	Not connected	0	−100	0	−100
P13	13	936	−74.46	762	−79.21
P27	27	1506	−58.91	1304	−64.42
P40	40	2101	−42.68	1842	−49.75
P53	53	2671	−27.14	2388	−34.86
P67	67	3212	−12.38	2855	−22.12
P80	80	3581	−2.31	3366	−8.18
P93	93	3665	−0.02	3655	−0.30
P30	100 (full)	3666	0	3666	0

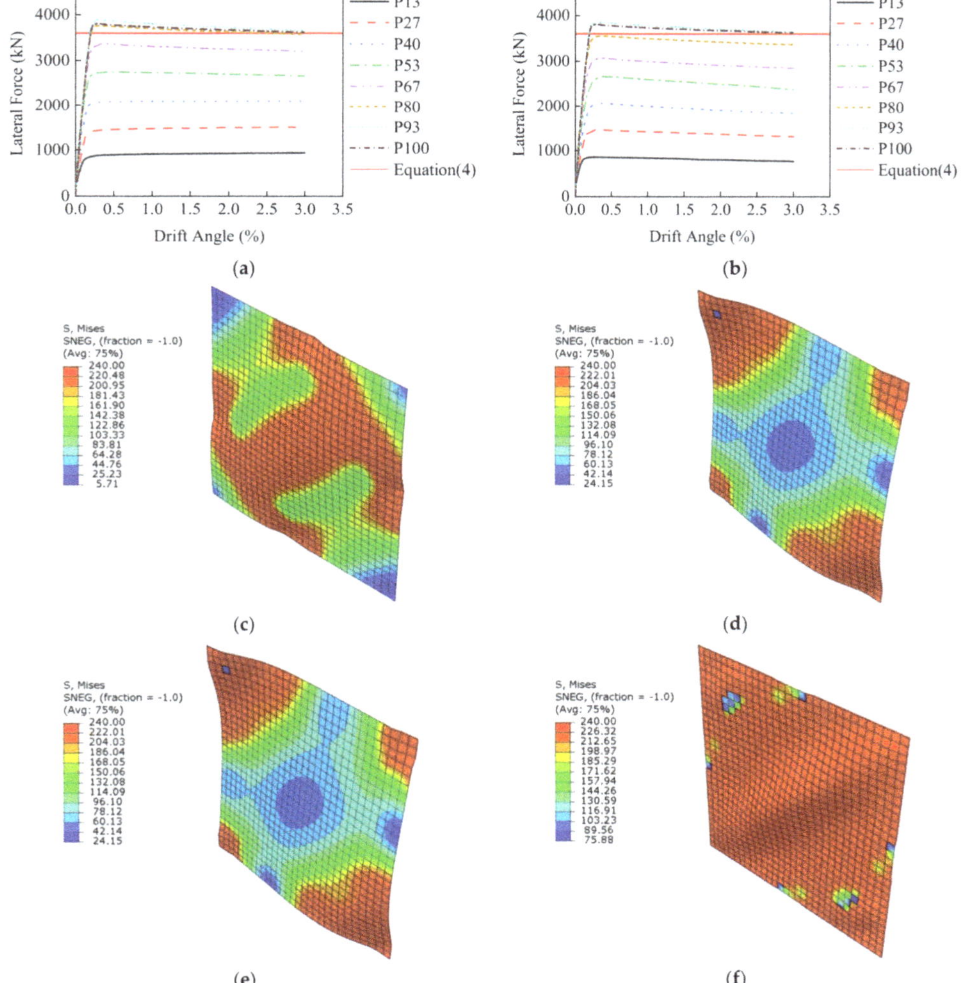

Figure 10. Load–displacement curves and stress distributions at 3% drift of S3 and S3R models: (**a**) S3; (**b**) S3R; (**c**) S3-P13; (**d**) S3R-P13; (**e**) S3-P80; (**f**) S3R-P80.

3.4. Structural Performance of S4 and S4R Models

S4 has a two-segment plate-to-frame connection, while S4R has a three-segment connection. Table 6 presents the shear strength of S4 and S4R. In Figure 11c, the S4-P20 has a large area yielding in the middle of the infill plate, whereas the corners are low stress or unstressed. In Figure 11d, the tension field exists on two sides of the S4R-P20 diagonal, yet the diagonal line does not yield. When the connectivity ratio of the S4-P80 and S4R-P80 is 80%, the shear strength is close to the full connection model. When the connection rate was 80% in Figure 11e,f, S4 and S4R almost fully yielded.

Table 6. Results of shear strength for S4 and S4R models.

Model	Connectivity Ratio γ (%)	S4		S4R	
		Value (kN)	Decrease Ratio (%)	Value (kN)	Decrease Ratio (%)
P0	Not connected	0	−100	0	−100
P20	20	1589	−56.65	1413	−61.45
P40	40	2463	−32.81	2238	−38.95
P60	60	3298	−10.03	3054	−16.69
P80	80	3638	−0.76	3587	−2.15
P100	100 (full)	3666	0	3666	0

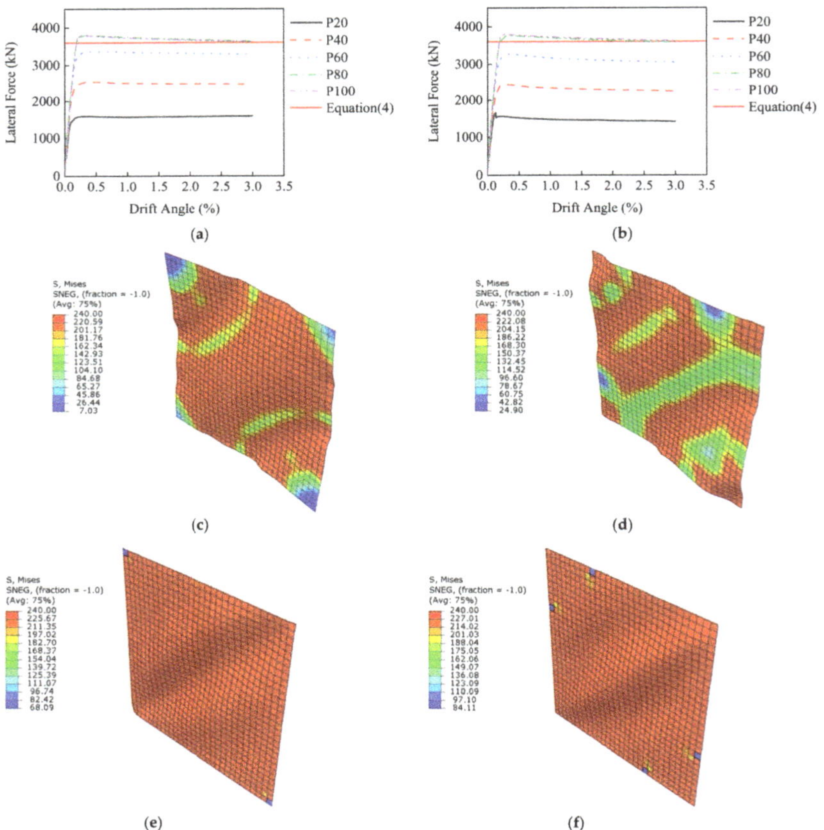

Figure 11. Load–displacement curves and stress distributions at 3% drift of S4 and S4R models: (**a**) S4; (**b**) S4R; (**c**) S4-P20; (**d**) S4R-P20; (**e**) S4-P80; (**f**) S4R-P80.

4. Discussion of Shear Strength of PC-SPSW

4.1. Influence of Partial Connection on Shear Strength

The shear capacities of partial connection steel plate shear walls are compared in Figure 12.

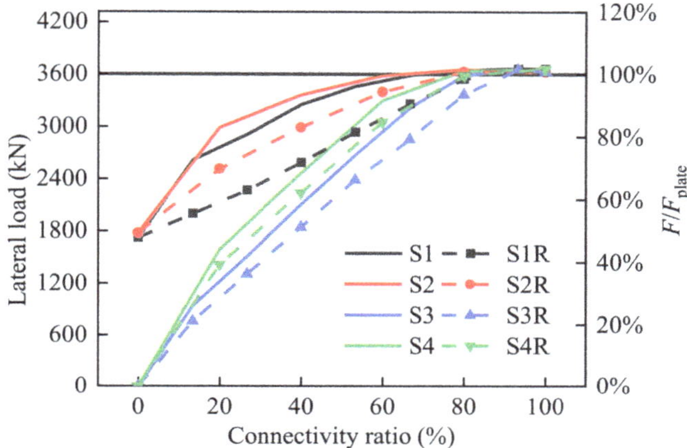

Figure 12. Relationship between shear strength and connectivity ratio.

Firstly, the effect of the connectivity ratio on the shear strength is examined. When the connectivity ratio is less than 50%, the shear strength of the fully beam-connected and partially column-connected types is greater than the four-edge partial connection because being fully beam-connected provides anchorage for the development of the infill plate tension field. Moreover, the low connectivity ratio prevents the infill plate from developing a significant area of the tension field, limiting the behavior of the infill plate material. The shear strength of the four-edge partial connection type increases with increasing the connectivity ratio. When the connectivity ratio hits 50%, even the shear strength of S4 exceeds that of S1R. Next, when the connectivity ratio exceeds 80%, the shear strength of both the two-edge partial connection and the four-edge partial connection is close to the case of the full connection. This is due to the infill plate's ability to produce a relatively complete tension field and fully develop the material properties of the steel plate.

Secondly, the influence of the connection types on the shear strength is analyzed. The fully beam-connected and partially column-connected types, such as S1 and S2, can still have approximately 80% of the shear strength even at the connectivity ratio of 20%. When the connectivity ratio is 60%, S1, S2, S2R, and S4 can retain more than 90% of the shear strength. Especially, S4 is partially connected on four edges, which means that the S4 model can utilize most of the material properties of the infill plate at a lower connectivity ratio. Therefore, the connection type of the S4 can significantly reduce the material and construction time required for plate-to-frame connection, and the S4 model is easy to construct and is a great connection type.

Next, the number of connected segments influences the shear strength. S1 and S3 are one-segment connections, while S2 and S4 are multi-segment connections. The shear strength of the steel plate shear wall with a multi-segment connection is higher than that of the one-segment connection for the same connectivity ratio. This is due to the fact that the multi-segment connection extends the connection range, making the area of the tension field more than that of the one-segment connection. Therefore, the distance of the segments can be selected appropriately.

Finally, the different effects of the connections from centers or corners on the shear strength are analyzed. For example, S1 and S3 are the connection from the centers, while S1R and S3R are the connection from the corners. Under the same connectivity ratio, the shear strength of the connection from corners is basically lower than that from the centers. This is because the tension field is firstly generated along the diagonal when the infill plate buckles. The type of connection from the centers shows an obvious tension field in the diagonal direction, such as in Figure 10c. In the case of the connection from the corners, the tension field generated along the diagonal cannot develop effectively because there is no restraint in the middle of the infill plate. Therefore, most of the infill plate does not enter the yielding stage, and the shear strength of the structure is small (in Figure 10d). In summary, the influence of the central connection on the structural shear strength is higher than that of the corner connection.

4.2. Shear Strength Equations of S1 and S1R

The shear strength of the infill plate with different connectivity ratios is fitted according to the finite element model, as shown in Figure 13. The shear strength of the S1 model increases rapidly with increasing connectivity ratio when the connectivity ratio is low. By changing the width (b = 1500, 3000, 4500 mm) of the steel plate shear wall to change the width-to-height (aspect) ratio of the infill plate, the prediction Equation (10) still agrees well with the finite element results. The bearing capacity and the connectivity ratio of the S1R model can be simplified to a linear relationship. The prediction Equation (11) under different width-to-height ratios is also in good agreement with the finite element results. Equations (10) and (11) can accurately predict the shear strength. The parameter $h/(h + b)$ in Equations (10) and (11) can reflect the changes in the width-to-height (aspect) ratio of the infill plate. Furthermore, this parameter indicates the relationship between the length of partially connected models to the total length.

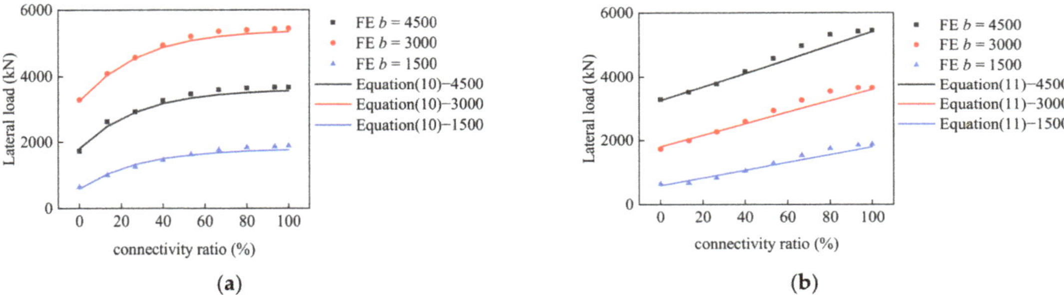

Figure 13. Comparison results of prediction equation and FE method for: (**a**) S1, and (**b**) S1R models.

When the plate is fully connected to the beams, the shear strength of the infill plate that is partially column-connected from the centers, such as S1, is shown in Equation (10). Equation (11) shows the shear strength of the infill plate that is partially column-connected from the corners, such as S1R.

$$F_{S1} = [1 - f_{S1}(\gamma)] \cdot F_{\text{plate}} = \left[1 - \frac{h}{h + b}e^{3.5\gamma}\right] \cdot \frac{1}{2}\sigma_y bt \sin(2\theta) \tag{10}$$

$$F_{S1R} = [1 - f_{S1R}(\gamma)] \cdot F_{\text{plate}} = \left[1 - \frac{h}{h + b}(1 - \gamma)\right] \cdot \frac{1}{2}\sigma_y bt \sin(2\theta) \tag{11}$$

4.3. Shear Strength Equations of S2 and S2R

Figure 14 represents the relationship between shear strength and connectivity ratio for different width-to-height ratios of S2 and S2R. The predicted results from Equations (12) and (13) match the finite element values. The parameter $h/(h + b)$ reflects the changes in width-to-height ratios with the length of partially connected models.

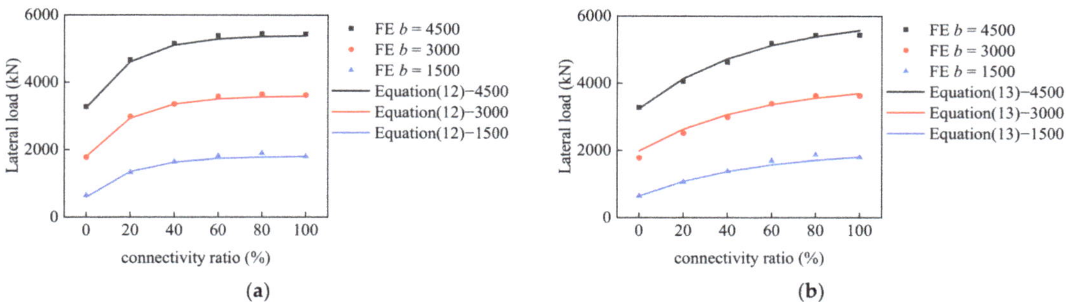

Figure 14. Comparison results of prediction equation and FE method for (**a**) S2 and (**b**) S2R models.

S2 and S1 are similar in connection types, therefore, Equation (12) is similar to Equation (10). However, the coefficients for the parameter of connectivity ratio are different. In the S2 example, the coefficient of the connectivity ratio is 5 in Equation (12), which is larger than the value of 3.5 in Equation (10). This coefficient reflects the influence of the number of connection segments on the shear strength. S1 is one segment connected, and S2 is two segments connected. The more segments, the greater the range of connection action.

$$F_{S2} = [1 - f_{S2}(\gamma)] \cdot F_{\text{plate}} = \left[1 - \frac{h}{h+b}e^{5\gamma}\right] \cdot \frac{1}{2}\sigma_y bt \sin(2\theta) \tag{12}$$

The equation for predicting the shear strength of the S2R type is shown as follows. In Equation (13), there is a constant of 1.1, which indicates the enhancement factor of the structure when the three segments of pate are connected from the centers and corners.

$$F_{S2R} = [1 - f_{S2R}(\gamma)] \cdot F_{\text{plate}} = \left[1.1 - \frac{h}{h+b}(1.1e^{\gamma})^2\right] \cdot \frac{1}{2}\sigma_y bt \sin(2\theta) \tag{13}$$

4.4. Shear Strength Equations of S3 and S4

The relationship between the shear strength and connectivity ratio of S3 and S4 is shown in Figure 15. Both S3 and S4 are the four-edged models partially connected to the beams and columns. Assuming the same connectivity ratio of beams and columns, here we only discuss the case of the width-to-height ratio of 1, that is, the parameter $(h + b)/(h + b)$ is 1. The prediction equations are consistent with the finite element results. The connection type of S3 is similar to S1, but the difference is that they are two-edge or four-edge partial connection models. Comparing Equations (14) and (10), the coefficient for the parameter of connectivity ratio is reduced by half to be 1.75, and the enhancement factor is 1.2. Similar conclusions can be drawn from Equations (15) and (12) for S4 and S2, where the coefficient for the parameter of connectivity ratio is also reduced by half to be 2.5 and the enhancement factor is 1.1.

$$F_{S3} = [1 - f_{S3}(\gamma)] \cdot F_{\text{plate}} = \left[1.2\left(1 - \frac{h+b}{h+b}e^{1.75\gamma}\right)\right] \cdot \frac{1}{2}\sigma_y bt \sin(2\theta) \tag{14}$$

$$F_{S4} = [1 - f_{S4}(\gamma)] \cdot F_{\text{plate}} = \left[1.1\left(1 - \frac{h+b}{h+b}e^{2.5\gamma}\right)\right] \cdot \frac{1}{2}\sigma_y bt \sin(2\theta) \tag{15}$$

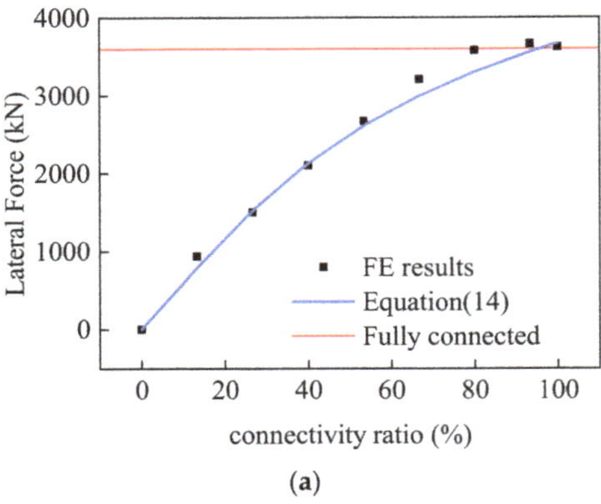
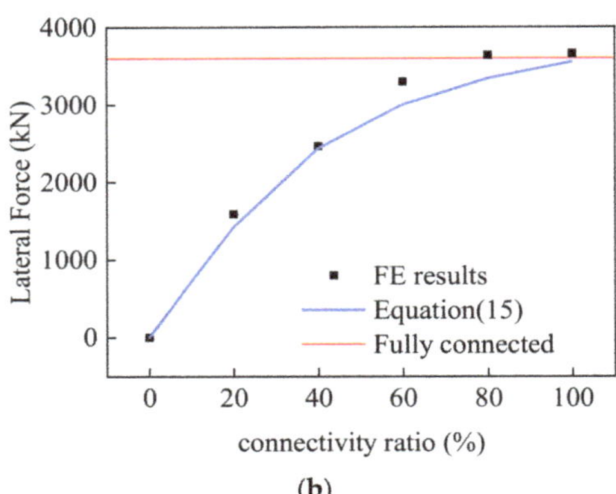

Figure 15. Comparison results of prediction equation and FE method for: (**a**) S3, and (**b**) S4 models.

5. Conclusions

1. This research proposes eight different types of plate-to-frame connections and investigates the effect of the connectivity ratio on the shear strength of the steel plate shear wall. When the infill plate is only connected to beams, the shear strength decreases by nearly 50% compared to the full connection. However, when the central part of the plate is connected to the column, such as S1 and S2, the load-bearing capacity of the structure is significantly higher than that of the beam-only connection. When the connectivity ratio of plate-to-columns hits 67%, the shear strength of the structure is almost equivalent to the full connection.

2. Considering the connectivity ratio and width-to-height ratio, the formulae of shear strength with different connection types are presented. The prediction formula correlates with the finite element results well. The shear strength increases rapidly with an increase in the connectivity ratio. When the connection ratio reaches 80%, the shear strength of the structure reaches more than 95% of the full connection type.

3. Multi-segment connections obtain a higher shear strength than one-segment connections at the same connectivity ratio. Increasing the connectivity ratio of the plate from the centers can improve the shear strength of the structure more than from the corners.

Author Contributions: Conceptualization, Y.Y. and Z.M.; methodology, Y.Y.; software, Y.Y.; validation, Y.Y.; formal analysis, Y.Y.; investigation, Y.Y.; resources, Y.Y., Z.M. and B.Z.; data curation, Y.Y.; writing—original draft preparation, Y.Y.; writing—review and editing, Z.M. and B.Z.; visualization, Y.Y.; supervision, Z.M. and B.Z.; project administration, Z.M.; funding acquisition, Z.M. All authors have read and agreed to the published version of the manuscript.

Funding: This research was funded by the National Natural Science Foundation of China, grant number 51578064 and the Natural Science Foundation of Beijing Municipality grant number 8172031.

Institutional Review Board Statement: Not applicable.

Informed Consent Statement: Not applicable.

Data Availability Statement: All the data supporting the results were provided within the article.

Conflicts of Interest: The authors declare no conflict of interest.

References

1. Elgaaly, M. Thin steel plate shear walls behavior and analysis. *Thin-Walled Struct.* **1998**, *32*, 151–180. [CrossRef]
2. Wagner, H. *Flat Sheet Metal Girder with Very Thin Metal Web: Part I General Theories and Assumptions*; Technical Memorandum, No. 604; National Advisory Committee for Aeronautics: Moffett Field, CA, USA, 1931.
3. Basler, K. Strength of plate girders in shear. *J. Struct. Div. ASCE* **1961**, *87*, 151–180. [CrossRef]

4. Thorburn, L.J.; Kulak, G.L.; Montgomery, C.J. *Analysis of Steel Plate Shear Walls*; Structural Engineering Report. No. 107; Department of Civil Engineering, University of Alberta: Edmonton, AB, Canada, 1983.
5. Jin, S.; Yang, S.; Bai, J. Numerical and experimental investigation of the full-scale buckling-restrained steel plate shear wall with inclined slots. *Thin-Walled Struct.* **2019**, *144*, 106362. [CrossRef]
6. Mu, Z.; Yang, Y. Experimental and numerical study on seismic behavior of obliquely stiffened steel plate shear walls with openings. *Thin-Walled Struct.* **2020**, *146*, 106457. [CrossRef]
7. Farahbakhshtooli, A.; Bhowmick, A.K. Nonlinear seismic analysis of perforated steel plate shear walls using a macro-model. *Thin-Walled Struct.* **2021**, *166*, 108022. [CrossRef]
8. Tong, J.; Guo, Y.; Zuo, J.; Gao, J. Experimental and numerical study on shear resistant behavior of double-corrugated-plate shear walls. *Thin-Walled Struct.* **2020**, *147*, 106485. [CrossRef]
9. Azandariani, M.G.; Gholhaki, M.; Kafi, M.A. Experimental and numerical investigation of low-yield-strength (LYS) steel plate shear walls under cyclic loading. *Eng. Struct.* **2020**, *203*, 109866. [CrossRef]
10. De Matteis, G.; Brando, G.; D'Agostino, M.F.M. Experimental analysis of partially buckling inhibited pure aluminium shear panels. In *Behaviour of Steel Structures in Seismic Areas*; CRC Press: Boca Raton, FL, USA, 2012; pp. 809–814. [CrossRef]
11. De Matteis, G.; Brando, G. Comparative analysis of dual steel frames with dissipative metal shear panels. *Key Eng. Mat.* **2018**, *763*, 735–742. [CrossRef]
12. Yu, J.; Huang, J.; Li, B.; Feng, X. Experimental study on steel plate shear walls with novel plate-frame connection. *J. Constr. Steel Res.* **2021**, *180*, 106601. [CrossRef]
13. Paslar, N.; Farzampour, A.; Hatami, F. Infill plate interconnection effects on the structural behavior of steel plate shear walls. *Thin-Walled Struct.* **2020**, *149*, 106621. [CrossRef]
14. Azandariani, G.M.; Rousta, A.M.; Mohammadi, M.; Rashidi, M.; Abdolmaleki, H. Numerical and analytical study of ultimate capacity of steel plate shear walls with partial plate-column connection (SPSW-PC). *Structures* **2021**, *33*, 3066–3080. [CrossRef]
15. Wei, M.; Liew, J.Y.R.; Fu, X. Panel action of novel partially connected buckling-restrained steel plate shear walls. *J. Constr. Steel Res.* **2017**, *128*, 483–497. [CrossRef]
16. Wei, M.; Liew, J.Y.R.; Yong, D.; Fu, X. Experimental and numerical investigation of novel partially connected steel plate shear walls. *J. Constr. Steel Res.* **2017**, *132*, 1–15. [CrossRef]
17. Du, Y.; Zhang, Y.; Zhou, T.; Chen, Z.; Zheng, Z.; Wang, X. Experimental and numerical study on seismic behavior of SCFRT column frame-buckling restrained steel plate shear wall structure with different connection forms. *Eng. Struct.* **2021**, *239*, 112355. [CrossRef]
18. Mirsadeghi, H.M.R.; Fanaie, N. Steel plate shear walls with partial length connection to vertical boundary element. *Structures* **2021**, *32*, 1820–1838. [CrossRef]
19. ABAQUS. *Analysis User's Manual IV. Version 6.10*; ABAQUS, Inc.: Palo Alto, CA, USA; Dassault Systèmes: Vélizy-Villacoublay, France, 2010.
20. Asl, M.H.; Safarkhani, M. Seismic behavior of steel plate shear wall with reduced boundary beam section. *Thin-Walled Struct.* **2017**, *116*, 169–179. [CrossRef]
21. Machaly, E.B.; Safar, S.S.; Amer, M.A. Numerical investigation on ultimate shear strength of steel plate shear walls. *Thin-Walled Struct.* **2014**, *84*, 78–90. [CrossRef]
22. Wang, M.; Zhang, X.; Yang, L.; Yang, W. Cyclic performance for low-yield point steel plate shear walls with diagonal T-shaped-stiffener. *J. Constr. Steel Res.* **2020**, *171*, 106163. [CrossRef]
23. Shekastehband, B.; Azaraxsh, A.A.; Showkati, H.; Pavir, A. Behavior of semi-supported steel shear walls: Experimental and numerical simulations. *Eng. Struct.* **2017**, *135*, 161–176. [CrossRef]
24. Sigariyazd, M.A.; Joghataie, A.; Attari, N.K.A. Analysis and design recommendations for diagonally stiffened steel plate shear walls. *Thin-Walled Struct.* **2016**, *103*, 72–80. [CrossRef]
25. Timoshenko, S.P.; Gere, J.M. *Theory of Elastic Stability*; McGraw-Hill Publishing Company: New York, NY, USA, 1961.
26. Sabelli, R.; Bruneau, M. *Steel Design Guide of Steel Plate Shear Walls*; No. 20; American Institute of Steel Construction (AISC): Chicago, IL, USA, 2007.
27. Ozcelik, Y.; Clayton, P.M. Strip model for steel plate shear walls with beam-connected web plates. *Eng. Struct.* **2017**, *136*, 369–379. [CrossRef]
28. Nascimbene, R. Towards Non–standard numerical modeling of thin-shell structures: Geometrically linear formulation. *Int. J. Comput. Methods Eng. Sci. Mech.* **2014**, *15*, 126–141. [CrossRef]
29. BS-EN 1993-1-6:2007; Eurocode 3: Design of Steel Structures Part 1–6: Strength and Stability of Shell. European Committee for Standardization: Brussels, Belgium, 2007.
30. Habashi, H.R.; Alinia, M.M. Characteristics of the wall–frame interaction in steel plate shear walls. *J. Constr. Steel Res.* **2010**, *66*, 150–158. [CrossRef]

Article

Fatigue Analysis of Long-Span Steel Truss Arched Bridge Part II: Fatigue Life Assessment of Suspenders Subjected to Dynamic Overloaded Moving Vehicles

Peng Liu [1], Hongping Lu [1], Yixuan Chen [1], Jian Zhao [2], Luming An [2], Yuanqing Wang [3] and Jianping Liu [1,*]

[1] School of Architecture and Civil Engineering, Shenyang University of Technology, Shenyang 110870, China; pliu@sut.edu.cn (P.L.); lhp19971008@163.com (H.L.); cyx980201@163.com (Y.C.)
[2] China Railway Bridge Engineering Group Co., Ltd., Beijing 100039, China; zhaojianll@126.com (J.Z.); anluming@stdu.edu.cn (L.A.)
[3] Department of Civil Engineering, Tsinghua University, Beijing 100190, China; wang-yq@mail.tsinghua.edu.cn
* Correspondence: liujianping024@sut.edu.cn

Abstract: In a half-through steel arched bridge, the suspenders are the critical load transfer component that transmits the deck system and traffic load to the arch rib. These suspenders are subjected to traffic and environmental vibrations and are prone to fatigue failure, especially for overloaded moving vehicles. This paper aims to study the impact of moving vehicles' overloaded rate on the fatigue performance of suspenders in a long-span three steel truss arch bridge. Based on the Mingzhu Bay steel arch bridge, a 3D finite element bridge model was first established and seven types of moving fatigue vehicle models were considered. Then the stress amplitude and dynamic response of the suspenders on the middle steel truss arch were studied under a standard, 25%, and 50% overloaded moving vehicles load. Following that, the Miner fatigue cumulative damage theory was employed to evaluate the fatigue life of the suspenders. The results show that the short suspenders in the middle steel truss arch have the shortest fatigue life but can still meet the design requirements under the standard load. However, the fatigue life of the suspenders decreases by 20% and 30% when the overloading rate reaches 25% and 50%. The fatigue life cannot meet the design requirement when the overload rate is 50%.

Keywords: suspender; steel arch bridge; fatigue life; overloaded rate; finite element

Citation: Liu, P.; Lu, H.; Chen, Y.; Zhao, J.; An, L.; Wang, Y.; Liu, J. Fatigue Analysis of Long-Span Steel Truss Arched Bridge Part II: Fatigue Life Assessment of Suspenders Subjected to Dynamic Overloaded Moving Vehicles. *Metals* **2022**, *12*, 1035. https://doi.org/10.3390/met12061035

Academic Editors: Zhihua Chen, Hanbin Ge and Siu-lai Chan

Received: 30 April 2022
Accepted: 14 June 2022
Published: 17 June 2022

Publisher's Note: MDPI stays neutral with regard to jurisdictional claims in published maps and institutional affiliations.

1. Introduction

The half-through steel truss arch bridge has been widely used in modern bridges due to its advantage of having excellent load capacity and a simple construction method [1]. The suspenders of the steel truss arch bridge are the key load-bearing component that transmit the deck system load, moving vehicle load, and wind load to the upper arch ribs. These suspenders are subjected to repeated cyclic stresses with a varying magnitude as a result of the dynamic vehicle loading during their lifecycle [2]. The short suspender of the arch bridge has a high amount of stiffness and is subject to high vibration frequency. The anchorage end of the arch bridge is in a frequent state of bending and shear stress and is vulnerable to rain and other corrosion effects [3]. For the dynamic load of the overloading vehicles in particular, the dynamic amplification effect can accelerate the fatigue damage as a result of the large stress ranges induced [4]. The accumulated fatigue damage can initialize the crack and may cause the bridge's failure. However, the fatigue design may underestimate the effect of overloading vehicles on the fatigue life of steel bridges [5]. For the fatigue life analysis of metal structures, Carpinteri et al. [6] reformulated frequency-domain critical-plane criteria to improve the accuracy of the fatigue life estimation of smooth metal structural members under multiaxial random loads. Maercek et al. [7] provided methods for the determination of energy-based characteristics of the fatigue life of materials and a new approach for the determination of strain energy density parameters. Pejkowski et al. [8] studied small cracks on the surface of fatigue specimens under multiaxial loading and found that there was a correlation between the observed damage mechanism and the quality of fatigue life prediction.

Many efforts have been made to investigate the effect of dynamic vehicle loading on the fatigue life of suspenders. Zhang et al. [9,10] established the finite element model of suspenders, and the fatigue reliability of a long-span bridge was analyzed under the combined dynamic load of vehicles and the wind load. The results showed that wind load significantly increased the stress on the suspender, but had little impact on the fatigue damage of the suspender. The dynamic effect of vehicles on long-span bridges is little and the impact of vehicle speed and road roughness can be ignored. Yang et al. [11] analyzed the impact of five-axis vehicles on simply supported and continuous beam bridges; the parametric studies were conducted and an equation for continuous beams subjected to moving vehicle loads was proposed. The results showed that the deflection, bending moment, and shear force at the midpoint of simply supported beams are linearly proportional to the velocity parameters. Based on the field monitoring data, Kwon et al. [12,13] used the equivalent probability density function to evaluate the fatigue reliability of steel bridges and proposed a method to predict the fatigue reliability evaluation of bridges in terms of the probability of equivalent stress amplitude distribution. To improve fatigue life estimation, the bilinear S–N method was integrated into a probabilistic model to analyze the uncertainties related to the fatigue deterioration process, and a probabilistic method based on bilinear stress life was proposed. Raju et al. [14] proposed a fatigue evaluation and design method to assess the remaining life of existing bridges; the results showed that the proposed method can evaluate the safety index more accurately. In addition, fatigue design methods have been included in the AASHTO specification.

Due to the randomness and complexity of moving vehicle loads, many scholars have established numerical vehicle models to simplify moving vehicle loads, and the fatigue performance of bridges has been investigated. Mohammadi et al. [5,15] proposed a bridge rated model considering the influence of overloading fatigue damage. The model was applied to five bridges, and it was shown that the fatigue damage caused by overloading is significant for old bridges, Furthermore, frequent overloading accelerates fatigue damage and shortens the fatigue life of bridges. E. J. Obrient et al. [16,17] proposed a semi-parametric fitting program that can simulate the traffic load effect. A micro-simulation algorithm was proposed to predict more accurately the applied traffic load on the long-span bridge when two vehicles were mixed between lanes in a traffic jam. Schilling et al. [18–21] investigated the design value of the number of stress cycles caused by vehicles passing through the span of various types of steel-frame highway bridges and a numerical method was proposed. The results showed that vibration response and vehicle spacing have little impact on the design value of cyclic stress under standard conditions. Wang et al. [22,23] established a mathematical model for counting typical vehicles, compared the static effect of heavy vehicles with that of standard designed vehicles, and studied the influence of the road roughness correlation on dynamic influencing factors. The results showed that the tandem axle load of heavy vehicles exceeded the limit in the guidelines. Zhao et al. [24] conducted a statistical analysis of the maximum bending moment and shear force of simply supported, 2-span, and 3-span continuous beam bridges; the top 5% of heavy vehicles in each vehicle category were selected and the results showed that a 5-axle, short, single vehicle produces greater shear forces in bridge girders than standard Permit truck models, and the 5-axle vehicle was proposed for the standard Permit vehicle and bridge ratings. Zhang et.al. [25,26] established the three-dimensional vehicle model and 3D dynamic bridge model, and the effect of vehicle speed and road roughness conditions on existing bridges was investigated. A fatigue reliability assessment framework for existing bridges in their service life considering the random effects of vehicle speed and road roughness conditions was proposed. This method can be used to modify the fatigue damage of multiple stress ranges with different amplitudes into one equivalent stress range cycle and shows more accuracy in a lognormal distribution to describe the modified equivalent stress range.

However, many studies mainly focused on investigating how the load affects the fatigue behavior of in-service bridges, and little attention has been paid to considering the effect of overloaded moving vehicles on the fatigue life of steel arch bridges. This paper aims to study the impact of moving vehicles' overloaded rate on the fatigue performance of suspenders in a long-span three steel truss arch bridge. Based on the Mingzhu Bay steel arch bridge, a 3D finite element bridge model was established and seven types of moving vehicle models were considered. The stress amplitude and dynamic response of the middle steel truss arch suspenders were studied under standard, 25%, and 50% overloaded moving vehicles. In addition, the Miner fatigue cumulative damage theory was employed to evaluate the fatigue life of suspenders subjected to standard and overloaded moving vehicles. The flow chart of this paper is shown in Figure 1.

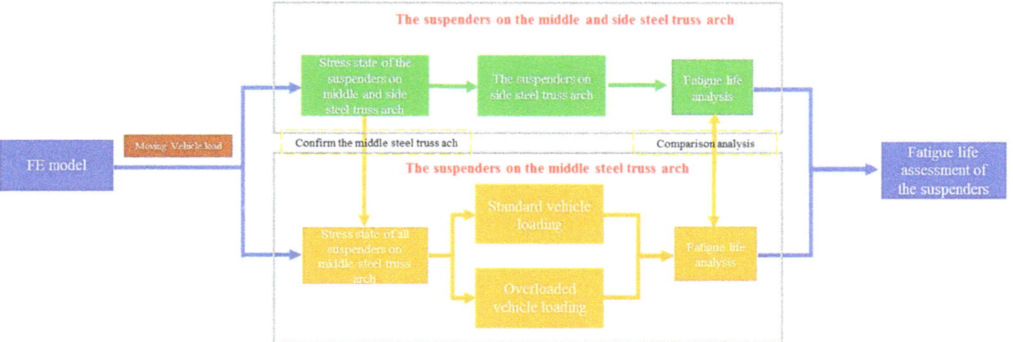

Figure 1. Flow chart of analysis.

2. Project Summary

The main bridge of Mingzhu Bay is (96 + 164 + 436 + 164 + 96 + 60 m) a six-span continuous three-steel truss arch bridge with a total length of 1016 m. Double orthotropic steel decks are used. The upper deck is a two-way eight-lane highway and there are sidewalks are on both sides. The total width of the deck is 43.2 m [24,27]. Both sides of the lower deck are reserved for the walking lanes. Figure 2 shows the bridge elevation and the schematic diagram of the side span and secondary side span of the main bridge.

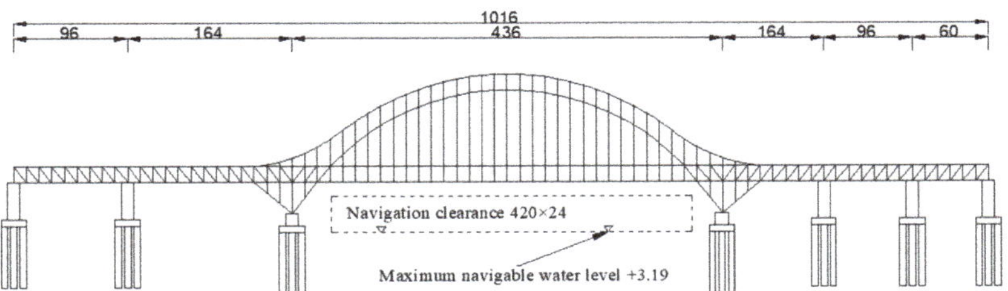

Figure 2. Elevation of Guangzhou Mingzhu Bay bridge (Unit: m).

3. Finite Element Analysis

3.1. FE Model of the Bridge

The 3D finite element bridge model was established by Midas/Civil [28], as shown in Figure 3. The whole model has 6676 nodes and 10,768 elements in total, which consist of 81 truss elements, 8067 beam elements, and 2620 shell elements.

Figure 3. FE model of Guangzhou Mingzhu Bay bridge.

3.2. Suspenders' Parameters

Truss elements were used to simulate the suspender of the steel truss arch bridge. Since the steel truss arch bridge is symmetrical, the half-span suspender was selected for analysis. For the transverse direction of the bridge, the short suspenders of the middle truss arch and the side truss arch were selected for the comparative analysis of fatigue life, as shown in Figure 4a, while the numbers of the suspenders are shown in Figure 4b. Figure 5 illustrates the geometry of the suspender.

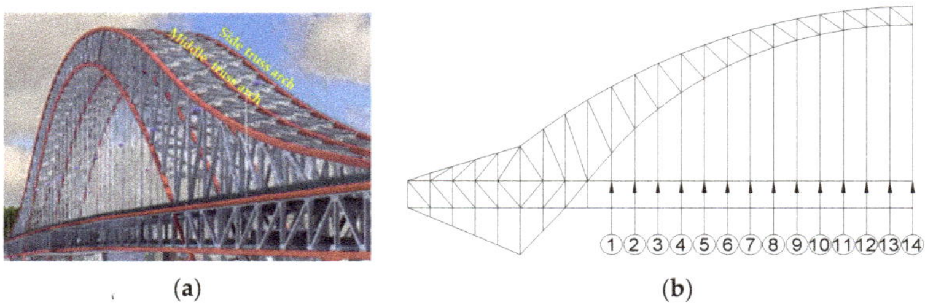

(**a**)   (**b**)

Figure 4. The steel truss arch and suspender number. (**a**) Steel truss arch; (**b**) the suspender numbers.

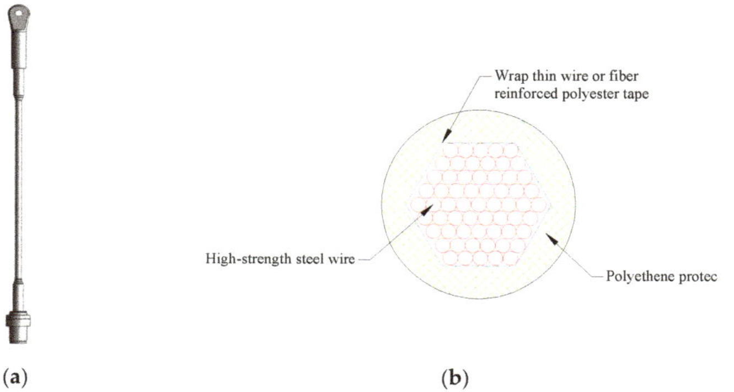

(**a**)   (**b**)

Figure 5. The geometry of the suspender. (**a**) The schematic of suspenders; (**b**) cross-section.

3.3. Chemical Composition and Mechanical Properties of Suspender Cable

The chemical composition and mechanical properties of suspenders are listed in Tables 1 and 2.

Table 1. Chemical composition of suspenders [29].

Chemical Composition	C	Si	Mn	Cr	S	Cu
Mass fraction (%)	0.85~0.90	0.12~0.32	0.60~0.90	0.10~0.25	≤0.025	≤0.10

Table 2. The mechanical properties of suspenders [30].

Nominal Diameter (mm)	Tensile Strength (MPa)	Elongation	Modulus of Elasticity (MPa)
7.0 ± 0.07	1670	≥ 4.0	$(2.0 \pm 0.1) \times 10^5$

4. Movie Vehicle Load and Fatigue Life Assessment Method

4.1. Moving Vehicle Load

The dynamic load of the vehicle is regarded as a spring-mass system with a uniform speed, while the design speed of the bridge is 80 km/h as simulated in the FE model. The vehicle load is considered as the harmonic force of uniform motion on the bridge deck. The vehicle is simplified to a uniformly moving biaxial mass. Based on the traffic volume statistics of Guangzhou Mingzhu Bay Bridge [31], the daily traffic flow can be predicted. According to the axle number and axle load, vehicles are divided into M1, M2, M3, M4, M5, M6, and M7 fatigue vehicle models. The vehicle load statistics and traffic flow statistics are shown in Table 3.

Table 3. Fatigue traffic model and traffic flow statistics.

Vehicle Model Type	Vehicle Model (Axle Weight, kN, Axle Spacing, mm)	Total Weight (kN)	Daily Traffic Flow	The Proportion of Total Traffic
M1	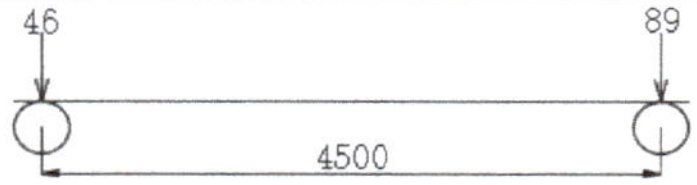	135	1543	14.78
M2	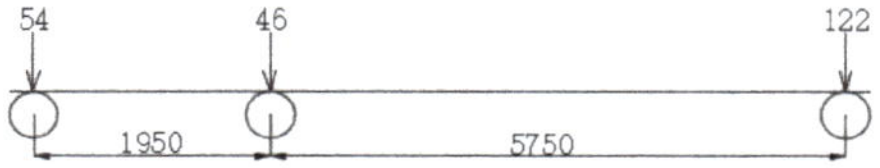	222	223	2.14
M3	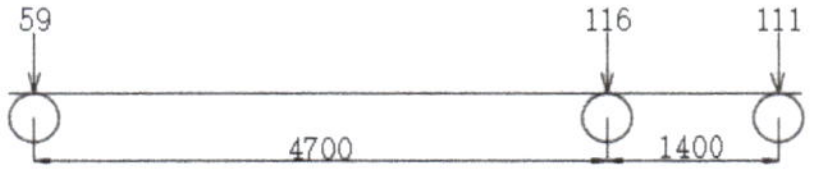	286	73	0.70
M4	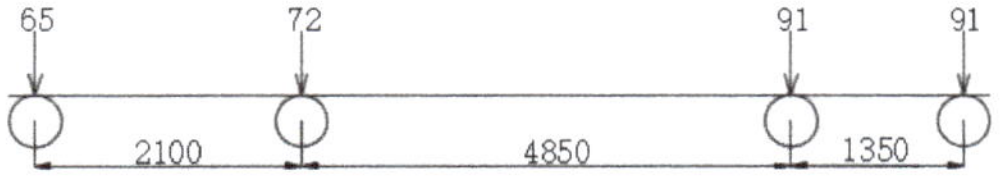	319	460	4.14
M5	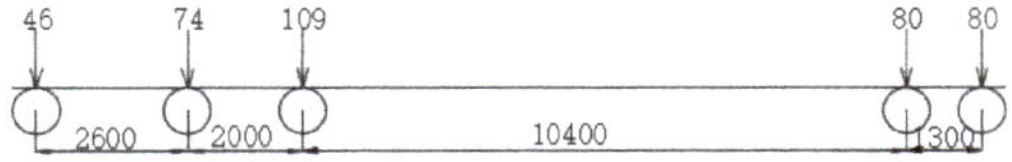	389	13	0.13
M6	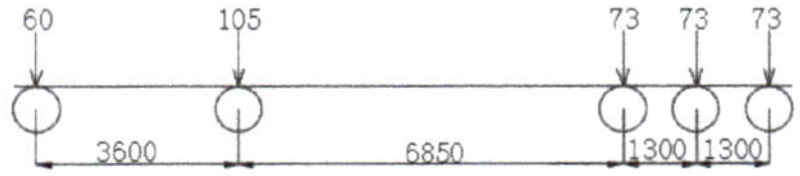	387	135	1.29
M7	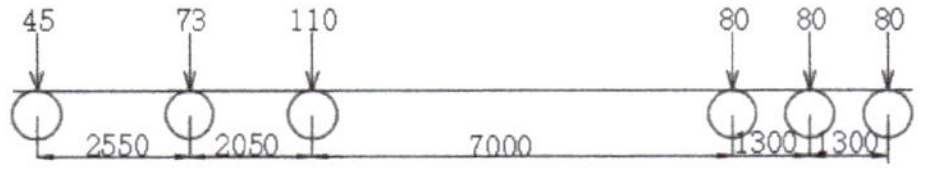	468	1453	13.92

4.2. Fatigue Life Assessment Method

Miner's linear cumulative fatigue damage theory [32] was adopted to evaluate the fatigue life of the suspenders. The fatigue damage of the suspender could be obtained from the *S-N* curve [33]. The fatigue life equation is listed in Equation (1):

$$\lg N = 15.1 + 13.5 \cdot \left(\frac{1}{\sigma_a} - \frac{\sigma_m}{\sigma_a - \sigma_b} \right) \tag{1}$$

where N is the number of cycles, σ_a, σ_b are the tensile strength of the suspenders and the mean stress of the suspenders, respectively, and σ_m is the mean stress value of σ_a and σ_b.

5. Results and Discussion

In order to investigate the stress state of the suspenders under a moving vehicle load, the unit moving vehicle load was applied to the bridge deck in the FE model. The vehicle started at 10 s and passed through the long-span arch bridge deck at 30 s. The stress amplitude of each suspender in the middle steel truss arch was studied, as shown in Figure 6.

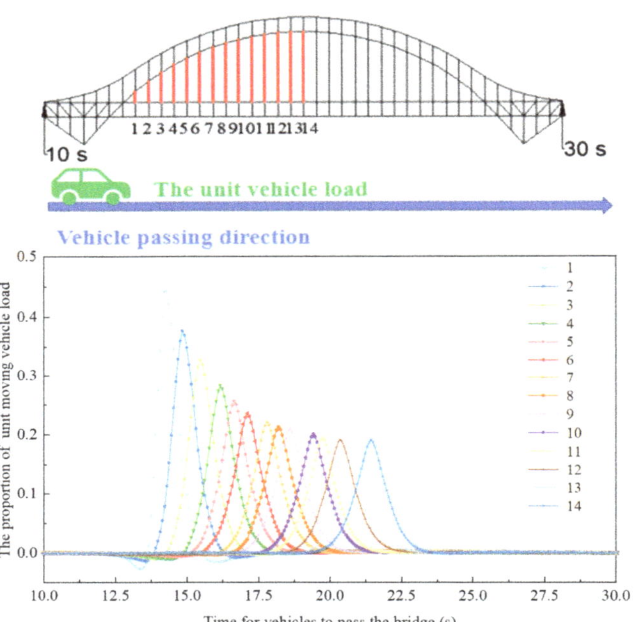

Figure 6. The suspender stress amplitude in middle steel truss arch.

It can be seen from Figure 6 that for the half arch suspenders from No. 1 to No. 14, the tensile stress amplitude decreased gradually when the vehicle passed on the bridge. The No. 1 suspender at the arch foot position achieved the largest stress amplitude.

5.1. Impact of Steel Truss Arch on the Short Suspender

There are three steel truss arches in the Mingzhu Bay bridge. The stress state of the short suspender in the middle and the side steel truss arch was investigated in the FE model. A number of stress cycles occur in the bridge members as the vehicle moves on the bridge. The dynamic responses of seven types of moving vehicles were considered and the stress amplitude of the suspender was obtained from the FE model. In addition, when the vehicle passes through the main span at a speed of 80 km/h, the time starts from 10 s and ends at 30 s. The stress of the suspender is shown in Figure 7 as the M1–M7 vehicle models passed the main span.

The Miner's cumulative fatigue damage theory indicates that all stress cycle amplitudes and their corresponding times of action need to be obtained for the fatigue assessment of the suspender. Therefore, it is necessary to process the fatigue stress spectrum of the suspender, extract all the cyclic amplitudes, and count the numbers of each cycle. This paper takes the short suspender of the steel truss arch bridge as the research object to evaluate the fatigue performance and predict the fatigue life. In terms of the rain-flow counting method [34,35], the stress amplitude of the short suspender of the middle truss and the side truss was counted for each cycle and accumulated by MATLAB [36] to generate the stress spectrum, as shown in Figure 8.

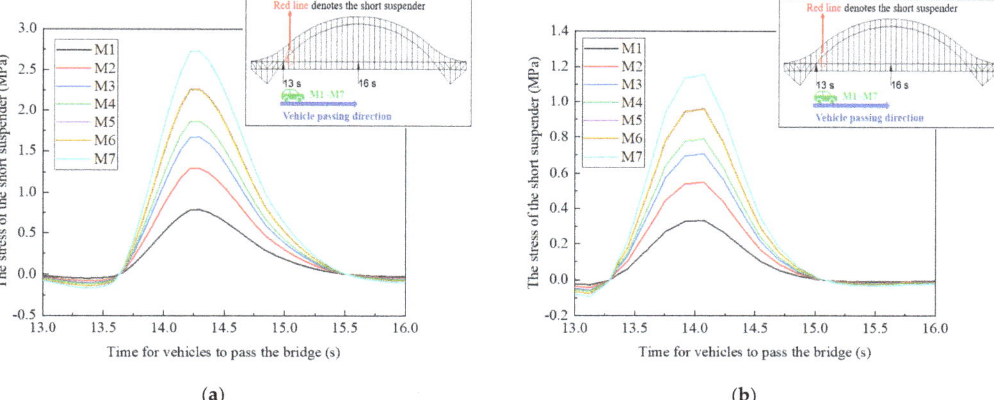

(**a**)

(**b**)

Figure 7. Dynamic response of the short suspenders under single vehicle load. (**a**) The middle steel truss arch; (**b**) the side steel truss arch.

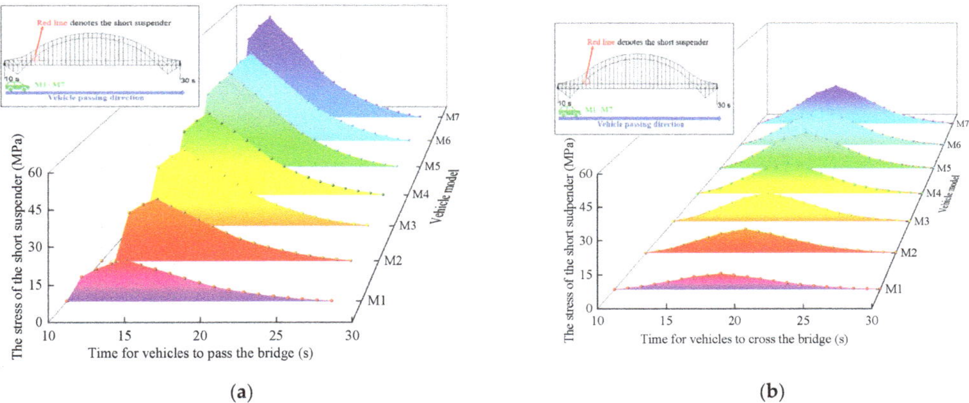

(**a**)

(**b**)

Figure 8. Dynamic response of the short suspenders under traffic load. (**a**) The middle steel truss arch; (**b**) the side steel truss arch; M1–M7 are vehicle model types.

For the middle steel truss arch, the stress on the suspender is about 50% higher than the suspender in the side steel truss arch with the same vehicle loading. Under the moving vehicle of M7, the maximum stress on the middle and the side suspender is 46.7 MPa and 23.6 MPa, respectively. Moreover, the stress amplitude and the number of cycles were extracted and the data are shown in Table 4.

The fatigue damage of the suspender can be obtained from the *S-N* curve Equation (1) to evaluate the fatigue life of the suspender. From the stress spectrum in Figure 8, the average stress of the short suspender of the middle truss is 274.16 MPa, the ultimate tensile stress of the suspender steel wire is 1670 MPa [37], and the *S-N* curve Equation (1) can be simplified into:

$$\lg N = 14.82 - 3.5\lg\sigma_a \tag{2}$$

Table 4. Stress amplitude statistics of the short suspender per day.

Stress Amplitude σ_a (MPa)	Number of Cycles (Times)	Stress Amplitude σ_a (MPa)	Number of Cycles (Times)
0~1	3900	25~30	0
1~5	1839	30~35	0
5~10	2061	35~40	0
10~15	0	40~45	0
15~20	0	45~50	0
20~25	0	>50	3900

The fatigue damage degree d under a single stress cycle can be deduced as

$$\text{d} = \frac{1}{\text{N}} = \frac{\sigma_a^{3.5}}{14.82^{10}} \tag{3}$$

With Miner's linear cumulative fatigue damage theory, the total fatigue damage degree D can be summed by the damage degree d

$$\text{D} = \sum d = \sum \frac{1}{N} = \sum \frac{\sigma_a^{3.5}}{14.82^{10}} \tag{4}$$

$$\sigma_{ae} = \left(\frac{\sum_{i=1}^{x} \sigma_{ai}^{m}}{x} \right)^{\frac{1}{m}} = \left(\frac{\sum \sigma_a^{3.5}}{n} \right)^{\frac{1}{3.5}} \tag{5}$$

where n is the total number of cyclic stresses; σ_{ae} is the equivalent stress amplitude of the suspender. Based on the data in Table 4 and Figure 8 the fatigue life of the short suspender can be calculated. Moreover, the fatigue life of the short suspender is 140 years, which meets the design life of 100 years.

5.2. Impact of Overloaded Rate

From the stress analysis of the suspenders from the middle truss and the side truss, it was found that the stress on the suspender in the middle truss arch was always larger than the suspenders in the side truss arch. Hence, the fatigue life assessment was conducted on the middle truss arch suspenders. The overload moving vehicles rate was investigated by a standard, 25%, and 50% overload with seven fatigue vehicle types; the overweight data and traffic flow are shown in Table 5.

Table 5. Overloaded vehicle data.

Vehicle Model	Standard	Overweight Rate		Traffic Flow Per Day
	Weight (kN)	25% (kN)	50% (kN)	
M1	135	168.75	202.5	1543
M2	222	277.5	333	223
M3	286	357.5	429	73
M4	319	398.5	478.5	460
M5	389	486.25	583.5	13
M6	387	483.75	580.5	135
M7	486	585	729	1453

The fatigue life of the suspenders on the half-span arch were evaluated under moving vehicle standard, 25%, and 50% overloaded rates (Figure 9).

It can be seen from Figure 9 the fatigue life of the suspender can meet the design life of 100 years under the standard fatigue load. With the overload rate up to 50%, the fatigue life of the suspender is lower than the design service life. For the short suspender along the bridge direction, the suspender decreases significantly, and its amplitude is about 50 years, which cannot meet the requirement of the design. The fatigue life of the suspender shows a non-linear decreasing trend with the increasing overloading ratio. Furthermore, the fatigue life of the suspender decreases by about 30%, which shows that the overload condition has a significant impact on the fatigue life of the suspender in the steel truss arch bridge.

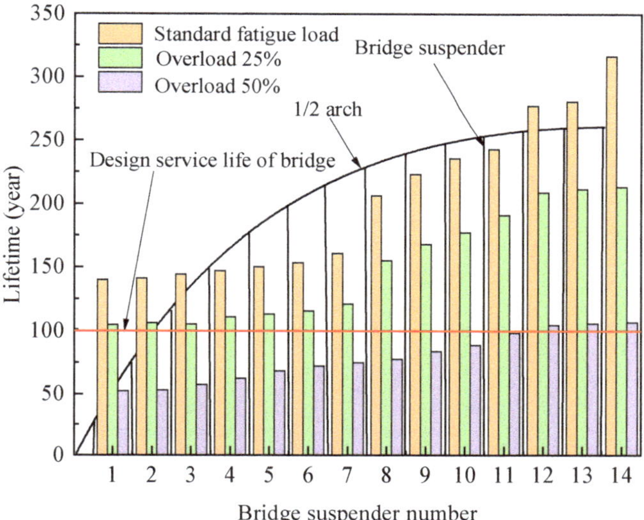

Figure 9. Fatigue life comparison of the suspenders in middle truss.

6. Conclusions

Based on the Guangzhou Mingzhu Bay bridge, the whole bridge model was established by the finite element software Midas Civil. The fatigue life of the suspender was evaluated under a standard fatigue load, and 25% and 50% overloading; the conclusions were drawn as follows:

1. The short suspenders near the arch foot show a larger stress amplitude than others, while the stress on the suspenders in the middle steel truss arch is greater than the suspenders in the side truss arch. However, the fatigue life of the short suspender under a standard traffic load is 140 years, which meets the design requirement of 100 years.
2. The fatigue life of the suspenders decreases by 20% and 30% under the overloaded rate of 25% and 50%, respectively. With a vehicle design speed of 80 km/h, the fatigue life of the short suspender on the mid-truss arch is about 50 years under the 50% overloaded rate, which is significantly lower than the bridge design life of 100 years.
3. The fatigue life shows a non-linear decreasing trend with the increasing overloading ratio. Overloading has a significant effect on the suspender life of steel truss arch bridges. Herein, the overloaded vehicles should be strictly controlled to ensure the safety of the bridge during its service.

The conclusions are based on the FE model analysis and Miner's cumulative fatigue damage theory. If we suppose that, at the specified stress level, the accumulation of fatigue damage is independent of the previous loading history of the material, then the damage from a single loading cycle should be the same, which means the fatigue life is not affected by the loading sequence regardless of whether the stress amplitude is from high to low or from low to high. However, the actual situation is that the loading sequence has a certain impact on the fatigue life of the components. Because the fatigue failure mechanism is under the same stress level, the damage degree with the microcrack formation and main crack growth is different. Therefore, the theoretical analysis results are usually not all the same as the test data in the construction.

Author Contributions: P.L., H.L. and Y.C. wrote the manuscript together; J.Z. and L.A. conducted data processing and drawings; Y.W. and J.L. revised the manuscript. All authors have read and agreed to the published version of the manuscript.

Funding: Science and Technology Research and Development Project of China Railway Construction Bridge Engineering Bureau Group Co., LTD. (DQJ-2018-A01), Tianjin Science and Technology Development Plan Project (19YDLZSF00030), National Natural Science Foundation of China (51038006).

Institutional Review Board Statement: Not applicable.

Informed Consent Statement: Not applicable.

Data Availability Statement: Not applicable.

Acknowledgments: The authors would like to thank China Railway Construction and Bridge Bureau Group Co., LTD. for the experiment support.

Conflicts of Interest: The authors declare no conflict of interest.

Nomenclature

N	number of fatigue cycles,
σ_m	mean stress value of σ_a and σ_b,
σ_a, σ_b	tensile strength of the suspenders and the mean stress of the suspenders,
d, D	fatigue damage per single cycle and fatigue damage per day,
m	coefficient of the fatigue damage,
x	number of cycles to reach a stress level,
n	number of vehicle cycles,
σ_{ae}	the equivalent stress amplitude of the suspender.

References

1. Yoda, T. *Classifications, Designloading, and Analysis Methods*; Butterworth-Heinemann: Oxford, UK, 2017; pp. 84–93.
2. Polepeddi, R. *The Effect of Fatigue Damage in Rating Bridges under Overload Conditions*; Illinois Institute of Technology ProQuest Dissertations Publishing: Chicago, IL, USA, 1997.
3. An, Y.; Guan, D.; Ding, Y.; Ou, J. Fast Warning Method for Rigid Hangers in a High-Speed Railway Arch Bridge Using Long-Term Monitoring Data. *J. Perform. Constr. Facil.* **2017**, *31*, 04017103. [CrossRef]
4. Wang, W.; Deng, L.; Shao, X. Fatigue Design of Steel Bridges Considering the Effect of Dynamic Vehicle Loading and Overloaded Trucks. *J. Bridg. Eng.* **2016**, *21*, 04016048. [CrossRef]
5. Mohammadi, J.; Polepeddi, R. Bridge Rating with Consideration for Fatigue Damage from Overloads. *J. Bridg. Eng.* **2000**, *5*, 259–265. [CrossRef]
6. Carpinteri, A.; Fortese, G.; Ronchei, C.; Scorza, D.; Spagnoli, A.; Vantadori, S. Fatigue life evaluation of metallic structures under multiaxial random loading. *Int. J. Fatigue* **2016**, *90*, 191–199. [CrossRef]
7. Macek, W.; Lagoda, T.; Mucha, N. Energy-based fatigue failure characteristics of materials under random bending loading in elastic-plastic range. *Fatigue Fract. Eng. Mater. Struct.* **2017**, *41*, 249–259. [CrossRef]
8. Pejkowski, Ł.; Seyda, J. Fatigue of four metallic materials under asynchronous loadings: Small cracks observation and fatigue life prediction. *Int. J. Fatigue* **2020**, *142*, 105904. [CrossRef]
9. Zhang, B.; Chen, W.; Xu, J. Load effect and fatigue damage of bridges under combined actions of traffic and wind: A case study. *Stahlbau* **2016**, *85*, 281–291. [CrossRef]
10. Zhang, W.; Cai, C.S.; Pan, F. Fatigue Reliability Assessment for Long-Span Bridges under Combined Dynamic Loads from Winds and Vehicles. *J. Bridg. Eng.* **2013**, *18*, 735–747. [CrossRef]
11. Yang, Y.-B.; Liao, S.-S.; Lin, B.-H. Impact formulas for vehicles moving over simple and continuous beams. *J. Struct. Eng.* **1995**, *121*, 1644–1650. [CrossRef]
12. Kwon, K.; Frangopol, D.M. Bridge fatigue reliability assessment using probability density functions of equivalent stress range based on field monitoring data. *Int. J. Fatigue* **2010**, *32*, 1221–1232. [CrossRef]
13. Kwon, K.; Frangopol, D.; Soliman, M. Probabilistic fatigue life estimation of steel bridges by using a bilinear S-N approach. *J. Bridge Eng.* **2012**, *32*, 121–1232. [CrossRef]
14. Raju, S.; Moses, F.; Schilling, C. Reliability calibration of fatigue evaluation and design procedures. *J. Struct. Eng.* **1990**, *5*, 1356–1369. [CrossRef]
15. Mohammadi, J.; Shah, N. Statistical Evaluation of Truck Overloads. *J. Transp. Eng.* **1992**, *118*, 651–665. [CrossRef]
16. Obrien, E.J.; Enright, B.; Getachew, A. Importance of the Tail in Truck Weight Modeling for Bridge Assessment. *J. Bridg. Eng.* **2010**, *15*, 210–213. [CrossRef]

17. Obrien, E.J.; Hayrapetova, A.; Walsh, C. The use of micro-simulation for congested traffic load modeling of medium- and long-span bridges. *Struct. Infrastruct. Eng.* **2012**, *8*, 269–276. [CrossRef]
18. Schilling, C.G. Stress Cycles for Fatigue Design of Steel Bridges. *J. Struct. Eng.* **1984**, *110*, 1222–1234. [CrossRef]
19. Schilling, C.G.; Klippstein, K.H. Fatigue of Steel Beams by Simulated Bridge Traffic. *J. Struct. Div.* **1977**, *103*, 1561–1575. [CrossRef]
20. Schilling, C.G.; Klippstein, K.H. New Method for Fatigue Design of Bridges. *J. Struct. Div.* **1978**, *104*, 425–438. [CrossRef]
21. Schilling, C.G.; Klippstein, K.H.; Barsom, J.M.; Blake, G.T. *Fatigue of Welded Steel Bridge Members under Variable-Amplitude Loadings*; National Cooperative Highway Research Program Rep. 188; Transportation Research Board: Washington, DC, USA, 1978; Volume 4.
22. Wang, T.-L.; Liu, C. *Influence of Heavy Trucks on Highway Bridges*; Rep. No. FL/DOT/RMC/6672-379; Florida Department of Transportation: Tallahassee, FL, USA, 2000.
23. Wang, T.-L.; Liu, C.; Huang, D.; Shahawy, M. Truck Loading and Fatigue Damage Analysis for Girder Bridges Based on Weigh-in-Motion Data. *J. Bridg. Eng.* **2005**, *10*, 12–20. [CrossRef]
24. Zhao, J.; Tabatabai, H. Evaluation of a Permit Vehicle Model Using Weigh-in-Motion Truck Records. *J. Bridg. Eng.* **2012**, *17*, 389–392. [CrossRef]
25. Zhang, W.; Cai, C.S. Fatigue Reliability Assessment for Existing Bridges Considering Vehicle Speed and Road Surface Conditions. *J. Bridg. Eng.* **2012**, *17*, 443–453. [CrossRef]
26. Zhang, W.; Cai, C.S. Reliability-Based Dynamic Amplification Factor on Stress Ranges for Fatigue Design of Existing Bridges. *J. Bridg. Eng.* **2013**, *18*, 538–552. [CrossRef]
27. Zhao, J.; Yin, G.S.; An, L.M.; Liu, Y.T.; Ren, Y.L. Middle span closure technology of main bridge of Guangzhou Mingzhu Way Bridge. *Bridge Constr.* **2021**, *51*, 127–133. (In Chinese)
28. MIDAS 2019. Computer Software. Midas: Beijing, China.
29. Jiang, C.; Wu, C.; Jiang, X. Uniform corrosion and pitting corrosion of high strength steel wires for bridge cables. *J. Tongji Univ.* **2018**, *46*, 1615–1621. (In Chinese) [CrossRef]
30. *Suspender of Highway Suspension Bridge: JT/T 449-2021*; Ministry of Transport of the People's Republic of China: Beijing, China, 2021. (In Chinese)
31. Hu, H.Y.; Zhao, J.; Ren, Y.L.; An, L.M.; Liu, Y.T. Overall Design of Main Bridge of Guangzhou Mingzhu Wan Bridge. *Bridge Constr.* **2021**, *51*, 93–99. (In Chinese)
32. *Traffic Volume Report of Guangzhou Mingzhu Wan Bridge*; China Railway Construction Bridge Bureau Group Co., Ltd. Postdoctoral Workstation: Guangzhou, China, 2019.
33. *Standard for Design of Steel Structures: GB 50017—2017 Beijing*; China Architecture and Building Press: Guangzhou, China, 2017. (In Chinese)
34. Xia, T.X.; Yao, W.X.; Liu, X.M.; Ji, Y.F. Study on the Applicability of Miner's Theory in multi-axis two-step Spectrum considering material dispersion. *Chin. J. Mech. Eng.* **2015**, *51*, 38–45. [CrossRef]
35. Zhang, S.Y.; Wu, H. Multi-axial Low-cycle Fatigue Life Evaluation based on Sequential Rain Flow Method and Critical Plane Method. *Mech. Q.* **2020**, *9*, 465–476. (In Chinese)
36. *MATLAB [Computer software]*; MathWorks: Natick, MA, USA, 2020.
37. *Guangzhou Mingzhu Wan Bridge (96+164+436+164+96+60) m Six-Span Continuous Steel Truss Arch Construction Calculation Book*; China Railway Construction Bridge Bureau Group Co., Ltd. Postdoctoral Workstation: Guangzhou, China, 2019.

Mechanical Properties of L-Shaped Column Composed of RAC-Filled Steel Tubes under Eccentric Compression

Tengfei Ma [1,2]**, Zhihua Chen** [1,3]**, Yansheng Du** [1,3,*]**, Ting Zhou** [4] **and Yutong Zhang** [1]

[1] School of Civil Engineering, Tianjin University, Tianjin 300072, China; mtf@nciae.edu.cn (T.M.); zhchen@tju.edu.cn (Z.C.); zhangyt@tju.edu.cn (Y.Z.)
[2] Department of Civil Engineering, North China Institute of Aerospace Engineering, Langfang 065000, China
[3] Key Laboratory of Coast Civil Structure Safety of Ministry of Education, Tianjin University, Tianjin 300072, China
[4] School of Architecture, Tianjin University, Tianjin 300072, China; zhouting1126@126.com
[*] Correspondence: duys@tju.edu.cn

Abstract: In this paper, the eccentric compression test of seven specimens was conducted to explore the application possibility of recycled aggregate concrete (RAC) in L-shaped columns composed of concrete-filled steel tubes (CFST). The main parameter is the replacement ratio of recycled coarse aggregates (RCA), and an L-shaped column of composed hollow steel tubes was set as a control. The test results indicated that the bearing capacity and stiffness of the L-shaped column composed of RAC-filled steel tubes (RACFSTs) are better than those of the L-shaped column composed of hollow steel tubes. The compressive strength of concrete is reduced by 73.1% as the replacement ratio of RCA increases from 0 to 100%, while that of the column is merely reduced by 23.9%. The strength disadvantage of RAC is compensated by the confinement of steel tubes. Besides, the result of the eccentric compression test (80 mm eccentricity) was compared with the axial compression test (0 mm eccentricity). The increase in eccentricity reduced the bearing capacity and ductility due to additional bending moments. The finite element model (FEM) was established by software ANSYS to compare with the experimental results. The bearing capacity deviation of FEM is 4.23~6.56%. The parametric analysis was carried out to summarize the influence of parameters such as eccentricity, material strength, and steel tube thickness. With the increase of eccentricity, the bearing capacity of the RACFST decreases gradually. In engineering design, the bearing capacity of the RACFST can be improved by increasing the strength and thickness of the steel.

Keywords: L-shaped column; recycled aggregate concrete; concrete-filled steel tube; eccentric compression; finite element model

Citation: Ma, T.; Chen, Z.; Du, Y.; Zhou, T.; Zhang, Y. Mechanical Properties of L-Shaped Column Composed of RAC-Filled Steel Tubes under Eccentric Compression. *Metals* **2022**, *12*, 953. https://doi.org/10.3390/met12060953

Academic Editor: Ezio Cadoni

Received: 27 April 2022
Accepted: 30 May 2022
Published: 31 May 2022

Publisher's Note: MDPI stays neutral with regard to jurisdictional claims in published maps and institutional affiliations.

1. Introduction

With the development of urbanization, the construction space has put forward higher requirements for the traditional structure. In the traditional structure, the external corner of the frame column not only reduces the construction space but is also difficult to handle in terms of architectural aesthetics, see Figure 1a. The special-shaped columns are proposed to solve the above problems, which can completely hide the structural column in the infill wall. The special-shaped column avoids the negative impact of the external corner on the building function, as shown in Figure 1b. Due to the architectural advantages of special-shaped columns, researchers have investigated the mechanical properties of various special-shaped columns. Li et al. [1] carried out an axial compression test on the steel-reinforced concrete short column with a T-section, and the effect of shaped steel ratio and shear-span ratio on bearing capacity was explored. Tokgoz et al. [2] presented the eccentric compression test of L-shaped high strength reinforced concrete and concrete-encased composite columns, which indicated that the addition of steel fibers to high-strength concrete improved the mechanical properties of the columns under eccentric loading. Xue et al. [3–5] conducted seismic damage tests on steel-reinforced concrete irregular section columns. The effects of different loading schemes, axial compression ratios, and steel distribution ratios on the seismic performance of specimens were revealed.

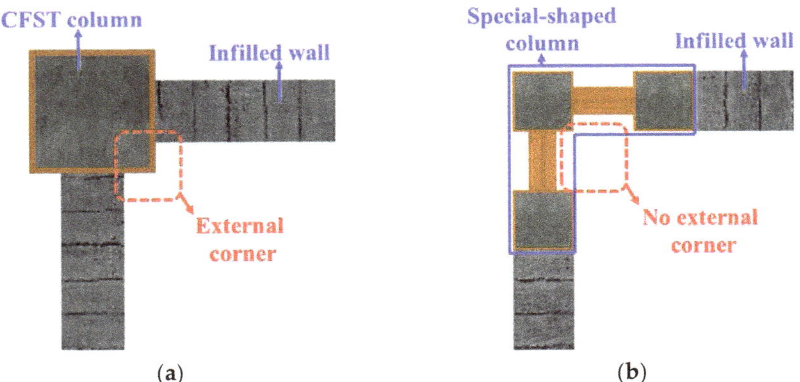

Figure 1. The comparison of the traditional column and the special-shaped column. (**a**) Traditional column. (**b**) Special-shaped column.

However, the steel-reinforced concrete columns require complex formwork in construction. The construction of concrete-filled steel tube (CFST) columns can avoid the use of complex formwork. The different types of special-shaped steel tubes are more convenient to prefabricate in the factory. Han [6–8], Tu [9,10], and Zhou [11,12] carried out mechanical tests of special-shaped CFST columns, respectively. The influence of different parameters on special-shaped CFST columns was explored in the tests. The appropriate design formula was also proposed. Li et al. [13] conducted a quasi-static test on the joint of T-shaped CFST columns and H-shaped steel beams to study the effects of the beam-column connection form and the axial compression ratio on its seismic performance. To further facilitate the prefabrication and construction of special-shaped mono CFST columns, special-shaped columns with small cross-sections were proposed by Chen et al. [14,15], as presented in Figure 2. The cross-section form of the column can be divided into L-shaped, T-shaped, and cross-shaped according to the structural requirements, corresponding to corner columns, side columns, and middle columns, respectively. This special-shaped column is formed by welding the mono CFST columns through the connection plates, which can ensure the collaborative work of the CFST columns. Rong et al. [15–18] conducted the axial compression tests of the special-shaped columns composed of the square CFST with different slenderness ratios. The test results summarized the strength failure and instability failure modes. The L-shaped column composited of the mono column connected with the perforated steel plate was proposed by Zhou et al. [19–22]. The axial compression test indicated that the L-shaped column connected by the perforated steel plate can still ensure cooperative deformation of the mono column, but the bearing capacity is reduced. By evaluating mechanical properties and processing complexity, the special-shaped columns connected with non-perforated steel plates are more suitable for the application. The previous research has demonstrated that the special-shaped column has high overall stiffness and strong energy dissipation capacity, which is very suitable for prefabricated high-rise buildings.

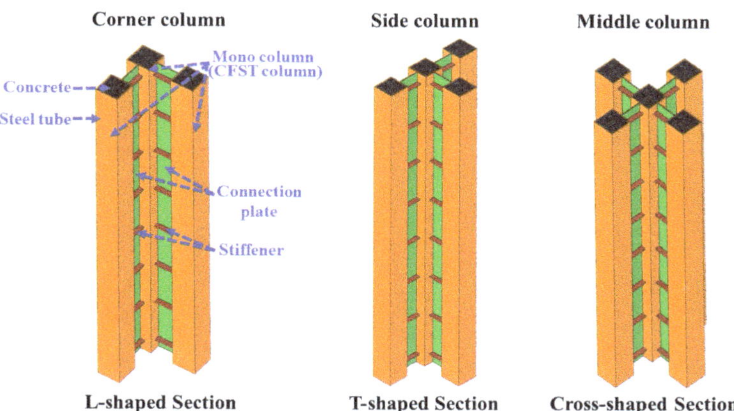

Figure 2. Special-shaped columns composed of mono columns.

Due to the continuous deterioration of the current global natural environment, the use of recyclable green building materials is an inevitable choice for the construction industry [23]. Recycled aggregate concrete (RAC) made from solid waste can not only reduce the mining of natural aggregates but also absorb construction waste. In China, construction waste accounts for 40% of the waste produced annually [24]. RAC has a great development space and application value. However, the lower strength and poorer elastic modulus of the RAC impeded the engineering application compared with ordinary concrete [25–27]. The reasons for the strength reduction of recycled concrete are: recycled aggregate has poor properties, the effect of the production process, the effect of the strength of parent concrete, and the effect of the repeated use of recycled aggregate [28]. To improve the mechanical properties of RAC, many scholars have studied the mechanical properties of RAC-filled steel tube (RACFST) columns by utilizing the confining effect of steel tubes [29]. In compression, the RAC undergoes lateral deformation due to Poisson's ratio effect. The steel tube produces passive confinement on the RAC, which can improve the compressive strength of the RAC. Meanwhile, the buckling of the steel tube is inhibited by the inner RAC. This is similar to ordinary CFST. In the application of recycled concrete-filled steel tubes, the initial defects caused by the strength, compactness, and pouring defects of recycled concrete should be reduced as much as possible. Wu et al. [30] conducted the quasi-static test of 15 thin-walled RACFST columns produced by large waste concrete blocks as recycled coarse aggregate (RCA). The test results presented that the bearing capacity of RACFST varies less when the replacement ratio of RCA varies between 0–40%. The bond behavior of CFST and RACFST were summarized by Lyu [31] based on the push-out test of 56 RACFST columns. The calculation formula was proposed for the bond strength between the concrete and steel tubes. Wang et al. [32] displayed that the reduction in compressive strength of RACFST is less than 10% compared to CFST for the same concrete mix ratio. The research on the mechanical properties of RACFST indicated that the confinement of steel tubes could compensate for the shortcomings of RAC. Therefore, it is feasible to apply RAC to the special-shaped column composed of mono CFST columns.

At present, the research on the innovative special-shaped columns composed of RACFST is still in shortage. To study the mechanical properties of shaped columns under the combined action of axial pressure and bending moments. In this paper, the eccentric compression test of the special-shaped columns composed of RACFST was conducted. Six specimens were designed with the different RCA replacement ratios of RAC as the parameter. A special-shaped column composed of hollow steel tubes was designed as a control specimen. The effects of the RCA replacement ratios of RAC on the mechanical properties of specimens were evaluated by comparing the failure modes and load-displacement curves of the tests. The strain analysis was applied to investigate the correspondence between the overall deformation and the local strain of the columns. The finite element model (FEM) was compared with the experimental results. Parametric analysis was carried out based on the validated FEM. The effect of eccentricity, material strength, and steel tube wall thickness on the bearing capacity of the specimens was investigated.

2. Experimental Program

2.1. Specimens Design

The L-shaped columns composed of CFST were created in Cangzhou, China [33]. The dimensions of the specimens in this paper are 1/2 scale L-shaped columns used in actual engineering. The mono column is an RAC-filled steel tube. The height of the mono column was 1500 mm. The wall thickness of the steel tubes was 3.75 mm. The L-shaped column was composed of three mono columns connected together by two connection plates. To enhance the stiffness of the connection plate, stiffener ribs with an interval of 200 mm were welded on the connection plate. The dimensions of the stiffeners were 100 mm × 20 mm × 5.75 mm. The 20 mm thick cover plate was welded on both ends of the L-shaped column to prevent local damage at the column end during compression. The details are presented in Figure 3. The concrete used in the steel tube was RAC. Six specimens were fabricated with different RCA replacement ratios of RAC, 0%, 20%, 40%, 60%, 80%, and 100%. An L-shaped column composed of hollow steel tubes was set as the control specimens. All the parameters are shown in Table 1.

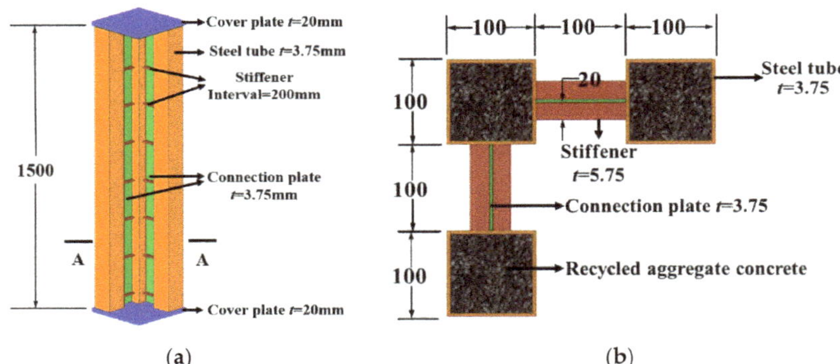

Figure 3. Dimensions of the L-shaped column (in mm). (**a**) General dimension. (**b**) Cross-section A-A.

Table 1. The detailed parameters of specimens.

Specimens	Steel Tube $H \times l \times l \times t$ (mm)	Connection Plate $H \times l \times t$ (mm)	$f_{c,cube}$ (MPa)	Replacement Ratio of RAC	Eccentric Distance (mm)
P80-0	1500 × 100 × 100 × 3.75	1500 × 100 × 3.75	33.36	0%	80
P80-20	1500 × 100 × 100 × 3.75	1500 × 100 × 3.75	29.00	20%	80
P80-40	1500 × 100 × 100 × 3.75	1500 × 100 × 3.75	17.59	40%	80
P80-60	1500 × 100 × 100 × 3.75	1500 × 100 × 3.75	13.30	60%	80
P80-80	1500 × 100 × 100 × 3.75	1500 × 100 × 3.75	11.94	80%	80
P80-100	1500 × 100 × 100 × 3.75	1500 × 100 × 3.75	8.99	100%	80
P80-K	1500 × 100 × 100 × 3.75	1500 × 100 × 3.75	-	No RAC	80

Note: H is the height of the component, l is the side length of the section, t is the thickness of the steel plate, $f_{c,cube}$ is the cubic compressive strength of RAC.

2.2. Material Properties

2.2.1. Recycle Aggregate Concrete

The mix ratio of the ordinary concrete used in this paper was C30 grade according to specification JGJ55 [34]. The concrete was fine aggregate concrete. Medium sand with a particle size of 0.3 mm~0.5 mm was used. The fineness module was 2.6 mm. The particle size of natural aggregate and recycled coarse aggregate was 5 mm~15 mm. The crushing index was 10%. The water absorption was 2.3%. Recycled aggregate concrete was prepared by replacing natural aggregates with different proportions of recycled coarse aggregates. The specific mix ratio is presented in Table 2. Three cubic blocks (100 mm × 100 mm × 100 mm) and three prism blocks (150 mm × 150 mm × 300 mm) for each RCA replacement ratio of RAC were tested to reduce the test error according to specification GB/T50081 [35]. The compressive strength and elastic modulus of RAC were obtained in Table 3. The standard compression test is presented in Figure 4.

Table 2. The detailed mix proportion of RAC (kg/m^3).

Replacement Ratio	0%	20%	40%	60%	80%	100%
Water	44.74	40.21	35.80	31.34	26.87	22.40
Cement	138.25	138.25	138.25	138.25	138.25	138.25
Sand	149.40	149.40	149.40	149.40	149.40	149.40
Coarse aggregate	267.60	214.08	160.56	107.04	53.52	0
RCA	0	58.00	115.98	173.96	231.95	289.94

Table 3. The material properties of RAC.

Replacement Ratio	$f_{c,cube}$ (Mpa)	$f_{c,prism}$ (Mpa)	E_c (Mpa)	Curing Age (Day)
0%	33.36	26.7	32,964	55
20%	29.00	25.8	23,064	55
40%	17.59	14.0	13,722	28
60%	13.30	12.2	14,369	28
80%	11.94	10.0	10,345	28
100%	8.99	8.4	9912	28

Note: $f_{c,cube}$ is the cubic compressive strength of RAC; $f_{c,prism}$ is the Prismatic compressive strength of RAC; E_c is the elasticity modulus of RAC. The concrete test block test was carried out on the same day as the specimen loading.

(a)

(b)

Figure 4. The compression test of the concrete. (**a**) Cube compression test. (**b**) Prism compression test.

2.2.2. Steel

According to specification GB/T228 [36], tensile tests were conducted for steel coupons taken from steel tubes, connection plates, and stiffener ribs. The dimension of the steel coupon and the test machine are illustrated in Figure 5. All the steel used in the specimens was Q235 grade. The mechanical properties of the steel, such as yield strength, ultimate strength, and elastic modulus, are shown in Table 4.

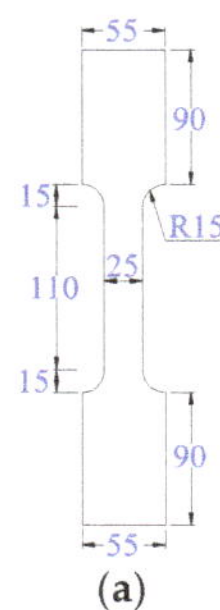

(a)

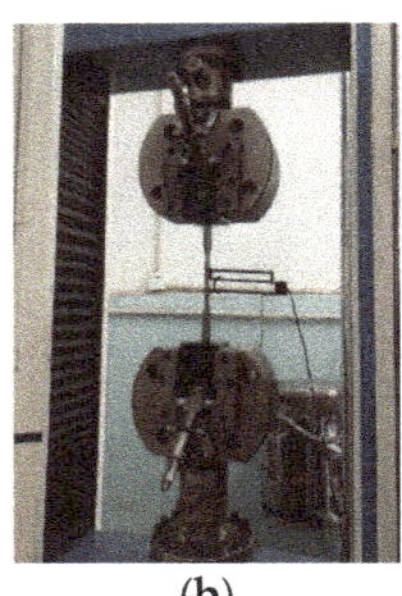

(b)

Figure 5. Material test of steel. (**a**) Dimension (in mm). (**b**) Test machine.

Table 4. The properties of the steel coupons.

Material	Steel Grade	t (mm)	f_y (MPa)	f_u (MPa)	E_s (MPa)
Steel tube	Q235	3.75	269	445	208,305
Connection plate	Q235	3.75	258	402	186,208
Stiffener	Q235	5.75	372	467	187,355

Note: t is the thickness of the steel; f_y is the yield strength of the steel; f_u is the ultimate strength of the steel; E_s is the elastic modulus of the steel.

2.3. Loading Device and Scheme

The loading device was a 500 T electro-hydraulic servo pressure test machine (Jilin guanteng Automation Technology Co., Ltd, Jilin, China). Both ends of the L-shaped column were hinged. The spherical hinge device was used to release the restraint, as presented in Figure 6.

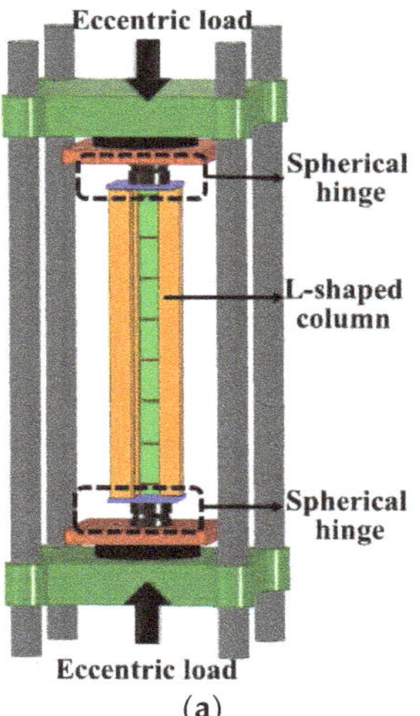

(a)

(b)

Figure 6. Loading device. (**a**) 3D model. (**b**) On-site photo.

To identify the loading point of the specimen, it was necessary to obtain the cross-sectional centroid of the L-shaped column. In this paper, the geometric cross-section method was used to calculate the centroid of L-shaped columns based on the T/CECS825 [37]. As shown in Figure 7a, the corner point of column B is the coordinate origin O. The direction along the edge of the column limb is defined as two perpendicular coordinate axes in the plane. The coordinates of the cross-sectional centroid can be obtained by calculating the static distances S_x and S_y of the cross-section to the x-axes and y-axes. The formula for calculating the coordinates of the centroid is presented in (1)–(2). To apply the bidirectional bending moment in the X-axis and Y-axis directions, the loading point of the specimen is 80 mm away from the centroid along the X' axis, see Figure 7b.

$$x_c = \frac{S_y}{A} = \frac{\int_A x \, dA}{A} \tag{1}$$

$$y_c = \frac{S_x}{A} = \frac{\int_A y \, dA}{A} \tag{2}$$

where x_c, y_c is the coordinates of the centroid from the y-axis and x-axis, respectively. S_x, S_y is the static distance of the cross-section to the x-axis and y-axis, respectively. A is the cross-sectional area enclosed by the geometric outer contour of the steel tubes and connection plates. dA is the micro area taken at the coordinates (x, y) within the cross-section.

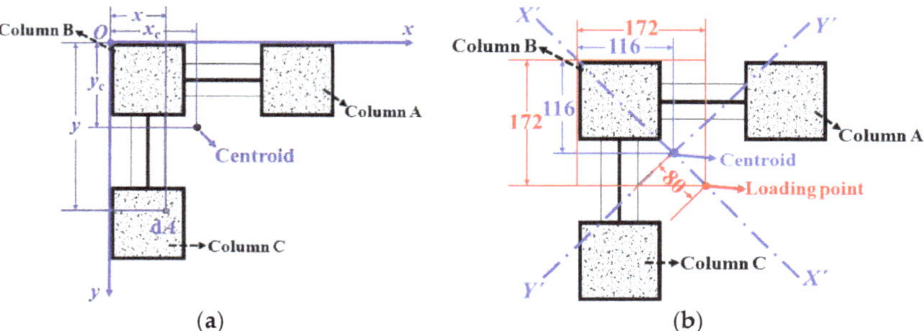

Figure 7. Position of the centroid and loading point. (**a**) Calculation of the centroid. (**b**) Loading point (in mm).

The loading scheme contains pre-loading and formal loading. During pre-loading, the load was applied at a rate of 20 kN/min to 10% of the yield load estimated by the numerical model. The operating status of the measurement equipment needs to be checked. Besides, the main purpose of pre-loading is to confirm that the spherical hinge coincides with the loading point of the specimen. The formal loading contains three stages. During the elastic stage, the load is gradually applied at intervals of 200 kN. During the yield stage, the load is applied at intervals of 100 kN. During the failure stage, the force control is replaced using displacement control with a loading rate of 0.3 mm/min. The test stopped when the load dropped to 85% of the peak load.

2.4. Measurement Arrangement

The linear variable displacement transducers (LVDTs) (Tjweekend, Tianjin, China) were used to monitor the displacement variation of the specimens, as shown in Figure 8. Twelve lateral LVDTs (H1–H12) were used to monitor the bending deformation of the specimen at both ends and in the middle of the column. Four vertical LVDTs (V1–V4) were arranged to monitor the compressive deformation of the specimen. To monitor the yielding and micro-deformation of the steel tubes, lateral and vertical strain gauges were arranged at the measurement points of the three mono columns and the connecting plates. The measurement points are divided into three zones along with the height of the column, see Figure 8b. Figure 8c provided details of the measurement points of the strain gauges and the orientation numbers of the columns.

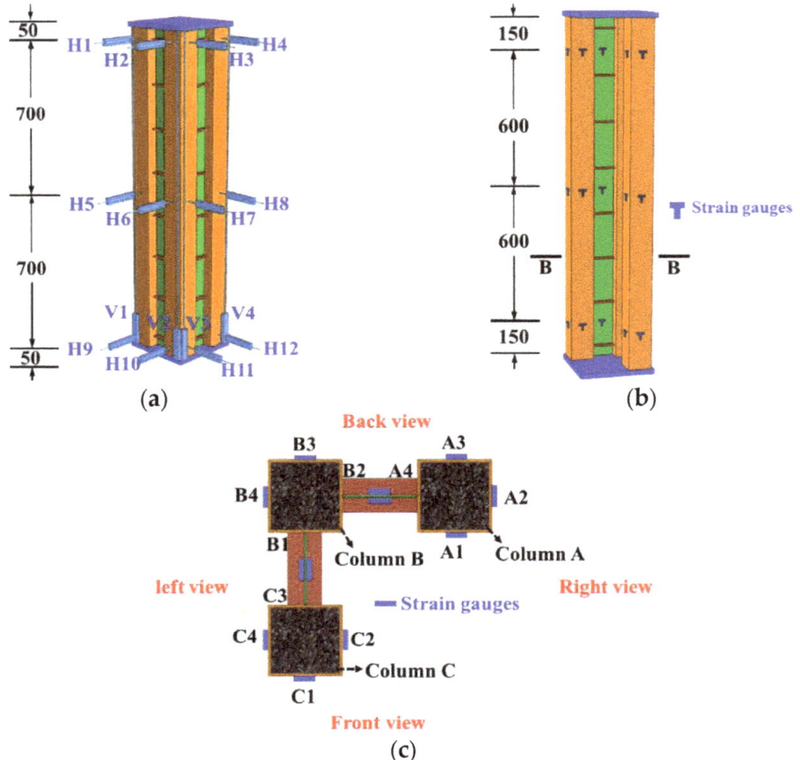

Figure 8. The layout of the measurement. (**a**) LVDT. (**b**) Strain gauges. (**c**) B-B cross-section.

3. Experimental Phenomenon

The coordinates and orientation numbers of the column cross-section are marked in Figures 7b and 8c to facilitate the presentation of the experimental phenomenon. According to the deformation process of the L-shaped column under eccentric compression, the failure mode can be divided into three stages: elastic stage, elastoplastic stage, and failure stage. There was no significant experimental phenomenon of the specimens in the elastic stage due to the micro elastic deformation of steel and RAC.

The specimen entered the elastoplastic stage as the load increased. The mono columns of specimens P80-0~P80-100 were slightly bent around the Y' axis due to the bending moment. The specimen deformed slowly as the load gradually increased to the peak load. The local buckling of column A and column C began to occur at the middle or upper end of the L-shaped column, as presented in Figure 9a,b. There was basically no local buckling of the steel tube in column B. This is because the section of column B was subjected to the tensile force under the eccentric load. Besides, the connection plates on both sides formed the confinement on column B. For specimen P80-K, the hollow steel tube began to exhibit inward buckling due to the absence of RAC inside, as shown in Figure 9c.

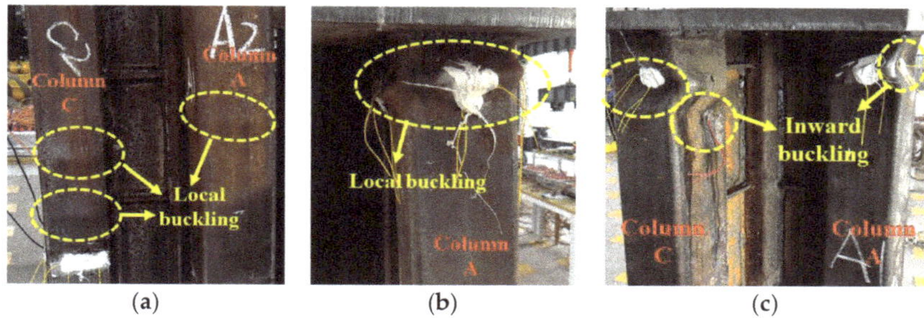

Figure 9. Typical phenomenon in elastoplastic stage. (**a**) Middle of the column. (**b**) Upper end of the column. (**c**) Inward buckling of steel tube.

In the failure stage, the specimen bent and deformed rapidly in a short time due to the eccentric load. Two compression failure modes appeared in the specimens P80-0~P80-100. In one of the failure modes, the failure section appeared in the middle of the column, including specimens P80-0, P80-20, and P80-100. The column section was bent around the Y' axis. Columns A and C presented multiple local bucklings in the middle of the column, as shown in Figure 10a,e,f. There was no obvious buckling deformation on the connection plate, indicating that the mono columns could be cooperatively deformed. In another failure mode, the failure section occurred at the upper end of the column. The obvious local buckling of columns A and C was concentrated at the upper end, as shown in Figure 10b–d. For specimen P80-K, the inward buckling occurred at the upper end of columns A and C. According to the comparison of the failure mode, the deflection of the specimen with the failure section at the column end was less than that of the specimen with the failure section in the middle of the column. Besides, the comparison of the failure mode between the L-shaped columns filled with RAC and without RAC indicated that the core concrete could effectively inhibit the inward buckling of steel tubes.

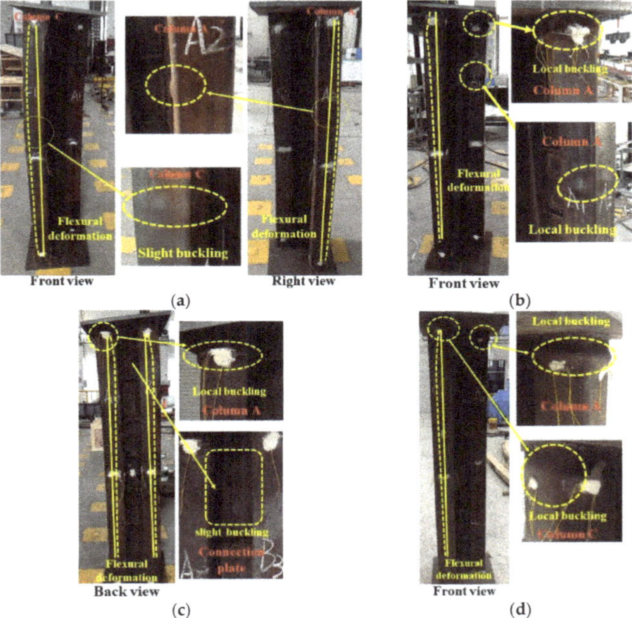

Figure 10. *Cont.*

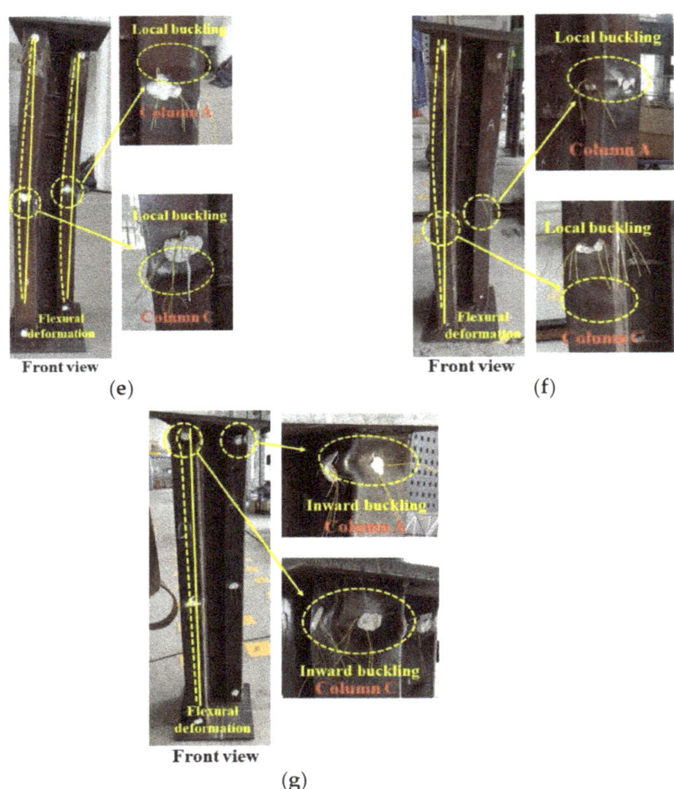

Figure 10. Failure phenomenon of the specimens. (**a**) P80-0. (**b**) P80-20. (**c**) P80-40. (**d**) P80-60. (**e**) P80-80. (**f**) P80-100. (**g**) P80-K.

All the L-shaped columns filled with RAC exhibited integral flexural deformation. The failure mode of the specimens presented the compression-bending damage of columns A and C. The effect of different replacement ratios on the failure mode was not significant. The L-shaped column without RAC was subject to premature loss of load capacity due to inward buckling of the steel tubes.

4. Test Results and Analysis

To facilitate the comparison of the mechanical properties of the specimens at different stages, the loading-displacement curves were divided into the elastic stage, elastoplastic stage, and failure stage. The critical point between the elastic stage and the elastoplastic stage is the yield point, which can be obtained by the "farthest point method" [38]. The critical point between the elastoplastic stage and the failure stage is the peak point, which corresponded to the peak load. The point corresponding to 85% of the peak load in the failure is the ultimate point. Besides, the ductility index (*DI*) was defined as Equation (3) to characterize the ductility of the L-shaped column. The secant stiffness (S_y) of the yield point was utilized to characterize the stiffness of the L-shaped column, which was defined as in Equation (4).

$$DI = \frac{\Delta_u}{\Delta_y} \tag{3}$$

$$S_y = \frac{N_y}{\Delta_y} \tag{4}$$

where Δ_u is the ultimate displacement, Δ_y is the yield displacement, N_y is the yield load of the specimen.

The data of the yield point, peak point, and ultimate point are shown in Table 5. All the loading-displacement curves of specimens are presented in Figure 11. The comparison of the loading-displacement curves indicated that specimen P80-0 had the highest stiffness, and the stiffness of the L-shaped columns decreased significantly as the replacement ratio of RAC increased. Compared to the yield point, secant stiffness Sy of specimen P80-0, the specimens P80-20, P80-40, P80-60, P80-80, P80-100, and P80-K decreased by 18.4%, 43.4%, 51.2%, 57.1%, 68.5%, and 72.7%, respectively. The above data indicated that the stiffness of the specimens decreased linearly with the RCA replacement ratio of the in-filled RAC increased from 0 to 40%. The stiffness of the specimens decreased slowly as the replacement ratio of RCA increased from 40% to 100%. By comparing P80-K with other specimens, the filling of RAC in the steel tube effectively improved the stiffness of the L-shaped column.

Table 5. The characteristic points of the specimens.

Specimens	Yield Point		Peak Point		Ultimate Point		DI	S_y (kN/mm)
	Δ_y (mm)	N_y (kN)	Δ_p (mm)	N_p (kN)	Δ_u (mm)	N_u (kN)		
P80-0	1.46	1450	5.75	1904	16.68	1618	11.42	993
P80-20	1.54	1248	6.28	1820	13.42	1547	8.71	810
P80-40	1.93	1085	6.80	1652	10.98	1404	5.69	562
P80-60	2.14	1038	7.16	1580	11.54	1343	5.39	485
P80-80	2.31	985	7.59	1510	12.22	1284	5.29	426
P80-100	2.90	909	8.36	1449	13.45	1232	4.64	313
P80-K	2.63	713	7.92	1222	13.14	1039	5.01	271

Note: Δ_y is the yield displacement, N_y is the yield load, Δ_p is the peak displacement, N_p is the peak load, Δ_u is the ultimate displacement, N_u is the ultimate load, DI is the ductility index, S_y is the secant stiffness of the yield point.

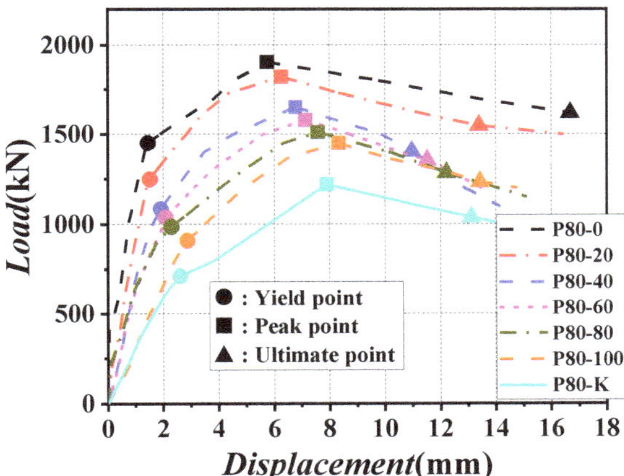

Figure 11. Axial load-displacement curve.

Besides, the variation of bearing capacity with the replacement ratio is displayed in Figure 12. The bearing capacity gradually decreased with the increase of the RCA replacement. When the RCA replacement ratio increased from 0% to 40%, the yield load and peak load decreased by 25.2% and 13.2%, respectively. When the RCA replacement ratio increased from 40% to 100%, the yield load and peak load decreased by 16.2% and 12.3%, respectively. This indicated that the bearing capacity decreased more rapidly when the RCA replacement ratio was below 40%. When the RCA replacement ratio exceeded 40%, the bearing capacity decreased at a slower rate. Besides, the yield load and peak load of specimen P80-K were 21.6–50.8% and 15.7–35.8% lower than other specimens, respectively. This indicated that the filling of concrete can increase the bearing capacity by preventing the steel tube from buckling inward and contributing to the vertical compressive strength.

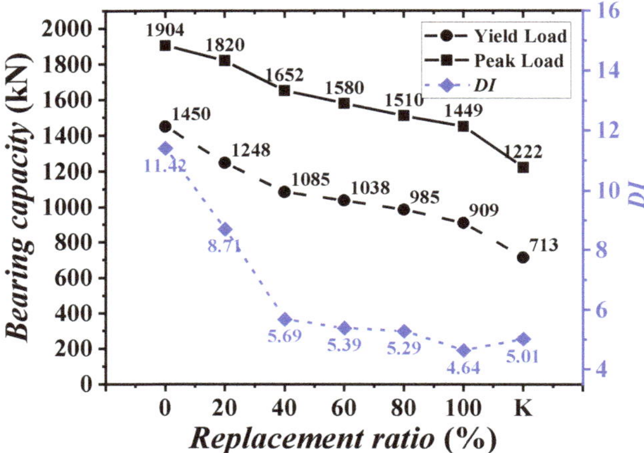

Figure 12. Bearing capacity and ductility.

From the variation curve of *DI* in Figure 12, the ductility of the L-shaped column decreased significantly with the increase of the RCA replacement ratio. When the RCA replacement ratio increased from 0% to 40%, the DI decreased by 50.2%. When the RCA replacement ratio increased from 40% to 100%, the DI decreased by 18.5%. The above data indicated that the ductility of the specimens decreased significantly when ordinary concrete was replaced by RAC. This phenomenon was especially obvious when the replacement ratio of RCA was less than 40%. When the replacement ratio of RCA exceeded 40%, the effect of the elevated replacement ratio on the ductility of the specimens decreased significantly. Besides, The *DI* of specimen P80-K was close to that of specimen P80-80, which indicated that the ductility of specimens with a higher RAC replacement ratio was closer to that of the L-shaped column composed of hollow steel tubes. Therefore, RAC with less than 40% RCA replacement ratio is recommended to be applied in L-shaped columns.

5. Finite Element Analysis

5.1. Finite Element Models

In this paper, the finite element analysis software ANSYS was used to simulate the test process. The peak load and failure mode of the RACFST obtained by finite element simulation were compared with the test results to verify the effectiveness of the numerical simulation method. The steel, connecting plate, and stiffener were modeled by SHELL181 elements, and the RAC was simulated by SOLID65 elements. Surface–surface contact between the steel tube and the inner concrete was used to simulate the interaction between steel and concrete. The concrete side as the "target surface" was simulated by the TARGE170 element, and the steel tube side as the "contact surface" was simulated by the CONTA173 element. Constraining the *X*, *Y*, and *Z* degrees of freedom of all nodes at the bottom of the RACFST and all nodes at the top of the RACFST was coupled with the load application position and constrained *X* and *Y* degrees of freedom. This is consistent with the experimental boundary conditions. The modeling method referred to previous literature [20]. The finite element model is presented in Figure 13.

In the modeling, the constraint effect of steel tubes on concrete was considered [39]. For the constitutive relationship of recycled concrete, the reader is referred to [40].

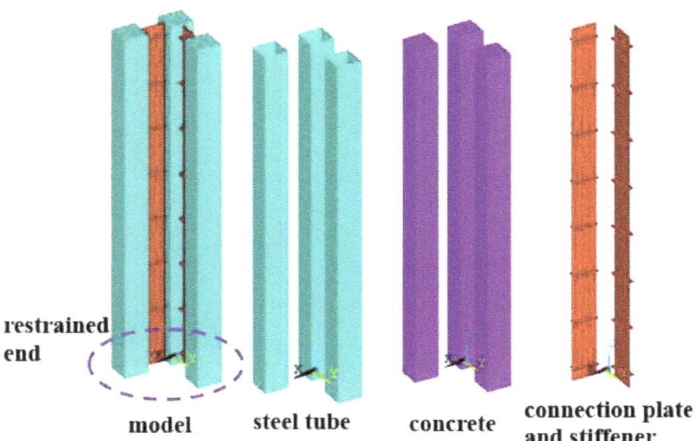

Figure 13. Finite element models.

5.2. The Comparison of the Axial Load-Displacement Curve

The load-vertical displacement comparison curve obtained from numerical simulation and test results is shown in Figures 14 and 15. It can be seen that the stiffness of the two curves agreed well. The comparison bearing capacity of the yield load and peak load of RACFST are listed in Table 6. For RACFST with different RCA replacement ratios, the average yield load from numerical simulation and test results was 1.10, and the average peak load was 1.05. It can be seen that the bearing capacity by numerical simulation was in good agreement with the experimental results. In conclusion, the numerical simulation method used in this paper can effectively simulate the eccentric bearing capacity of the RACFST.

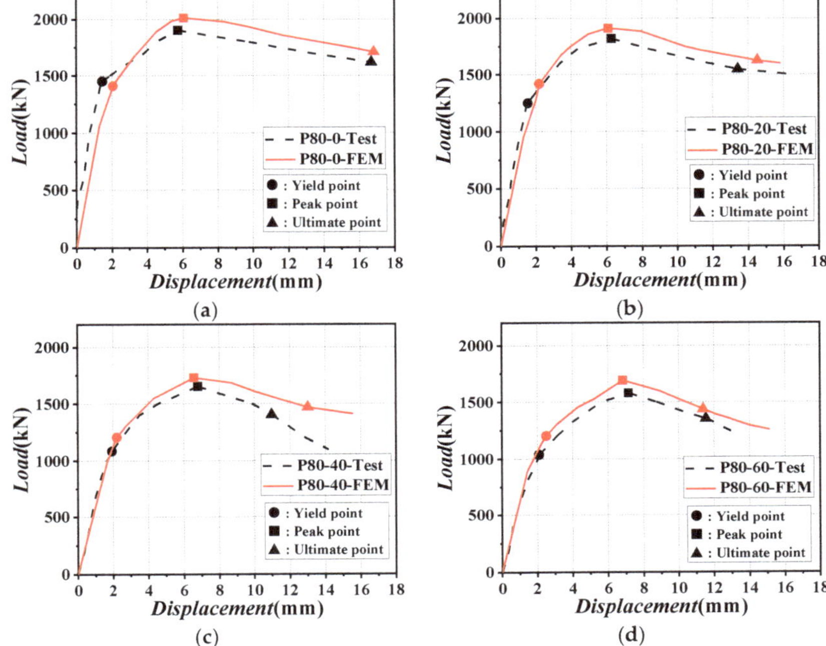

Figure 14. *Cont.*

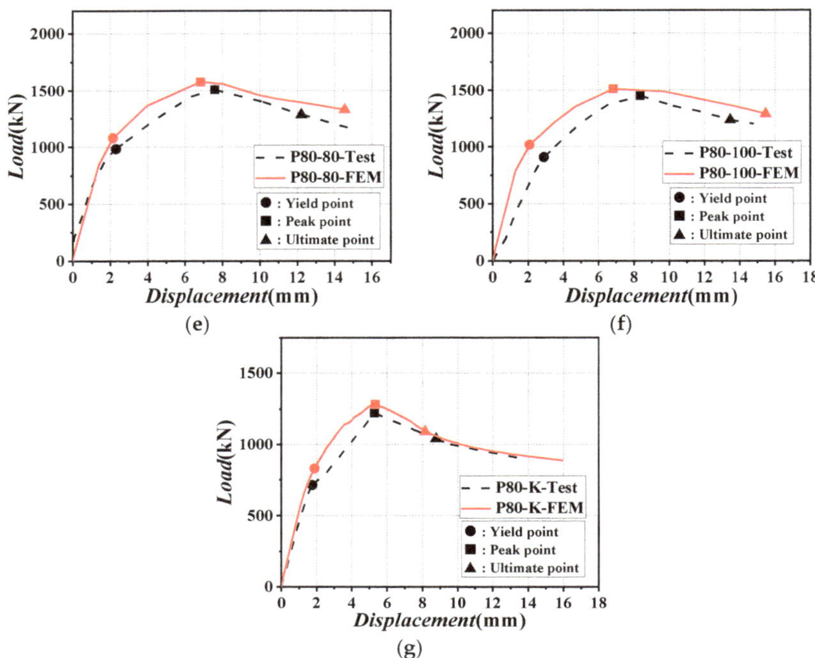

Figure 14. Comparison of the axial load-displacement curve. (**a**) P80-0. (**b**) P80-20. (**c**) P80-40. (**d**) P80-60. (**e**) P80-80. (**f**) P80-100. (**g**) P80-K.

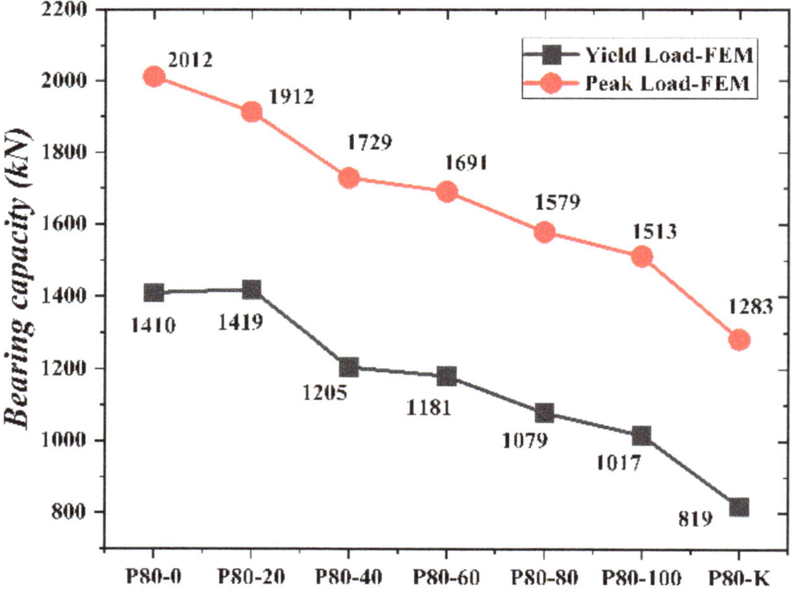

Figure 15. Bearing capacity of FEM.

Table 6. Comparison of the bearing capacity.

Specimens	Yield Point		$N_{\text{y-FEM}}/N_{\text{y-test}}$	Peak Point		$N_{\text{p-FEM}}/N_{\text{p-test}}$
	$N_{\text{y-FEM}}$ (kN)	$N_{\text{y-test}}$ (kN)		$N_{\text{p-FEM}}$ (kN)	$N_{\text{p-test}}$ (kN)	
P80-0	1410	1450	0.97	2012	1904	1.06
P80-20	1419	1248	1.14	1912	1820	1.05
P80-40	1205	1085	1.11	1729	1652	1.05
P80-60	1181	1038	1.14	1691	1580	1.07
P80-80	1079	985	1.10	1579	1510	1.05
P80-100	1017	909	1.11	1513	1449	1.04
P80-K	819	713	1.14	1283	1222	1.05
	MV		1.10	*MV*		1.05

Note: $N_{\text{y-FEM}}$ is the yield load of FEM, $N_{\text{y-test}}$ is the yield load of test results, $N_{\text{p-FEM}}$ is the peak load of FEM, $N_{\text{p-test}}$ is the peak load of test results, *MV* is the mean value.

5.3. The Verification of the Failure Mode

The comparison failure modes of each specimen are shown in Figure 16. The overall deformation of each specimen basically shows the bending deformation around the weak axis. The failure modes of each specimen in FEM and test were similar. The local buckling position of each specimen in FEM and test were similar. There was a slight difference between the FEM results and the test results, which may be caused by the failure to control the speed of applying the load during the test. It may also be caused by the fabrication deviation of the specimen or the non-uniformity of the filling concrete. In general, the failure modes of the FEM results were generally similar to the test. The above modeling method can provide a reference for exploring L-shaped columns composed of RACFST.

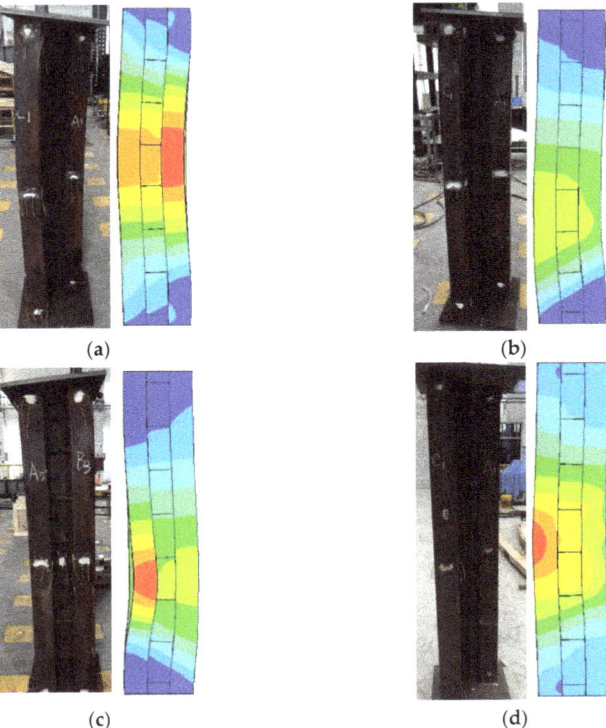

(a)

(b)

(c)

(d)

Figure 16. *Cont.*

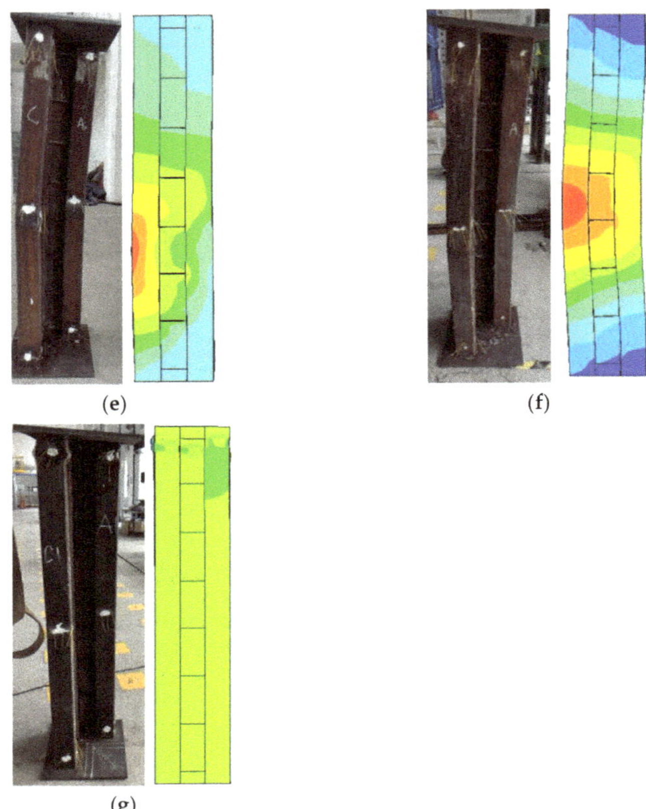

Figure 16. Comparison of the failure mode. (**a**) P80-0. (**b**) P80-20. (**c**) P80-40. (**d**) P80-60. (**e**) P80-80. (**f**) P80-100. (**g**) P80-K.

6. Parametric Analysis

6.1. The Effect of the Eccentricity

In this paper, five different eccentric loading positions were selected to study the eccentric mechanical properties of RACFST. The eccentric positions were all on the OX'axis, as shown in Figure 7. The eccentricities of loading points were 0 mm (centroid), 40 mm, 80 mm, 120 mm, and 160 mm. The load-eccentricity curve of RACFST is shown in Figure 17. It can be seen from Figure 17 that the peak load of RACFST decreased with the increase of eccentricity, and the greater the RAC ratio, the greater the decrease in peak load. When the replacement ratio of RAC was 0% and the loading position increased from 0 mm to 160 mm, the peak load of RACFST decreased by about 35%. When the replacement ratio of RAC was 100%, the peak load of RACFST decreased by about 45%.

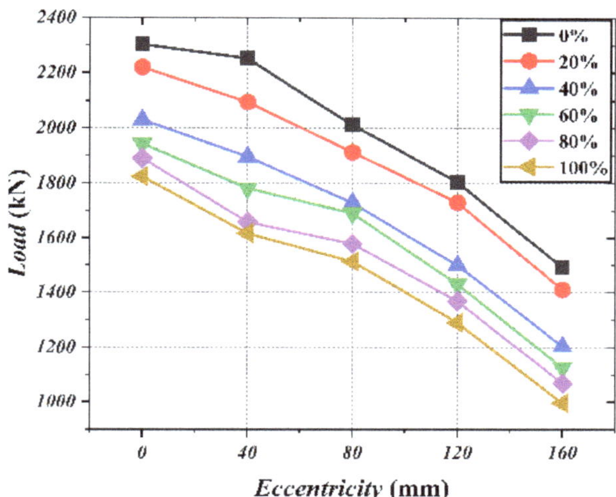

Figure 17. Load-eccentricity curve.

6.2. The Effect of the Steel Strength

In this paper, Q235, Q345, Q390, Q420, and Q460 were selected as variables to study the effect of steel strength on the eccentric mechanical properties of RACFST. The load-steel products curve of RACFST is shown in Figure 18. It can be seen from Figure 18 that with the improvement of steel strength, the peak load of RACFST decreased. When the steel strength was increased by one level, the amplitude of the peak load of RACFST increased; it could be increased by about 35%. Thus, the peak load of RACFST can be improved by improving the steel strength.

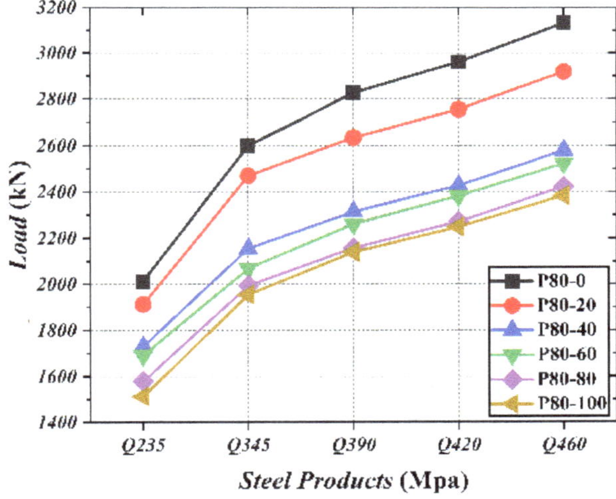

Figure 18. Load-steel products curve.

6.3. The Effect of the Thickness of the Steel

In order to study the influence of steel thickness on the eccentric mechanical properties of RACFST, nine different steel thicknesses of 3.75 mm, 5.75 mm, 7.75 mm, 9.75 mm, 11.75 mm, 13.75 mm, 15.75 mm, 17.75 mm, and 19.75 mm were selected to simulate the peak load. The steel thickness mentioned here refers to the thickness of the steel tube connecting the plate and stiffener. The load-steel thickness curve of RACFST is shown in Figure 19. Figure 19 shows that with the increase in steel thickness, the peak load of RACFST shows an upward trend. When the steel thickness increased from 3.75 mm to 19.75 mm, the increased amplitude of the peak load of RACFST under different RAC replacement ratios was 220%~317%, and the peak load increased with the RAC replacement ratio. This shows that the peak load of RACFST can be improved by increasing the steel thickness.

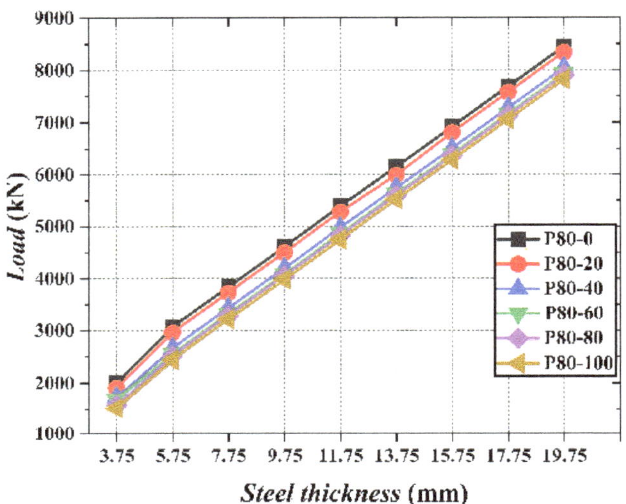

Figure 19. Load-steel thickness curve.

7. Conclusions

In this paper, the eccentric compressive performance of the L-shaped column composed of recycled aggregate concrete-filled steel tubes was investigated. The eccentric compression performance of RACFST with the different replacement ratios of recycled coarse aggregate was explored and compared with the hollow L-shaped column. In addition, a numerical simulation of eccentricity, steel strength, and steel thickness was conducted. The research conclusions are summarized as follows:

(1) The peak load of the hollow L-shaped column is lower than that of the L-shaped column composed of recycled aggregate concrete-filled steel tubes with different coarse aggregate replacement ratios, which shows that the recycled concrete improves the peak load of the RACFST. Compared with the hollow L-shaped column, the peak load of the RACFST with a 0% replacement ratio of RCA increased by 35.82%, and the peak load of the RACFST with a 100% replacement ratio of RCA increased by 15.67%.

(2) Under eccentric loading, with an increased RCA replacement ratio, the stiffness of the RACFST decreases. When the replacement ratio of RCA exceeds 40%, the reduction rate of stiffness slows down.

(3) Under eccentric loading, with an increased RCA replacement ratio, the ductility of the RACFST decreases. When the replacement ratio of RCA exceeded 40%, the effect of the elevated replacement ratio on the ductility of the specimens decreased significantly.

(4) With the increase of eccentricity, the bearing capacity of the RACFST decreases gradually.

(5) With the improvement of steel strength, the bearing capacity of the RACFST shows an upward trend; that is, the bearing capacity of the RACFST can be improved by strengthening the steel strength.

(6) With the increase in steel thickness, the bearing capacity of the RACFST shows an upward trend; that is, the bearing capacity of the RACFST can be improved by increasing the steel thickness.

Author Contributions: T.M.: conceptualization, testing, software, writing—review and editing; Z.C.: supervision, visualization; Y.D.: validation, formal analysis, methodology; T.Z.: investigation, resources; Y.Z.: data curation, writing—original draft and review. All authors have read and agreed to the published version of the manuscript.

Funding: This work is sponsored by The National Key R&D Program of China (Chen, Z: 2019YFD1101005), the Science and Technology Project of the Hebei Education Department (Ma, T: Grant No.QN2021003), and the Science and Technology Project of the North China Institute of Aerospace Engineering (Ma, T: ZD-2022-05).

Institutional Review Board Statement: Not applicable.

Informed Consent Statement: Not applicable.

Data Availability Statement: Not applicable.

Acknowledgments: The authors of the paper appreciate the support from the National Natural Science Foundation of China (Grant No. 51808182) and the China Postdoctoral Science Foundation (Grant No. 2020M670680).

Conflicts of Interest: The authors declare that they have no known competing financial interests or personal relationships that could have appeared to influence the work reported in this paper.

References

1. Li, Z.; Zhang, X.; Guo, Z.; Xu, K. Experimental study on the mechanical property of steel reinforced concrete short columns of t-shaped cross-section. *China Civ. Eng. J.* **2007**, *40*, 1–5.
2. Tokgoz, S.; Dundar, C. Tests of eccentrically loaded L-shaped section steel fibre high strength reinforced concrete and composite columns. *Eng. Struct.* **2012**, *38*, 134–141. [CrossRef]
3. Xue, J.; Zhou, C.; Liu, Z. Research on damage of solid-web steel reinforced concrete T-shaped columns subjected to various loadings. *Steel Compos. Struct.* **2017**, *24*, 409–423.
4. Xue, J.; Zhou, C.; Lin, J. Seismic performance of mixed column composed of square CFST column and circular RC column in Chinese archaized buildings. *Steel Compos. Struct.* **2018**, *29*, 451–464.
5. Xue, J.; Zhou, C.; Liu, Z.; Qi, L. Seismic response of steel reinforced concrete spatial frame with irregular section columns under earthquake excitation. *Earthq. Struct.* **2018**, *14*, 337–347.
6. Wang, F.-C.; Han, L.-H. Analytical behavior of special-shaped CFST stub columns under axial compression. *Thin-Walled Struct.* **2018**, *129*, 404–417. [CrossRef]
7. Ren, Q.-X.; Han, L.-H.; Lam, D.; Hou, C. Experiments on special-shaped CFST stub columns under axial compression. *J. Constr. Steel Res.* **2014**, *98*, 123–133. [CrossRef]
8. Zhang, Y.-B.; Han, L.-H.; Zhou, K.; Yang, S. Mechanical performance of hexagonal multi-cell concrete-filled steel tubular (CFST) stub columns under axial compression. *Thin-Walled Struct.* **2019**, *134*, 71–83. [CrossRef]
9. Li, T.; Tu, Y. Research on behavior of L-shaped multi-cell composite concrete-filled steel tubular short columns subjected to biaxial eccentric compression. *Ind. Constr.* **2018**, *48*, 156–161.
10. Sui, Y.; Tu, Y.; Zhang, J. Research on mechanical properties of T-shaped concrete filled steel tubular column subjected to biaxial eccentric compression. *Steel Constr.* **2017**, *9*, 41–46.
11. Li, Q.; Zhou, X.; Li, G.; Liu, Z.; Wang, Z.; Wang, X.; Xian, G. Experimental study on the behavior of special T-shaped composite columns with concrete-filled square steel tubulars under eccentric loads. *J. Civ. Environ. Eng.* **2021**, *43*, 102–111.
12. Wang, Z.; Zhou, X.; Wei, F.; Li, M. Performance of Special-Shaped Concrete-Filled Square Steel Tube Column under Axial Compression. *Adv. Civ. Eng.* **2020**, *2020*, 1–16. [CrossRef]
13. Li, B.-Y.; Yang, Y.-L.; Chen, Y.-F.; Cheng, W.; Zhang, L.-B. Behavior of connections between square CFST columns and H-section steel beams. *J. Constr. Steel Res.* **2018**, *145*, 10–27. [CrossRef]
14. Zhihua, C.; Zhenyu, L.; Bin, R.; Xiliang, L. Experiment of Axial Compression Bearing Capacity for Crisscross Section Special-Shaped Column Composed of Concrete-Filled Square Steel Tubes. *J. Tianjin Univ.* **2006**, *39*, 1275–1282.
15. Chen, Z.; Rong, B.; Apostolos, F. Axial compression stability of a crisscross section column composed of concrete-filled square steel tubes. *J. Mech. Mater. Struct.* **2009**, *4*, 1787–1799. [CrossRef]
16. Rong, B.; Chen, Z.; Fafitis, A.; Yang, N. Axial Compression Behavior and Analytical Method of L-Shaped Column Composed of Concrete-Filled Square Steel Tubes. *Trans. Tianjin Univ.* **2012**, *18*, 180–187. [CrossRef]

17. Chen, Z.; Rong, B. Research on axial compression stability of L-shaped column composed of concrete-filled square steel tubes. *Build. Struct.* **2009**, *39*, 39–43.

18. Rong, B.; Chen, Z.; Zhou, T. Research on axial compression strength of L-shaped short column composed of concrete-filled square steel tubes. *Ind. Constr.* **2009**, *39*, 104.

19. Zhou, T.; Xu, M.Y.; Chen, Z.H.; Wang, X.D.; Wang, Y.W. Eccentric loading behavior of L-shaped columns composed of concrete-filled steel tubes. *Adv. Steel Constr.* **2016**, *12*, 227–244.

20. Zhou, T.; Xu, M.; Wang, X.; Chen, Z.; Qin, Y. Experimental study and parameter analysis of L-shaped composite column under axial loading. *Int. J. Steel Struct.* **2015**, *15*, 797–807. [CrossRef]

21. Zhou, T.; Chen, Z.H.; Liu, H.B. Seismic behavior of special shaped column composed of concrete filled steel tubes. *J. Constr. Steel Res.* **2012**, *75*, 131–141. [CrossRef]

22. Wang, Y.; Chen, Z.; Zhou, T.; Li, Y.; Wang, X. Parametric analysis of special-shaped column composed of concrete-filled square steel tubes under cyclic loading. *J. Earthq. Eng. Eng. Vib.* **2014**, *34*, 126–132.

23. Turk, J.; Cotic, Z.; Mladenovic, A.; Sajna, A. Environmental evaluation of green concretes versus conventional concrete by means of LCA. *Waste Manag.* **2015**, *45*, 194–205. [CrossRef] [PubMed]

24. Zhang, X.; Gao, X. The hysteretic behavior of recycled aggregate concrete-filled square steel tube columns. *Eng. Struct.* **2019**, *198*, 109523.1–109523.13. [CrossRef]

25. Eguchi, K.; Teranishi, K.; Nakagome, A.; Kishimoto, H.; Shinozaki, K.; Narikawa, M. Application of recycled coarse aggregate by mixture to concrete construction. *Constr. Build. Mater.* **2007**, *21*, 1542–1551. [CrossRef]

26. Xiao, J.; Li, J.; Zhang, C. Mechanical properties of recycled aggregate concrete under uniaxial loading. *Cem. Concr. Res.* **2005**, *35*, 1187–1194. [CrossRef]

27. Khatib, M.J. Properties of concrete incorporating fine recycled aggregate. *Cem. Concr. Res.* **2005**, *35*, 763–769. [CrossRef]

28. Kim, J. Influence of quality of recycled aggregates on the mechanical properties of recycled aggregate concretes: An overview. *Constr. Build. Mater.* **2022**, *328*, 127071. [CrossRef]

29. Huang, L.; Liang, J.; Gao, C.; Yan, L. Flax FRP tube and steel spiral dual-confined recycled aggregate concrete: Experimental and analytical studies. *Constr. Build. Mater.* **2021**, *300*, 124023. [CrossRef]

30. Wu, B.; Zhao, X.-Y.; Zhang, J.-S. Cyclic behavior of thin-walled square steel tubular columns filled with demolished concrete lumps and fresh concrete. *J. Constr. Steel Res.* **2012**, *77*, 69–81. [CrossRef]

31. Lyu, W.-Q.; Han, L.-H. Investigation on bond strength between recycled aggregate concrete (RAC) and steel tube in RAC-filled steel tubes. *J. Constr. Steel Res.* **2019**, *155*, 438–459. [CrossRef]

32. Wang, Y.; Chen, J.; Geng, Y. Testing and analysis of axially loaded normal-strength recycled aggregate concrete filled steel tubular stub columns. *Eng. Struct.* **2015**, *86*, 192–212. [CrossRef]

33. Xu, M.; Zhou, T.; Chen, Z.; Li, Y.; Bisby, L. Experimental study of slender LCFST columns connected by steel linking plates. *J. Constr. Steel Res.* **2016**, *127*, 231–241. [CrossRef]

34. *JGJ55-2011*; Specification for Mix Proportion Design of Ordinary Concrete. China Build. Ind. Press: Beijing, China, 2011.

35. *GB/T 50081-2019*; Standard for Test Methods of Concrete Physical and Mechanical Properties. Stand. Adm. China: Beijing, China, 2019.

36. *GB/T 228.1-2010*; Metallic Materials-Tensile Testing-Part 1: Methods of Test at Room Temperature. Stand. Adm. China: Beijing, China, 2010.

37. *T/CECS825-2021*; Technical Specification for Special-Shaped Column Composed of Concrete-Filled Rectangular Steel Tubes. China Build. Ind. Press: Beijing, China, 2021.

38. Du, Y.; Zhang, Y.; Chen, Z.; Yan, J.B.; Zheng, Z. Axial compressive performance of CFRP confined rectangular CFST columns using high-strength materials with moderate slenderness. *Constr. Build. Mater.* **2021**, *299*, 123912. [CrossRef]

39. Yang, Y.; Han, L. Compressive and flexural behaviour of recycled aggregate concrete filled steel tubes (RACFST) under short-term loadings. *Steel Compos. Struct.* **2006**, *6*, 257–284. [CrossRef]

40. Li, Q.; Chen, Z.; Du, Y.; Wu, Y.; Liu, X. Study on constitutive model of core concrete of recycled aggregate concrete filled steel tubular columns under compression. *Ind. Constr.* **2021**, *51*, 108–115.

Article

A Reliability Analysis Framework of Ship Local Structure Based on Efficient Probabilistic Simulation and Experimental Data Fusion

Shuming Xiao [1], Yang Han [2], Yi Zhang [2], Qikun Wei [3], Yifan Wang [3], Na Wang [1], Haodong Wang [3], Jingxi Liu [3,4] and Yan Liu [3,4,*]

[1] Wuhan Rules and Research Institute, China Classification Society, Wuhan 430022, China; sm_xiao@ccs.org.cn (S.X.); nwang@ccs.org.cn (N.W.)
[2] First Engineering Co., Ltd., China Construction Third Bureau, Wuhan 430048, China; hanyanguzhou@foxmail.com (Y.H.); zhangyi_cctb@foxmail.com (Y.Z.)
[3] School of Naval Architecture and Ocean Engineering, Huazhong University of Science and Technology, Wuhan 430074, China; weiqikun@hust.edu.cn (Q.W.); wyff@hust.edu.cn (Y.W.); wanghaodong@hust.edu.cn (H.W.); liu_jing_xi@hust.edu.cn (J.L.)
[4] Hubei Key Laboratory of Naval Architecture and Marine Hydrodynamics (HUST), Wuhan 430074, China
*   Correspondence: yanliuch@hust.edu.cn

Citation: Xiao, S.; Han, Y.; Zhang, Y.; Wei, Q.; Wang, Y.; Wang, N.; Wang, H.; Liu, J.; Liu, Y. A Reliability Analysis Framework of Ship Local Structure Based on Efficient Probabilistic Simulation and Experimental Data Fusion. *Metals* **2022**, *12*, 805. https://doi.org/10.3390/met12050805

Academic Editors: Hihua Chen, Giovanni Meneghetti, Hanbin Ge and Siu-Lai Chan

Received: 28 February 2022
Accepted: 4 May 2022
Published: 6 May 2022

Publisher's Note: MDPI stays neutral with regard to jurisdictional claims in published maps and institutional affiliations.

Abstract: This paper presents a comprehensive framework for the reliability analysis of ship local structures. Existing reliability analysis of ship local structures relies on empirical analysis without experimental validation. The presented framework improves the probabilistic simulation process by combining finite element analysis and the Kriging surrogate model to increase the computational efficiency in uncertainty quantification. In addition, ultimate strength test data are introduced to update the prior distribution based on Bayesian data fusion. A cross-deck structure of a ship is studied in detail to present the application of this work. The framework provides a valuable reference for the reliability analysis of ship local structures and promotes the development of reliability-based design code. The novelty of this paper is that it introduces the combination of testing and probabilistic simulation into the reliability analysis of ship local structures.

Keywords: reliability analysis; ship local structure; probabilistic simulation; Kriging model; Bayesian data fusion

1. Introduction

Ship structural strength is a pre-requisite for vessels to avoid the risks brought by severe sea conditions and emergencies. It is necessary to conduct reliability analysis to assess the reliability level of ship structures. The reliability analysis method is an effective tool for safer engineering design [1]. Traditionally, in the marine industry, the empirical safety factor method [2,3] is applied to meet safety requirements, but it has limited effectiveness. This paper proposes a reliability analysis framework based on efficient probability simulation and Bayesian updating to address the issue of the reliability design of ship local structures. Structural reliability assessment with regard to ship local structures has been attracting growing attention. Zhu [4] investigated the ultimate strength and reliability assessment of ore carrier local structures. Zhang [5] studied the safety assessment of a plate structure considering damage caused by an underwater explosion. Zhang [6] investigated the ultimate strength of steel plates and stiffened panels and their employment in ship designs. However, the aforementioned studies lack testing validation of realistic ship local structures. The purpose of this paper is to investigate the reliability of a ship local structure through the combination of testing and efficient probabilistic simulation for practical use.

Reliability analysis is particularly useful in assessing ship structural integrity when facing uncertainties. Nordsenstrom [7] took the wave bending moment and the bearing capacity of the hull longitudinal bending as random variables, considered that the wave bending moment follows the Weibull distribution and the still water bending moment and the ship bearing capacity obey the normal distribution, and calculated the failure probability of the hull structure by the full probability method. Mansour [8,9] conducted a systematic study of the probabilistic model of the longitudinal strength of the hull. Probabilistic models are widely used to analyze the strength of ship structures. However, probabilistic simulations are subject to aleatory uncertainties associated with a lack of experimental data. Hence, it is necessary to introduce testing data for the modification of probabilistic simulation results.

The longitudinal strength of vessels has been widely researched by many scholars. Caldwell [10] firstly estimated the ultimate longitudinal strength of steel ships. Researchers explored the effect of initial imperfections or damage on longitudinal strength [11–13]. The ultimate strength of a ship hull girder with openings was investigated by Zhao [14]. Xu [15] and Xia [16] studied the relationship between the load forms and longitudinal strength. Thus far, researches of longitudinal strength have exhaustively explored many relative aspects in terms of reliability analysis. The design of ship structures is not always governed by longitudinal strength. Reliability analysis aiming at specific local structures has great potential to improve the overall performance of ship design. However, design rules based on the reliability analysis of ship local structures have not been well developed.

The computational demand of reliability analysis is tractable on analytical systems, but it can become impossible with complex computer codes such as finite element models [17]. With no analytical formula, the finite element method is essential for the reliability analysis of complex ship local structures, which leads to high computing costs. The surrogate model technique can handle the problem effectively. Due to the complexity of ship structures, the introduction of a surrogate model into the application has practical significance. Researchers proposed a method that combined the first-order reliability method and Kriging model as the response surface, assessed the prediction accuracy, and verified the usefulness and efficiency of the reliability analysis using the Kriging model [18]. Methods based on the Kriging method have been established to optimize the computing problem of the parameterized model [19,20]. The combination of the finite element model and Kriging surrogate model is adopted in this framework to improve the efficiency of probabilistic simulations.

The presented framework proposes an applicable approach to conduct reliability analysis with respect to real ship local structures. Aiming at ship local structural reliability design, this paper proposes a comprehensive framework, which conducts probability simulation using the Kriging method, combines test data fusion based on Bayesian update, and provides structural reliability design recommendations in practice. Considering the trend of larger-sized vessels, the midship structures are subjected to more transverse compressive loads, which is likely to lead to buckling failure. Therefore, this paper selects a cross-deck structure for tests to verify the effectiveness of the reliability analysis framework. The framework effectively overcomes several key problems: time-consuming and expensive probabilistic finite element simulations, testing of realistic ship local structures, and reliability design code development for specific ship structure design considerations. The main contribution of this paper is that it explores the efficient application of structural reliability analysis, with specific illustration on a typical ship local structure for simulation, testing, and practical design application.

2. Reliability Analysis Framework

2.1. Cross-Deck Structure

The cross-deck structure is designed to avoid tearing caused by the direct welding of the bulkhead to the deck. The cross-deck structure is usually located between the transverse bulkhead and hatch coaming, and its structure mainly includes inclined plates and vertical plates. The typical structural arrangement is illustrated in Figure 1a,b.

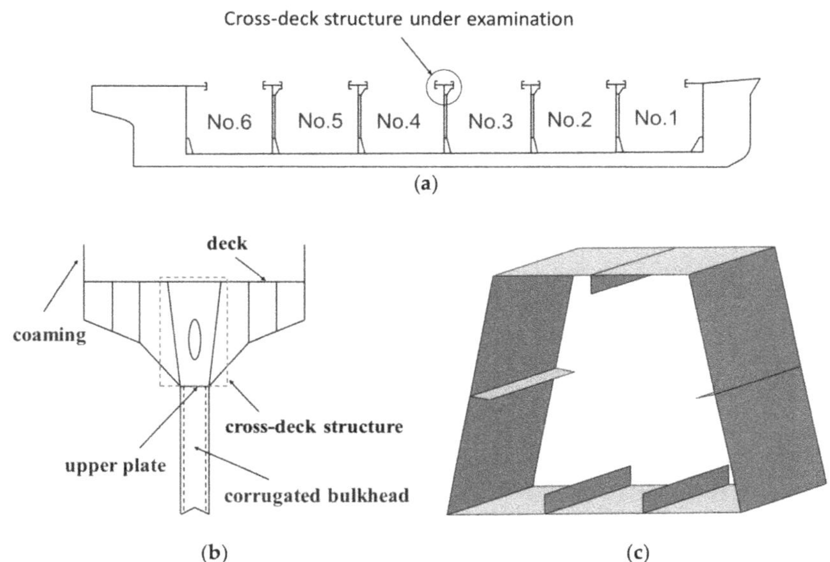

Figure 1. Cross-deck structure and its simplified model (**a**) The position of cross-deck structure in ships. (**b**) Schematic diagram of cross-deck structure (**c**) Simulation model of cross-deck structure.

In this paper, the cross-deck structure is simplified as a box girder structure for ultimate strength test consideration. The parametric model of the box girder is established for finite element simulation. The model of the box girder structure is shown in Figure 1c. This framework introduces tests to improve the precision of reliability analysis. A test specimen, shown in Figure 1c, is produced for the ultimate strength test, and the details of the test specimen are listed in Table 1.

Table 1. Main parameters of box girder structure.

Parameter	Unit	Data
Upper bottom width	mm	400
Bottom width	mm	600
Section height	mm	500
Overall length	mm	1500
Plate thickness	mm	3
Stiffener thickness	mm	3
Stiffener height	mm	30
Material	\	Q235

2.2. Efficient Probabilistic Simulation and Bayesian Update

An ultimate strength surrogate model based on the Kriging method of the cross-deck structure is introduced based on the response values of structural strength simulations. The response values are obtained by the finite element method, which is applied with parametric modeling. It is necessary to conduct a precision check for the Kriging surrogate model through random sampling. The stochastic distribution data of ultimate strength are obtained through random sampling by the Monte Carlo method of performance function. However, the result of probabilistic simulation is based on prior estimate simulation, leading to a deviation from actual performance. Hence, it is necessary to introduce Bayesian updating to modify the distribution parameters with the box girder test results. The flow chart of the framework is illustrated in Figure 2. As shown in the flow chart, the probability simulation part and experiment part start in parallel, and the experiment part provides data for Bayesian updating.

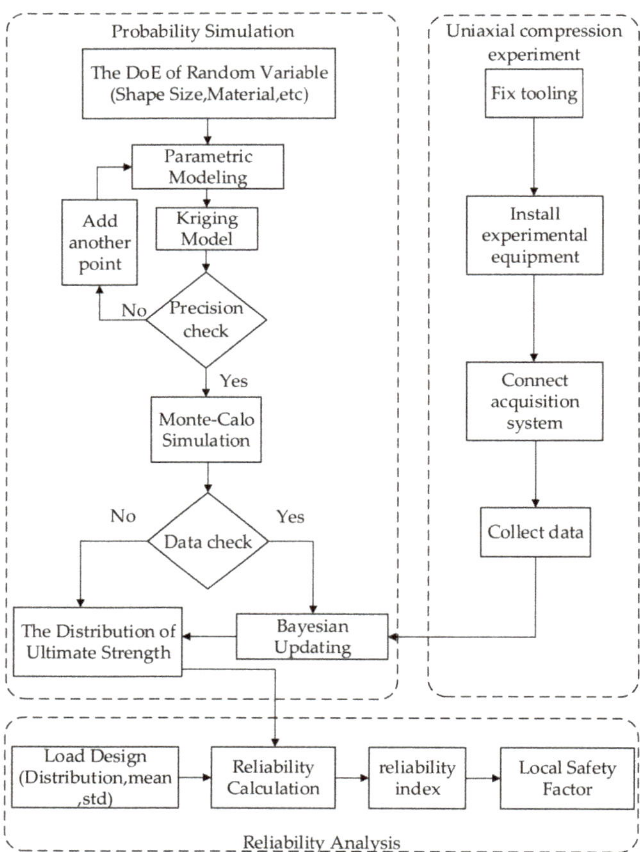

Figure 2. Flow chart of reliability analysis framework.

3. Efficient Probabilistic Simulation

With the improvement of simulation tools, the reliability analysis method has also changed from numerical analysis to simulation. For complex structures, it is difficult to obtain efficient solutions with probabilistic simulation. This section introduces the Kriging-assisted probabilistic simulation method for the box girder.

3.1. Probability Parametric Modeling

Referring to the research of Zhu et al. [21], in order to understand the influence of model parameters on the ultimate bearing capacity of the box girder, the plate thickness t_p, Young's modulus E_p, stiffener height h_s, and stiffener thickness t_s of the stiffened panels are studied as random variables. The ultimate strength distribution of the box girder model is obtained from probabilistic simulation. In Table 2, values of mean are given on the basis of the test specimen. The distribution type and coefficient of variation are selected according to the study of Zhu et al. [21].

Table 2. Distribution and parameters of random variables.

Random Variables	Distribution Type	Mean	Coefficient of Variation
stiffener height h_s	normal	30 (mm)	0.03
Young' modulus E_p	lognormal	210,000 (MPa)	0.03
plate thickness t_p	normal	3 (mm)	0.03
stiffener thickness t_s	normal	3 (mm)	0.03

There are some commonly used sampling methods, such as Monte Carlo sampling, optimal Latin hypercube sampling, uniform sampling, and Hammersley sequence sampling. The optimal Latin hypercube sampling is widely used because of its good spatial uniformity and projection uniformity. This study adopts this method to sample the random variables in Table 2 as experimental points and test points, respectively. The sampling interval of each variable is $[\mu - 3\sigma, \mu + 3\sigma]$, where μ is the mean of the variable and σ is the standard deviation of the variable. The number of sampling points is 50.

For a large number of experimental design points generated by sampling, manual finite element modeling will be inconvenient. When the finite element modeling process is the same but the characteristic parameters are different, the parametric modeling can be performed by modifying the characteristic parameters with code. Parametric modeling has a significant effect on reducing the workload. In this study, the development of Abaqus based on Python can realize the automation of a series of finite element analysis. Figure 3 shows the calculation result of the ultimate strength of the box girder model.

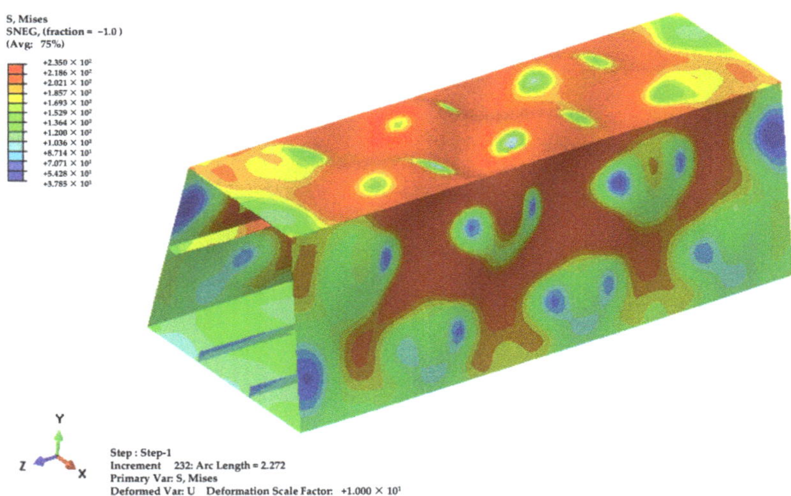

Figure 3. Finite element analysis of ultimate strength of box girder model.

3.2. Surrogate Model

The response data $y_1, y_2, \ldots, y_n$ of the ultimate strength obtained by the parametric modeling of the finite element are used as the experimental points to construct an approximate surrogate model for ultimate strength by the Kriging method. The surrogate model schematic diagram is presented in Figure 4. Referring to the works of Sacks et al. [22] or Simpson et al. [23], the Kriging method used is ordinary Kriging, where $f_f(x)$ is approximated by:

$$f_f(x) \approx \widetilde{f_f}(x) = F + Z(x) \tag{1}$$

where F is an estimator, $Z(x)$ is a local error term specific to each design point in the independent variable space, and x is a vector variable describing a design point in the independent variable space.

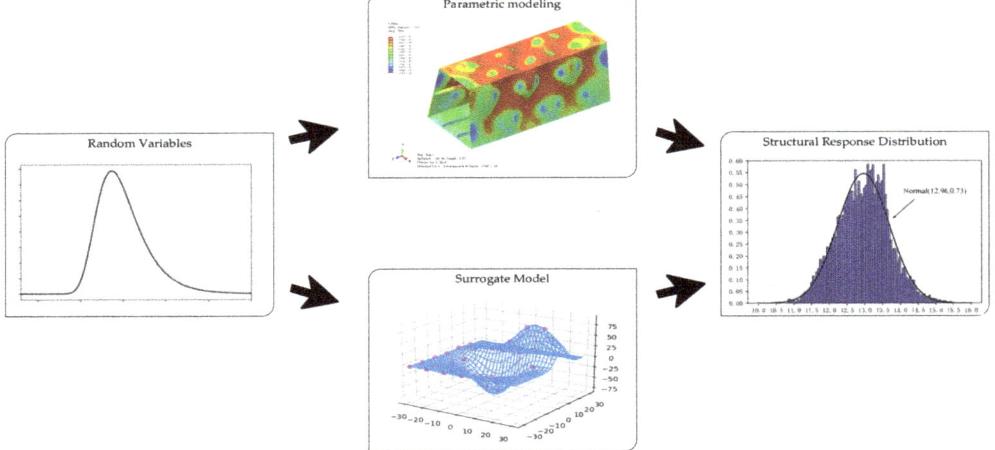

Figure 4. Surrogate model schematic diagram.

The relationship of the plate thickness t_p, Young's modulus E_p, stiffener height h_s, stiffener thickness t_s, and ultimate strength can be constructed through the surrogate model. In order to verify the accuracy of the surrogate model for calculating the ultimate strength, the experimental test points are used to carry out the accuracy analysis. Using experimental test points as input, we compare the calculation results based on the finite element and surrogate model and calculate the root mean square error. The construction of the surrogate model is completed only when the error is less than the specified value. Figure 5 presents the error distribution of the experimental test points. It can be seen from the figure that, compared with the results of the finite element calculation, the mean square error of the surrogate model is 0.170, and the maximum error is 0.56%. In summary, the accuracy of the surrogate model is sufficient.

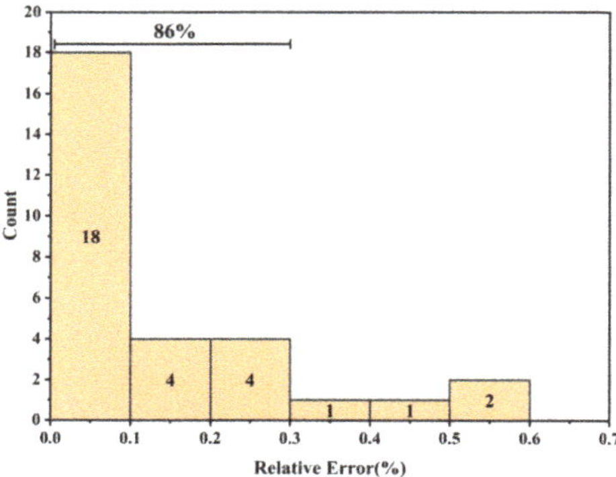

Figure 5. Error distribution of experimental test points.

3.3. Ultimate Strength Uncertainty Quantification

After establishing the ultimate strength surrogate model of the box girder structure and verifying its accuracy, the Monte Carlo method is applied to sample random variables and calculate the response value through the surrogate model. Finally, the sampled response values are fitted to a normal distribution. The mean of the fitted distribution is 8.578×10^5 N, and the standard deviation is 3.97×10^4 N.

4. Test of Box Girder Structures for Ultimate Strength

4.1. Test Specimen and Test Purpose

The test specimen is processed on the basis of the model shown in Figure 1, a box girder structure. The processing is performed according to specifications [24] formulated by the China Classification Society.

The failure mode is affected by the load characteristics, material mechanical properties, and structural size, and these factors have different distribution characteristics. Therefore, the structural reliability analysis of the cross-deck structure needs to explore the failure signature under the design load and the failure path of the structure firstly. Hence, the purposes of the test include the following:

- To obtain the failure path of the structure;
- To obtain the influence characteristics of uncertain factors;
- To verify the reasonability of the selection of reliability analysis parameters.

4.2. Test Process

4.2.1. Initial Defect Measurement

It is difficult to avoid initial defects during the cutting and welding process of steel plates. The initial defects have significant impacts on the ultimate strength. Hence, it is necessary to conduct initial defect measurement before the test. The traditional measurement method has the problems of low precision and complicated operation. Therefore, a handheld laser 3D scanner is introduced. The handheld laser 3D scanner positioning technology is able to provide high-precision 3D scanning results, not subjected to dimension, material, and color, etc. The scanning results of the specimen are shown in real time on the computer during the scanning process. The handheld scanner and scanning results are illustrated in Figure 6.

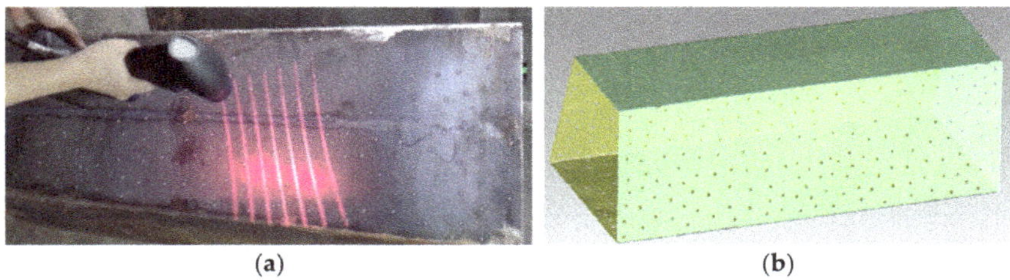

(a)   (b)

Figure 6. Three-dimensional scanning (**a**) Three-dimensional scanning with laser scanner (**b**) Three-dimension scanning point cloud map.

4.2.2. Test Equipment

The test equipment is shown in Figure 7. A hydraulic jack, displacement sensor, force sensor, and cover plate are arranged from top to bottom.

Two special upper and lower cover plates are designed, and the thickness of the plates is 50 mm. The cover plates are provided with grooves with a depth of 10 mm. The width of the grooves is slightly larger than the thickness of the box girder plates and stiffeners. The grooves of the upper and lower cover plates are used to hold both ends of the test specimen, and the force is evenly transmitted to the specimen through the cover plate at the upper end.

The force sensor is fixed between the top of the hydraulic jack and cover plate to measure the axial load on the model. Two laser displacement sensors are arranged symmetrically to measure the axial displacement of the specimen. At the same time, a camera is placed outside each surface of the model to record the deformation process of the specimen during the test.

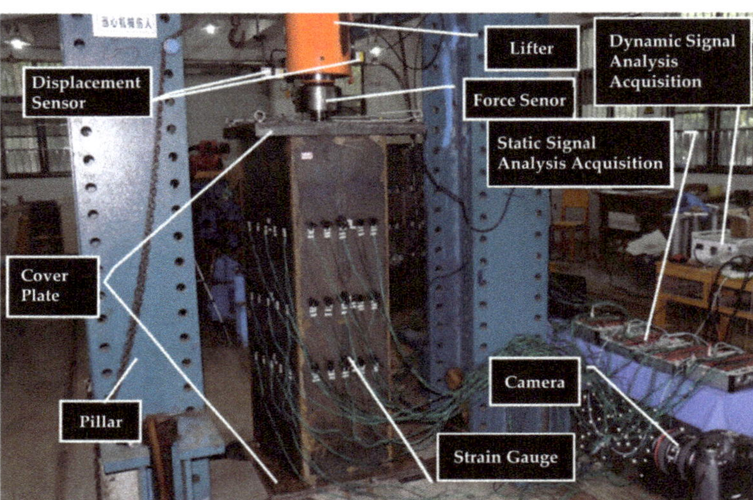

Figure 7. Arrangement of test equipment.

4.3. Credibility of Test

We apply an axial compressive load to the box girder structure and measure its structural displacement. When the load is increased to 603 KN, the box girder structure is obviously deformed, as shown in Figure 8. Meanwhile, the force sensor values drop, indicating that the structure has been buckling.

Figure 8. Buckling failure of the specimen.

A total of 8 sets of box girders are manufactured and tested. Load–displacement curves for a test specimen (No. 7) and the ultimate strength load of 8 sets are illustrated in Figure 9. The displacement takes the average of the two laser displacement sensors' values. The curves show that the test results are convincing, indicating the credibility of the test design, tooling, and loading process.

4.4. Data Fusion Using Bayesian Updating

The result of probabilistic simulation based on prior distribution is updated with test data. Test data fusion based on Bayesian updating improves the credibility of the results significantly. The Bayesian formula is the foundation of Bayesian updating, as shown below:

$$f(\theta|x) = \frac{f(x|\theta)f(\theta)}{f(x)} \tag{2}$$

where the parameter θ is a continuous variable:

$$f(x) = \int_{-\infty}^{+\infty} f(x|\theta)f(\theta)d\theta \tag{3}$$

The parameter $\theta = (\theta_1, \cdots, \theta_k)$ is a $k \times 1$ vector, and k is the dimension of θ. The parameter $x = (X_1, \cdots, X_n)$ is a $n \times 1$ vector, and n is the number of stochastic variables X_i. $f(\theta)$ in Equation (2) is the prior distribution, which is obtained by probabilistic simulation. $f(x|\theta)$ is the conditional probability distribution of X under given parameter θ, which can be obtained through test data. The formula of $f(x|\theta)$ is illustrated as follows [25]:

$$f(x|\theta) = \prod_{j=1}^{m} \frac{1}{\sqrt{2\pi\sigma}} \cdot \exp(-\frac{1}{2}(\frac{x_j - \hat{x}}{\sigma})^2) \tag{4}$$

where x_j and $\hat{x}$ are the jth experimental data and expected value, σ is the standard deviation of the statistical data, and m is the number of experimental data. As shown in Figure 10, the distribution after Bayesian updating is more centralized than the previous distribution.

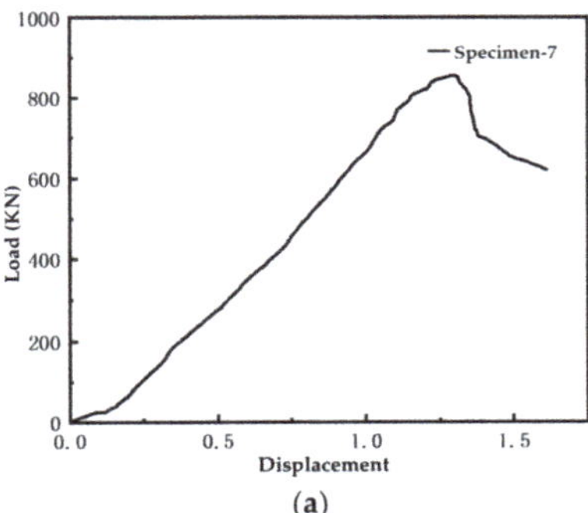

(a)

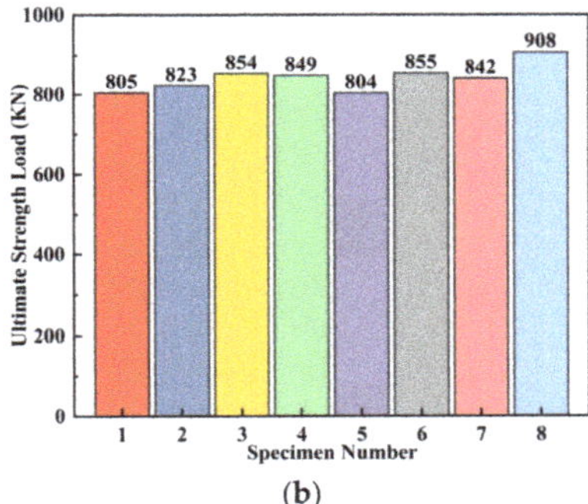

(b)

Figure 9. Load–displacement curve and histogram of ultimate strength load (**a**) Load-displacement curve of specimen-7 (**b**) Ultimate strength load of 8 sets of test specimen.

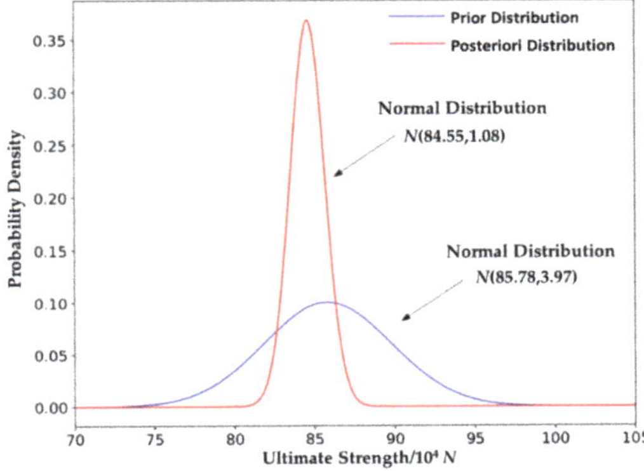

Figure 10. Probability density distribution plot before and after Bayesian update.

4.5. Analysis of Test and Comparison with Simulation

The initial defect has a great influence on the ultimate strength. The ultimate strength decreases with the increase in the initial defect. Therefore, the deformation caused by welding should be minimized as much as possible during processing. Insufficient stiffness of the upper and side plates is the main reason for the failure of the entire structure. Increasing the thickness of the upper and side plates appropriately is suggested when designing a trapezoidal section box girder. The buckling failure mode and ultimate strength of the test and simulation are anastomotic. The accuracy of the finite element method used in this framework is confirmed by the test.

5. Reliability Analysis

The purpose of this study is to assess the reliability of the cross-deck structures through reliability analysis. Generally speaking, the structural reliability analysis process can be divided into three parts. Firstly, we determine the distribution probability and related statistics of structural random variables $X = (x_1, x_2, \cdots, x_N)$. Secondly, we establish the limit state equation $G(X)$ of the structure. When $G(X) < 0$, it means that the structure fails. Finally, we calculate the probability that $G(X) < 0$ is the failure probability P_f of the structure. The main methods of reliability analysis are the first-order reliability method, the Monte Carlo simulation method, and Response Surface Methodology. In this work, the first-order reliability method will be used.

5.1. Limit State Equation

Considering the yield failure of the structure under lateral compression, the limit state equation of the box girder structure is defined as follows:

$$G(X) = F - F_{cd} \tag{5}$$

where F is the ultimate strength of the box girder structure, which is related to the shape, size, material properties, and boundary conditions of the box girder structure. It can be obtained by probabilistic simulation with the stochastic finite element method or Bayesian update with experimental data; F_{cd} is the lateral load force on the section of the box girder structure, which is affected by factors such as the hull structure, cargo load, still water load, and wave load. It can be obtained by formula derivation.

With the knowledge of the probabilistic distribution function of yield loads and resistance for the box girder structure, the probability of ultimate yield failure of the structure can be assessed as:

$$P_f(X) = P(G(X) < 0) = \Phi(-\beta) \tag{6}$$

where Φ is the standard cumulative normal distribution function, P_f is the probability of occurrence of the event $G(t) < 0$, and β is the reliability index.

5.2. Load Design

Referring to the article [21], the lateral load force F_{cd} of the box girder structure is affected by the hull structure, cargo load, still water, and wave load. However, due to the lack of information on cargo loads, still water load, and wave load, this paper adopts the method of assuming loads. Figure 11 shows the probability density distribution of three different distributed loads under consideration.

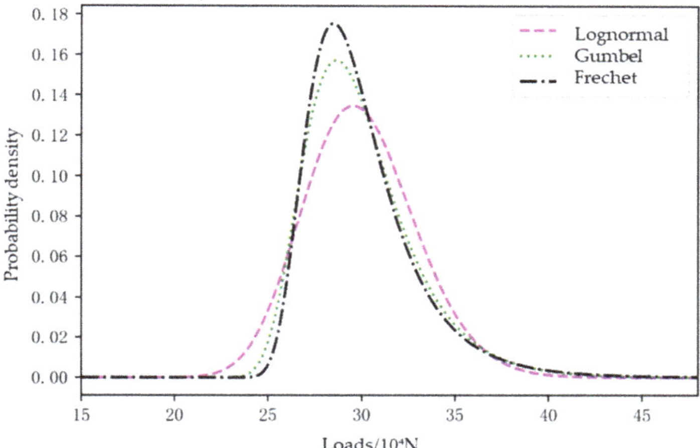

Figure 11. Probability density distribution of three different distributed loads.

5.3. Local Safety Factor Calculation

In this study, the local safety design criteria for the box girder are:

$$\gamma_s F_{cd} \leq \frac{F}{\gamma_r} \tag{7}$$

where γ_s, γ_r are the Partial Safety Factors for load and the Partial Safety Factors for resistance, respectively. F_{cd}, F are the average values of load and resistance, respectively. The relationship between these parameters is shown in Formulas (8)–(10):

$$K = \gamma_r \cdot \gamma_s \tag{8}$$

where K is the local safety factor of the box girder structure,

$$\gamma_r = \frac{1 - \frac{K_{R_u}}{\mu_{R_u}} C.O.V_{R_u}}{1 - 0.75\beta C.O.V_{R_u}} \tag{9}$$

$$\gamma_s = \frac{1 + 0.75\beta C.O.V_{S_e}}{1 + \frac{K_{S_e}}{\mu_{S_e}} C.O.V_{S_e}} \tag{10}$$

where β is the reliability index obtained by the reliability calculation method. The $C.O.V_{R_u}$ and $C.O.V_{S_e}$ are the coefficients of variation of resistance and load, respectively. K_{R_u} takes the value that makes the characteristic resistance R_u^k equal to the 5th percentile of the resistance distribution. K_{S_e} takes the value that makes the characteristic loading effect S_e^k equal to the 5th percentile of the loading distribution. Table 3 shows the various coefficients and local safety factors under different load distributions and parameters.

Table 3. Coefficients and local safety factors under different load distributions and parameters.

Statistical Parameters	Distribution Type	β	γ_s	$\frac{1}{\gamma_r}$	K
mean: $30 \times 10^4 N$ standard deviation: $3 \times 10^4 N$	Lognormal	10.351	1.993	0.847	2.353
	Gumbel	7.141	1.970	0.847	2.325
	Frechet	4.936	1.969	0.847	2.325
mean: $40 \times 10^4 N$ standard deviation: $4 \times 10^4 N$	Lognormal	7.491	1.600	0.907	1.764
	Gumbel	4.949	1.582	0.907	1.744
	Frechet	4.107	1.582	0.907	1.744
mean: $50 \times 10^4 N$ standard deviation: $5 \times 10^4 N$	Lognormal	5.273	1.337	0.947	1.411
	Gumbel	3.768	1.322	0.947	1.395
	Frechet	3.340	1.322	0.947	1.395

6. Discussion on Limitations of Framework

Although this framework presents a comprehensive reliability analysis of practical ship local structures through the combination of testing and probabilistic simulation, several limitations of the framework have to be recognized.

The test is built on a scaling model and not a real cross-deck structure because of the high cost and difficulties of a full-scale ship structure test. Secondly, reliability results can be affected by the choice of probabilistic distribution type and parameters, such as those listed in Table 2. Moreover, a probabilistic load is assumed in the framework. An axial load is considered in the test, while more complex loading conditions can arise in real situations. Thus, it must be noted that these limitations should be taken into consideration before applying this developed framework to ship local structures.

7. Conclusions

The ultimate strength experiment and FEM analysis indicated that the stiffness insufficiency of the upper and side plate is the main reason for the failure of the box girder structure. The thickness of the upper and side plate should be increased appropriately for more capable designs. The high fitness of results on the buckling failure mode and ultimate bearing capacity between the experimental and numerical models reveals the accuracy of the FEM. For model calculation, initial defects lead to a decline in the ultimate strength value. Hence, the deformation caused by welding should be controlled as far as possible during component processing.

The reliability study of a cross-deck structure shows a concrete application of the presented reliability analysis framework, which takes experimental data and FEM modeling into consideration. The Kriging surrogate model significantly reduces the required amount of probabilistic simulations by using parametric modeling. In addition, the Bayesian updating method decreases the uncertainty of the ultimate strength and rectifies errors between the test and modeling. According to the reliability results, the height of the stiffener and the thickness of the plate play a positive role in structural reliability levels.

This paper proposes a reliability analysis framework and explores the comprehensive combination of probabilistic simulation and testing. The framework improves the precision and decreases the computation cost and demands effectively. The study offers contributions to the development of the reliability-based design code for ship local structures.

Author Contributions: Conceptualization, S.X. and J.L.; methodology, Y.L.; software, H.W. and Y.W.; validation, H.W., Q.W. and Y.W.; formal analysis, H.W.; investigation, Y.H. and Y.Z.; resources, N.W.; data curation, S.X. and N.W.; writing—original draft preparation, Q.W. and Y.W.; writing—review and editing, Y.L.; visualization, Q.W.; supervision, J.L.; project administration, N.W.; funding acquisition, S.X. All authors have read and agreed to the published version of the manuscript.

Funding: This research was funded by the National Natural Science Foundation of China, Grant No. 52071150. The opinions and conclusions presented in this paper are those of the authors and do not necessarily reflect the views of the sponsoring organizations.

Data Availability Statement: Not applicable.

Acknowledgments: The authors acknowledge the Laboratory of Ship Structural Strength of the School of Naval Architecture and Ocean Engineering, Huazhong University of Science and Technology for the help with conducting the experimental study in this research.

Conflicts of Interest: The authors declare no conflict of interest.

References

1. Leveson, N.G. *Engineering a Safer World*; MIT Press: Cambridge, MA, USA, 2012.
2. Ayyub, B.; Assakkaf, I.; Sikora, J.; Adamchak, J.; Atua, K.; Melton, W.; Hess, P. Reliability-Based Load and Resistance Factor Design (LRFD) Guidelines for Hull Girder Bending. *Nav. Eng. J.* **2002**, *114*, 43–68. [CrossRef]
3. Deng, L.; Zou, L.; Tang, W.; Fang, C. Reliability analysis of longitudinal strength of inland river ships and optimization of sub-item safety factors. In Proceedings of the 2017 Outstanding Academic Papers of the Chinese Society of Naval Architects and Marine Engineers, Wuhan, China, 1 July 2016; pp. 13–20. (In Chinese).
4. Zhu, L.; Pan, M.; Zhou, H.-W.; Das, P.K. Transverse Strength and Reliability Assessment on Ore Carrier Cross-Deck Structures. In Proceedings of the 27th International Ocean and Polar Engineering Conference, San Francisco, CA, USA, 25–30 June 2017.
5. Yongkun, Z.; Xin, G. Research on the ship local structure damage subjected to underwater explosion. In Proceedings of the 2017 29th Chinese Control And Decision Conference (CCDC), Chongqing, China, 28–30 May 2017; pp. 6113–6117.
6. Zhang, S. A review and study on ultimate strength of steel plates and stiffened panels in axial compression. *Ships Offshore Struct.* **2015**, *11*, 81–91. [CrossRef]
7. Nordsenstrom, N. *Probability of Failure for Weibull Load and Normal Strength*; Det Norske Veritas: Berum, Norway, 1969.
8. Mansour, A. Methods of Computing the Probability of Failure Under Extreme Values of Bending Moment. *J. Ship Res.* **1972**, *16*, 113–123. [CrossRef]
9. Mansour, A.E. Probabilistic Design Concepts in Ship Structural Safety and Reliability. *Trans. SNAME* **1993**, *80*, 64–97.
10. Caldwell, J.B. Ultimate longitudinal strength. *Trans. Roy Inst. Naval. Arch.* **1965**, *107*, 411–430.
11. Vhanmane, S.; Bhattacharya, B. Estimation of ultimate hull girder strength with initial imperfections. *Ships Offshore Struct.* **2008**, *3*, 149–158. [CrossRef]
12. Kim, D.K.; Park, D.K.; Kim, H.B.; Seo, J.K.; Kim, B.J.; Paik, J.K.; Kim, M.S. The necessity of applying the common corrosion addition rule to container ships in terms of ultimate longitudinal strength. *Ocean. Eng.* **2012**, *49*, 43–55. [CrossRef]
13. Van Vu, H.; Sang-Rai, C. Generate the Ultimate Longitudinal Strength of Damaged Container Ship Considering Random Geometric Properties under Combined Bending Moment. In Proceedings of the 2nd Annual International Conference on Material, Machines and Methods for Sustainable Development, Cham, Switzerland, 12–15 November 2020; pp. 495–501.
14. Zhao, N.; Chen, B.-Q.; Zhou, Y.-Q.; Li, Z.-J.; Hu, J.-J.; Soares, C.G. Experimental and numerical investigation on the ultimate strength of a ship hull girder model with deck openings. *Marine Struct.* **2022**, *83*, 103175. [CrossRef]
15. Xu, M.; Song, Z.; Zhang, B.; Pan, J. Empirical formula for predicting ultimate strength of stiffened panel of ship structure under combined longitudinal compression and lateral loads. *Ocean. Eng.* **2018**, *162*, 161–175. [CrossRef]
16. Xia, T.; Yang, P.; Song, Y.; Hu, K.; Qian, Y.; Feng, F. Ultimate strength and post ultimate strength behaviors of hull plates under extreme longitudinal cyclic load. *Ocean. Eng.* **2019**, *193*, 106589. [CrossRef]
17. Echard, B.; Gayton, N.; Lemaire, M. AK-MCS: An active learning reliability method combining Kriging and Monte Carlo Simulation. *Struct. Saf.* **2011**, *33*, 145–154. [CrossRef]
18. Shi, X.; Teixeira, Â.P.; Zhang, J.; Soares, C.G. Kriging response surface reliability analysis of a ship-stiffened plate with initial imperfections. *Struct. Infrastruct. Eng.* **2014**, *11*, 1450–1465. [CrossRef]
19. De Baar, J.; Roberts, S.; Dwight, R.; Mallol, B. Uncertainty quantification for a sailing yacht hull, using multi-fidelity kriging. *Comput. Fluids* **2015**, *123*, 185–201. [CrossRef]
20. Bichon, B.; Eldred, M.; Swiler, L.; Mahadevan, S.; McFarland, J. Efficient Global Reliability Analysis for Nonlinear Implicit Performance Functions. *Aiaa J. AIAA J.* **2008**, *46*, 2459–2468. [CrossRef]
21. Pan, M.; Zhu, L.; Kumar, D.P. Structural reliability assessment on cross-deck structures of ore carrier (in Chinese). *J. Ship Mech.* **2020**, *24*, 118–126.
22. Sacks, J.; Welch, W.J.; Mitchell, T.J.; Wynn, H.P. Design and analysis of computer experiments. *Stat. Sci.* **1989**, *4*, 409–423. [CrossRef]
23. Simpson, T.W.; Poplinski, J.; Koch, P.N.; Allen, J.K. Metamodels for computer-based engineering design: Survey and recommendations. *Eng. Comput.* **2001**, *17*, 129–150. [CrossRef]
24. China Classification Society. *Materials and Welding Specifications*; China Classification Society: Beijing, China, 2021.
25. Perrin, F.; Sudret, B.; Pendola, M. Bayesian updating of mechanical models—Application in fracture mechanics. In *CFM 2007—18ème Congrès Français de Mécanique*; AFM, Maison de la Mécanique: Courbevoie, France, 2007.

Article

Intelligent Analysis for Safety-Influencing Factors of Prestressed Steel Structures Based on Digital Twins and Random Forest

Haoliang Zhu * and Yousong Wang

School of Civil Engineering and Transportation, South China University of Technology,
Guangzhou 510640, China; yswang@scut.edu.cn
* Correspondence: zhuhaoliang86@163.com

Abstract: The structure of a prestressed steel structure is complex, which can result in insufficient control accuracy and the low efficiency of the structural safety. The traditional analysis method only obtains the mechanical parameters of the structure and it cannot obtain the key factors that affect the structural safety. In order to improve the intelligence level of the structural safety performance analysis, this study proposes an intelligent analysis for the safety-influencing factors of prestressed steel structures that is based on digital twins (DTs) and random forest (RF). Firstly, the high-precision twin modeling is carried out by the weighted average method. The design parameters and the mechanical parameters of the structure are extracted in real time in the twin model, and the parameters are classified by the RF. The fusion mechanism of the DTs and RF is formed, and the intelligent analysis model of the structural safety factors is established. Driven by the analysis model, the correlation mechanism between the design parameters and the mechanical parameters is formed. The safety state of the structure is judged by the mechanical parameters, and the key design parameters that affect the various mechanical parameters are analyzed. Through the integration of the design parameters and mechanical parameters, the intelligent analysis process of the safety-influencing factors of prestressed steel structures is formed. Finally, an intelligent analysis of the importance of the safety-influencing factors is carried out with the string-supported beam structure as the test object. Driven by the integration of DTs and RF, the key design parameters that affect the various mechanical parameters are accurately obtained, which provides a basis for the intelligent control of the structural safety.

Keywords: digital twin; random forest; prestressed steel structure; influencing factors; intelligent analysis

Citation: Zhu, H.; Wang, Y. Intelligent Analysis for Safety-Influencing Factors of Prestressed Steel Structures Based on Digital Twins and Random Forest. *Metals* **2022**, *12*, 646. https://doi.org/10.3390/met12040646

Academic Editor: Zhihua Chen, Hanbin Ge and Siu-lai Chan

Received: 8 March 2022
Accepted: 9 April 2022
Published: 11 April 2022

Publisher's Note: MDPI stays neutral with regard to jurisdictional claims in published maps and institutional affiliations.

1. Introduction

Prestressed steel structures have the advantages of strong spanning capacities, beautiful shapes, light weights and short construction periods, and they are used in public buildings, such as in large stadiums [1]. At present, prestressed steel structures mainly include the cable dome structure [2]; the wheel spoke structure [3]; the cable truss structure, without the inner ring space [4]; the cable net structure [5]; the cable net shell structure [6] and so on. Prestressed steel cable is widely used in cable dome structures, and the compression rod is less and shorter, so that it can give full play to the tensile strength of the steel, and the structural efficiency is very high. The cable dome structure was applied to the gymnastics hall of the Seoul Olympic Games in South Korea. Radiated cable truss is mainly composed of an external pressure ring, cable truss and an inner pull ring. The Jabir Ahmed Stadium in Kuwait is a single-layer spoke cable truss. The ringless prestressed cable-supported structure system includes two categories: the ringless prestressed cable-supported structure system, and the ringless prestressed string-supported structure system. The Beijing Olympic Badminton Pavilion applies a cableless dome structure. The Beijing Winter Olympic Speed Skating Hall is the largest single-layer orthogonal cable-net structure in the world. The accurate safety maintenance of long-span spatial structures is also an important standard for the measurement of the national construction technology and level. The mechanical properties of the components directly determine the safety performance of the structure [7]. In the structure, different design parameters have also become important factors that affect the structural safety state. Since large-span spatial structures are mostly used in buildings of high importance, the safety performance of the structures is strictly required [8–10].

Guo et al. [11] investigated the effect of the initial cable length error in the prestressing state on the sensitivity of prestressed cables to the length error. By controlling the length error, the prestress level during the cable tensioning was effectively improved. In order to ensure the stability of the construction process of cable dome structures, Zhang et al. [12] propose a joint-square double-strut cable dome structure. This structure effectively improves the safety control accuracy of the structure. Wang et al. [13] analyzed the most active parameters (e.g., the cable force) in the construction process of spatial-structure prestressed cables in the whole process of prestressed cable tension. The safety of the construction process structure is ensured by analyzing the cable force. Arezki et al. [14] investigated the effect of temperature variation on the safety performance of cable truss structures and cables. Basta et al. [15] studied the quantitative evaluation of the decomposability of the cable-net structure on the basis of building information modeling (BIM).

By analyzing the research of the abovementioned scholars, new structural forms and efficient technical methods are studied for the safety maintenance of prestressed steel structures. With the development of a new generation of information technology and the promotion of industrial information systems, the application of intelligent technology to engineering construction has become a research hotspot. The application of DTs and intelligent algorithms in engineering practice can significantly improve the accuracy and intelligence of the structural performance analysis [16,17]. The integration of DTs and intelligent algorithms can realize the virtual simulation of the safety state of prestressed steel structures, and it can form the association mining between the design parameters and the mechanical parameters. Finally, the key factors that affect the structural safety performance were obtained in order to achieve the precise maintenance of structural safety.

DTs simulate and depict the state and the behavior of physical entities with a dynamic virtual model with high fidelity. As a link between the real physical world and the virtual digital space, it is the key enabling technology for the realization of intelligent construction [18,19]. Artificial intelligence has been applied in many disciplines and has formed a variety of intelligent algorithms [20], which can extract high-level features from the original data for perceptual decision making, and can improve the objectivity and accuracy of the information evaluation [21]. Liu et al. [22] propose a DT-driven dynamic guidance method for fire evacuation. The method integrates the Dijkstra algorithm to realize the real-time acquisition of environmental information, the three-dimensional visualization of the indoor layout and evacuation path planning. Lu et al. [23] integrated DTs, machine learning and data analysis to create a simulation model that represents and predicts the current and future conditions of physical counterparts. The integration of technology promotes the implementation and development of smart cities. Acharya et al. [24] propose a visual positioning method to achieve the real-time and accurate positioning of indoor buildings. The 3D indoor model is used to eliminate the image-based indoor environment reconstruction requirements, and the deep convolution neural network is fused to fine-tune the image. Random forest (RF) has strong advantages in data processing, especially data classification. Bhuiyan et al. [25] conducted a comprehensive assessment of 17 developed economies by using RF methods, a fuzzy decision-making test and assessment laboratory methods. For a structural health monitoring and reliability analysis, Liu et al. [26] studied an uncertain dynamic load identification strategy combined with the Kalman filter algorithm and the RF model. Soleimani [27] propose a machine learning algorithm (the RF ensemble learning method) to evaluate the importance of the modeling parameters for estimating the seismic demand. The results of the RF analysis are helpful to better understand the seismic performances of bridges. Therefore, the integration of

DTs and intelligent algorithms provides new ideas and methods for the intelligent transformation and upgrading of the construction industry. Driven by DTs, the high-fidelity twin model of the structure is established. By extracting the structural safety indicators and their corresponding influencing factors in the twin model, the importance of various factors is analyzed by RF, which provides a reliable basis for structural safety control and maintenance.

In view of the demand for the intelligent analysis of the construction safety of prestressed steel structures, the advantages of DTs and intelligent algorithms are combined. Prestressed steel structures are mostly used in large public buildings, and their dynamic behavior needs real-time simulation. Therefore, building a high-fidelity simulation model based on DTs is a key step. In this study, an intelligent analysis method for the safety-influencing factors of prestressed steel structures that is based on DTs and random forest (RF) is proposed. Firstly, the construction method of the high-precision twinning model of the structure is formed on the basis of the weighted average method. By analyzing the fusion mechanism of DTs and RF, an intelligent analysis model of the structural-safety-influencing factors is formed from five dimensions. Driven by the analysis model, the correlation mechanism between the design parameters and the mechanical parameters is established. Design parameters are the influencing factors of the structural safety. The safety performance of the structure is reflected according to the mechanical parameters. The key design parameters that affect the structural safety performance are obtained by the classification of the mechanical parameters. When the structural mechanical parameters exceed the limit, the key design parameters can be corrected to accurately formulate the safety maintenance measures. The resulting theoretical method is applied to the analysis of the influencing factors of the safety performance of beam string structures. By analyzing the influence of the change in the design parameters on the mechanical parameters in the test structure, the correction of the key influencing factors can significantly improve the safety performance of the structure. This study provides a reliable basis for structural health monitoring by analyzing the key factors that affect the structural safety performance.

2. Construction of Structural Twin Model Based on Weighted Average Method

In the process of the intelligent analysis of the safety-influencing factors of prestressed steel structures, the establishment of a high-precision twin model is the first step. Firstly, the finite element model of the structure is established, which can simulate the mechanical properties in real time. In order to strengthen the simulation ability of the finite element model, it is necessary to modify the virtual twin model of the structural construction process. The average weighting method [28] corrected the virtual model by combining the monitoring data of the real structure and the simulation data of the finite element model. The basic principle involves the selection of different weights for different location sensitivity indicators to achieve the goal of minimizing the sum of the squared Euclidean distances between the fusion results and each sensitivity indicator.

Assuming that the actual monitoring value of a certain mechanical parameter of each node of a cable in a prestressed steel structure under a certain working condition is $x_1, x_2, \dots, x_n$, then the average value ($\bar{x}$) of the actual monitoring value is expressed as Equation (1):

$$\bar{x} = \sum_{i=1}^{n} x_i / n \tag{1}$$

Then, the weight (ω_i) of each mechanical parameter is expressed by Equation (2), which is expressed as:

$$\omega_i = (1/d_i) / \sum_{i=1}^{n} (1/d_i) \tag{2}$$

In the equation, d_i is the Euclidean distance between the mechanical parameters and the average value of the mechanical parameters, which is expressed as Equation (3):

$$d_i = \|\bar{x} - x_i\| \tag{3}$$

Combined with the data of each node, the weighted average of the monitoring value of the mechanical parameters ($\hat{x}$) of the whole cable can be calculated, which is expressed as Equation (4):

$$\hat{x} = \sum_{i=1}^{n} \omega_i d_i \tag{4}$$

According to the weighted average of the monitoring values of the mechanical parameters of each cable, the overall monitoring value (D) of the structure can be calculated, which is expressed as Equation (5):

$$D = \mu + \alpha_c \sigma \tag{5}$$

where α_c is the confidence level of the mechanical parameter analysis, which is 1.5 in this study. μ and σ denote the mean and the standard deviation of the weighted average of the monitored values of the mechanical parameters ($\hat{x}$) for each cable, respectively.

Similarly, by using the above steps, the simulation values (D^*) of the mechanical parameters in the finite element model can be calculated. By using Equation (6) to judge the fidelity of the simulation model, the correction of the virtual model is completed, which ensures that the simulation data effectively represent the mechanical properties of the real structure:

$$E_D = \frac{|D^* - D|}{D} \times 100\% \tag{6}$$

In Equation (6), E_D is a metric for determining the fidelity of the twin model.

3. Analysis Framework of Influencing Factors Driven by Fusion of DTs and RF

DTs provide an important theoretical basis and technical support for the bidirectional connection and real-time interaction between virtual space and physical space. DTs use digital technology and virtual model simulation technology to explore and predict the running state of physical space. DTs are a link to realize the interaction and seamless connection between physical space and virtual space, and they provide more real-time, efficient and intelligent services. The strong nonlinear fitting ability of artificial intelligence is suitable for analyzing complex mapping relations, and its generalization ability, reliability and robustness have breakthrough advantages over performance prediction methods. Artificial intelligence can dispense with the dependence on a large number of signal-processing technologies and diagnostic experience, and can complete the adaptive feature extraction and security state analysis. The high-fidelity behavior simulation of the structure state that is driven by online data is completed by DT technology, and the real-time-state visualization of the structure is realized. On the basis of the intelligent algorithm, the analysis of the structural safety performance data is realized. The integration of DTs and the intelligent algorithm provides a theoretical basis for the intelligent analysis of the influencing factors of the structural safety performance.

3.1. Fusion Mechanism of DTs and RF

By combining DTs with RF [29], a highly robust analysis model with good performance can be obtained. By combining the data characteristics that are required for training the RF and the historical data structure in the DT model, the rules for RF to obtain twin data information can be obtained. On the basis of the prediction results of the DTs, the adjustment suggestions for the structural safety maintenance are provided. The collaborative interaction mechanism between the DT model and the RF model is formed so as to establish the fusion mechanism of the structural-safety-influencing factors on the basis of the DT model and the RF (DTs-RF).

The integration of DTs and RF forms an intelligent analysis system for the structural-safety-influencing factors. Its essence is as follows: Through the twin model of the virtual space to realize the real physical space structure to the virtual space synchronous mapping, realize the visualization and digitization of all of the elements of the structural safety maintenance. Through the virtual model to simulate the reality, the collaborative feedback makes it run synchronously with the physical construction, and it realizes the virtual mapping and virtual control. By integrating the data analysis method of RF, the intelligent analysis of the data-driven structural-safety-influencing factors is realized. Finally, the collaborative manipulation and interactive feedback of the physical dimension of the real maintenance system and the information dimension of the virtual maintenance system are formed. A new structural safety analysis and maintenance mode with iterative optimization and intelligent analysis is constructed.

As is shown in Figure 1, the fusion mechanism of DTs and RF is built. The basis of this fusion mechanism is the intelligent collection and dynamic perception of real structural safety information. In this study, intelligent sensing devices, such as 3D laser scanners and sensors, are used to intelligently collect and dynamically perceive the environmental and mechanical properties of the structure [30], thereby providing a basis for the establishment of the structural twin model and the intelligent analysis. Information management and the control platform, which are based on multisource heterogeneous data, are the core of the fusion of DTs and RF, which can analyze and predict the safety performance of the structure in real time. Finally, the unsafe state of the structure is corrected, and the feasibility of the correction is analyzed in the twin model so as to accurately guide and control the safety state of the structure over the whole life cycle.

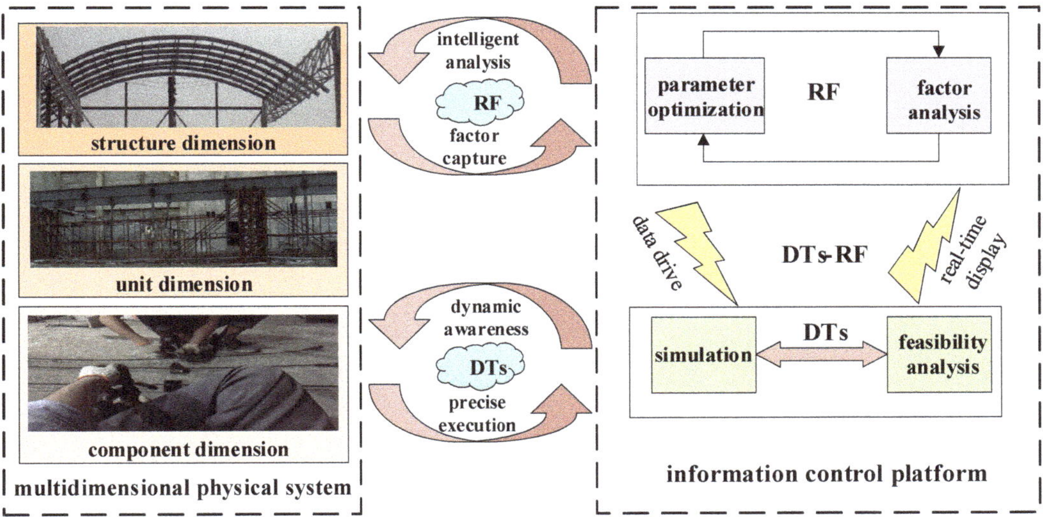

Figure 1. Fusion mechanism of DTs and RF.

Through the analysis of the fusion mechanism of DTs and RF, it can be concluded that the intelligent analysis of the structural-safety-influencing factors that are driven by the fusion of the two has the following characteristics:

1. Real-time perception reflects reality by virtuality. Through the information collection of the physical construction system, and the establishment of the virtual model for the whole life cycle of the structure, the visual monitoring of the whole process of the safety-state change is realized. The whole process of the full-factor multidimensional and multiscale information fusion is realized, which provides a synchronous operation model for the construction structure site;

2. Data-driven intelligent diagnosis. Through the information management and control platform, the DT model is combined with the RF so that the system can make full use of historical data, real-time data collection, simulation data and other multiple-construction information. The system excavates and analyzes all kinds of data, in-

telligently diagnoses the structural safety over the whole life period and avoids the construction risk in time;

3. Scientific prediction with virtual control. The data model and the intelligent scheme established by RF and other means. Through the simulation of the twin model, the maintenance measures are finally fed back into the actual maintenance process to realize the intelligent safety control of the real structure over the whole life cycle.

3.2. Intelligent Analysis Model of Structural Safety Factors

In the intelligent analysis method for the safety-influencing factors of prestressed steel structures, finding the key design parameters that affect the various mechanical parameters of the structure is the core step. Driven by DTs, a multidimensional model is formed. Under the interaction and cooperation of the various levels of the multidimensional model, the intelligent analysis of the structural safety factors is realized. In the physical space, the effect of the sensor on the structure and the mechanical parameters of the structure are collected in real time. At the same time, the virtual model of the solid structure is created in the process of the finite element analysis. The corresponding working conditions are set in the model so as to simulate the twin data of the role of the structure and the mechanical properties of the structure. Thus, the twin simulation of the real structure is realized. The model foundation is established for the intelligent analysis of the safety-influencing factors of prestressed steel structures driven by DTs, which provides a basis for the structural safety performance maintenance. The multidimensional model for the intelligent analysis of the structural-safety-influencing factors is established by Equation (7). The twin simulation model consists of five dimensions:

$$DT_M = \left(S_{pr}, \ S_{vm}, \ P_{td}, \ L_{fa}, \ C_n \right) \tag{7}$$

where DT_M is a multidimensional model for the analysis of the influencing factors of the structural safety; S_{pr} denotes a physical structure entity; S_{vm} refers to the virtual structure model; P_{td} is the twin data processing layer; L_{fa} equals the functional application layer; and C_n is a connection between the components. The twin model can realize the simulation mapping of the real structure, and it can process the structural parameters of real monitoring and virtual simulation. In the structure, the influence degrees of various design parameters are obtained by analyzing the change in the mechanical parameters. Thus, the design parameters of the structure can be modified when a mechanical parameter of the structure changes. Finally, the intelligent maintenance of the structure is realized. The multidimensional model for the analysis of the influencing factors of structural safety is shown in Figure 2. In the multidimensional model, the virtual–real interaction and the spatiotemporal evolution of the structural-safety-state changes are fully considered. In the perspective of the virtual–real interaction, the virtual structure model is established according to the physical structure entity to realize the real mapping of the security state, and it can provide feasibility verification for the final maintenance measures. From the perspective of the spatiotemporal evolution, with the change in the structural state, the design parameters and the mechanical parameters are constantly changing. Finally, the influencing factors of the safety performance are analyzed by analyzing the relationship between the mechanical parameters and the design parameters. Driven by DTs and RF, the real-time feedback of the structural safety state and the closed-loop control of the safety risk are realized.

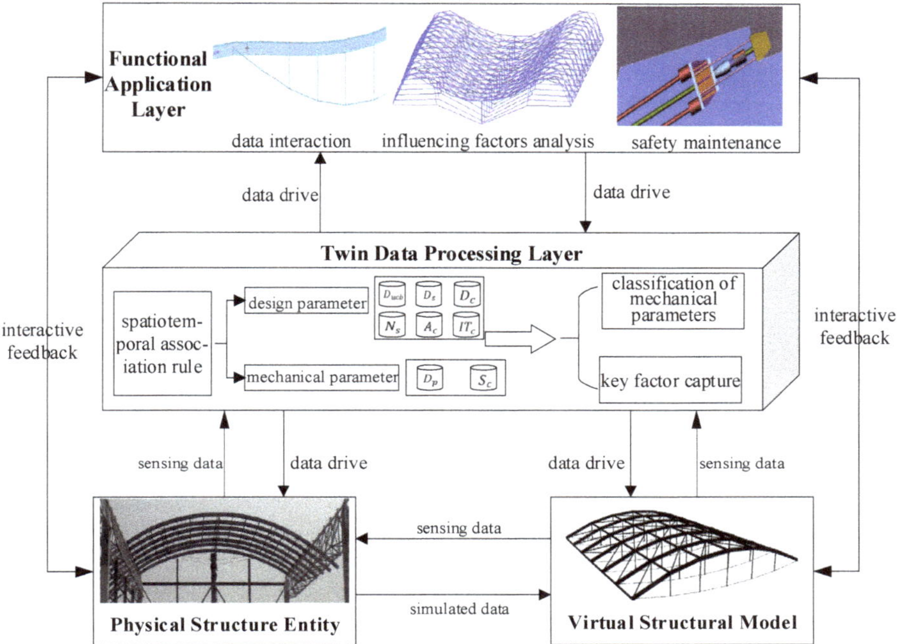

Figure 2. Multidimensional model for structural-safety-influencing factor analysis.

4. Importance Analysis of Structural-Safety-Influencing Factors under Multisource-Parameter Fusion

Driven by the multidimensional model for the structural safety factor analysis, the correlation mechanism between the structural design parameters and the mechanical parameters is first clarified by the twin model. By classifying the mechanical parameters, the key design parameters that affect the structural mechanical parameters are obtained under the driving force of the RF. Therefore, when a certain mechanical parameter of the structure exceeds the limit, the key mechanical parameters are corrected to realize the safety control of the structure.

4.1. Correlation Mechanism between Design Parameters and Mechanical Parameters

The structures of long-span prestressed steel structures are complex, and the safety state changes in real time. The structural safety risk involves multiple factors, such as the cable prestress, the component stress, the configuration, the tensioning machine, the construction parameters, the environmental change and the safety management level. The corresponding data and risk prediction control are often out of state, which also affects the efficiency of the structural safety risk state judgment and the control decision.

In the process of structural construction safety prediction, the corresponding relationship between the design parameters and the mechanical parameters is extracted according to the simulation results. With the help of the intelligent algorithm, association rules between the data are mined. According to the characteristics of prestressed steel structures, the correlation mechanism between the design parameters and the mechanical parameters is formed during the construction and maintenance stages of the structure, which is expressed as Equation (8):

$$f(a_1,a_2,\ldots,a_m) \overset{R}{\Leftrightarrow} g(b_1,b_2,\ldots,b_n) \tag{8}$$

In the equation, $f(a_1,a_2,\ldots,a_m)$ denotes the set of design parameters. $a_1,a_2,\ldots,a_m$ denote the specific design parameters, such as the size of the member, the initial tension of the cable, the number of the structural pieces, etc.; and $g(b_1,b_2,\ldots,b_n)$ represents the aggregate of the mechanical parameters. $b_1,b_2,\ldots,b_n$, respectively, equal the specific mechanical parameters, such as the cable force, the stress and the vertical displacement of the structure. $\overset{R}{\Leftrightarrow}$ denotes the corresponding relationship between the different design parameters and the mechanical parameters according to the technical standards and the data association rules of the structures. Driven by the association rules, the intelligent analysis of the influencing factors of the structural safety can be realized. The mechanical parameters of the structure are obtained by changing the design parameters, and the data set is formed to judge the key design parameters that are driven by RF.

4.2. Analysis Process of the Importance of Structural Safety Factors

By establishing the correlation mechanism between the design parameters and the mechanical parameters, the analysis set of the safety-influencing factors is formed. In the process of analysis, the mechanical parameters are classified, and the key factors are analyzed by RF. By constructing different sample training sets, the RF expands the differences among the classification models of the decision tree so as to improve the extrapolation prediction ability of the combined classification model [31–33]. A classification model sequence $(\{h_1(x),h_2(x),\ldots h_K(x)\})$ is obtained through K-round training. At this time, a multiclassification model system is formed. The final classification result is obtained by using a simple majority voting decision, and the final classification decision is expressed as Equation (9):

$$H(x) = arg \max_{Y} \sum_{i=1}^{K} I(h_i(x) = Y) \tag{9}$$

In the formula, $H(x)$ represents the combined classification model; h_i refers to the single decision tree classification model; Y denotes the output variable; and $I(\cdot)$ is the indicative function. In the process of constructing the RF classification algorithm, it is necessary to set the number of decision trees and to ensure the maximum number of features when the model is optimal.

According to the correlation mechanism between the design parameters and the mechanical parameters and the classification of the mechanical parameters, the data samples of the mechanical analysis are formed. In the sample, the design parameters are used as the input elements, and the security level is used as the output element. A sample acquisition can be expressed as Equation (10):

$$DP \overset{MP}{\Rightarrow} \begin{cases} \overset{D_p}{\rightarrow} (A,\ B,\ C,\ D) \\ \overset{S_c}{\rightarrow} (A,\ B,\ C,\ D) \end{cases} \tag{10}$$

In the formula, DP represents the design parameters of the structure, and MP represents the mechanical parameters of the structure. The safety performance is characterized by the mechanical parameters of the structure. There are mainly two types of mechanical parameters in the prestressed steel structure: the vertical displacement (D_p) and the stress of the cable (S_c). The two types of mechanical parameters are classified into four grades (A,B,C,D). On the basis of the obtained samples, the training set and the test set of the analysis model are divided. Driven by RF, a high-precision mechanical property analysis model is established by adjusting the super-parameters. The grades of the various mechanical parameters can be judged by the designing parameters, and the key factors that affect the changes in the various mechanical parameters can be obtained. The key influencing factors of the mechanical parameters are the corresponding design parameters. The importance analysis process of the safety-influencing factors of prestressed steel structures is shown in Figure 3.

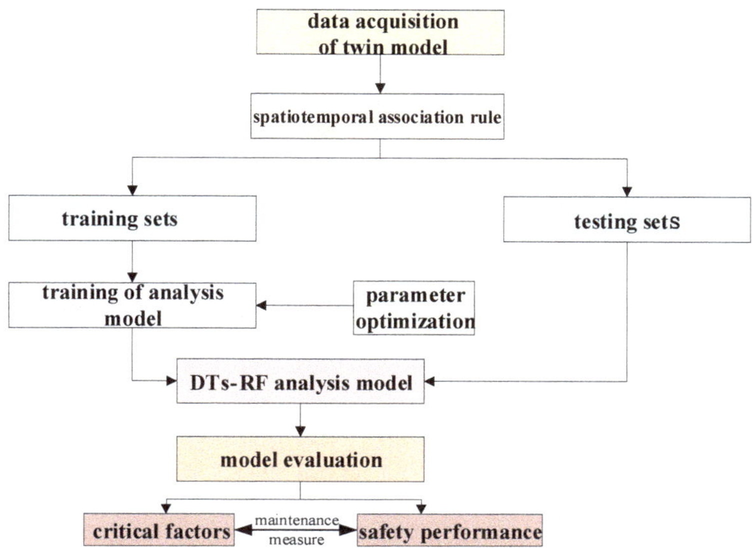

Figure 3. Importance analysis process of prestressed steel structure safety factors.

5. Analysis of Safety Factors of Beam String Structures

The data association mechanism of DT modeling is applied to RF to form an intelligent analysis method for the structural safety factors. The theoretical method is applied to the safety assessment of a beam string structure. The correlation mechanism between the design parameters and the mechanical parameters is established. Driven by this research method, the key design parameters are obtained through the classification of the structural mechanical parameters, and the intelligent analysis of the structural-safety-influencing factors is realized.

5.1. Test Structure Model

Taking the chord beam roof of a convention and exhibition center as the research object, one of the plane string structures is selected as the calculation model. One of the design parameters is taken in this study, and the model construction is shown in Figure 4. The span of the structure is $L = 48$ m, the rise height is 0 m, the sag is 3.5 m, and the number of struts is 9. The upper chord beam adopts H-shaped steel (Q345B). The pole adopts a round steel tube (Q235B). The lower chord cable adopts two types of 55 light-circle stress-relief steel wires.

In the process of the structural construction safety analysis, the size of the abovementioned chord beam, the size of the strut, the diameter of the lower chord cable, the number of struts, the arc of the lower chord cable and the initial tension of the lower chord cable are taken as the structural design parameters. The arc of the lower chord cable is used to reflect the length of each strut. In this test structure, the tangent value (tan α) of the radian is expressed by Equation (11):

$$\tan \alpha = \frac{h_s}{l_c} \tag{11}$$

In the equation, h_s represents the length of the mid-span brace, and l_c refers to the distance from the starting point of the lower chord cable to the midspan.

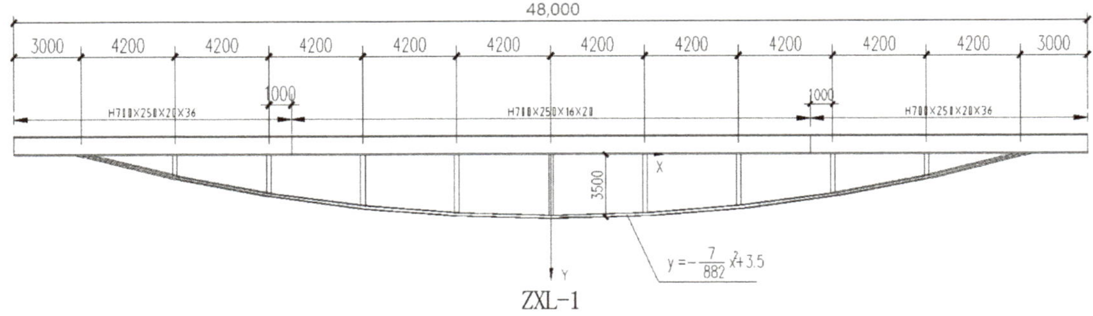

Figure 4. Structure model.

According to the design parameters, the mechanical parameters of the structure can be generated by setting the construction conditions in the finite element software. The main mechanical parameters that are analyzed in this study are the vertical displacement and the stress of the lower chord cable. The working condition that is considered in the construction process is: $0.9 \times$ constant load $+ 1.5 \times$ wind load.

5.2. Relationship between Design Parameters and Mechanical Parameters

In the analysis process of this structure, in order to accurately determine the mechanical parameters, two analysis models of the vertical displacement that correspond to the design parameters and the stress of the lower chord cable, respectively, are established. The values of the six design parameters, according to engineering experience, are shown in Table 1.

Table 1. Values of design parameters.

Parameter Type	Taking Values
Size of upper chord beam (mm)	H700 $\times$ 250 $\times$ 20 $\times$ 36, H700 $\times$ 250 $\times$ 16 $\times$ 20
Size of strut (mm)	$\varnothing$150 $\times$ 8, $\varnothing$160 $\times$ 8, $\varnothing$170 $\times$ 8
Diameter of lower chord cable (mm)	50, 55, 60, 65, 70
Number of struts	7, 9, 11
Arc of lower chord cable (°)	8, 9, 10
Initial tension of the lower chord cable (KN)	600, 700, 800, 900, 1000

A high-precision finite element twin model is established by the weighted average method. Taking the combination of one of the design parameters as an example, the specific parameters are shown in Table 2. Among them, there are nine struts, the radian of the lower chord cable is 8° and the initial tension is 700 kN. The twin model of the structure is established under the correction of the weighted average method. Under the condition of self-weight, the accuracy of the twin model is verified by comparing the internal forces of each component in the experimental model to the twin structure. In the test, the internal force of each component is collected by the column tension-and-compression sensor. The comparison between the simulation value of the internal force in the twin model and the acquisition value of the internal force in the test structure is shown in Table 3. The internal force error of the component is within 3%, which realizes the high-fidelity mapping of the test structure. The test structure is symmetrical. Therefore, under the condition of self-weight, the internal force of half of the components in the structure is compared. The numbers of the components are shown in Figure 5.

Table 2. Sectional dimensions of components.

Section of Top Chord			Size of Strut		Cable Size	
Left Section (mm)	Midportion (mm)	Right Section (mm)	Model	Area (m^2)	Model	Area (m^2)
H700 × 250 × 20 × 36	H700 × 250 × 16 × 20	H700 × 250 × 20 × 36	∅150 × 8	0.0038	$2SNS/S\varphi7 \times 55$	0.0021

Table 3. Comparison of internal force simulation values and acquisition values.

Component Number	Simulation Value (N)	Acquisition Value (N)	Error
1	140,377	140,253	0.088%
2	−6435	−6447	−0.186%
3	−6853	−6842	0.161%
4	−6907	−6893	0.203%
5	−6939	−6947	−0.115%
6	−6949	−6962	−0.187%

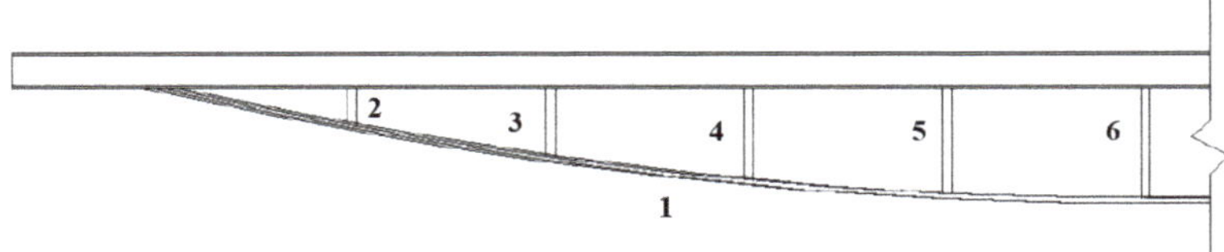

Figure 5. Numbering of components.

The mechanical parameters of the twin model structure are shown in Figure 6. By comparing them with the parameters in the real structure, the high fidelity of the twin model is ensured. The structure forms that correspond to the other design parameters are judged by the weighted average method, as described in Section 1. The mechanical parameters of the structure are formed under load.

5.3. Analysis of Structural Safety Factors

Driven by the correlation mechanism between the design parameters and the mechanical parameters, the mechanical properties of the structure are analyzed by integrating random forests. The mechanical properties of the structure are analyzed under the action of: 0.9 × constant load + 1.5 × wind load. In the high-fidelity twin model, the design parameters of the structure are set. The combination of the design parameters is selected according to Table 1. In this study, a total of 560 structural analyses were carried out in the combination of design parameters. Limited to the length of the article, Table 4 shows the design parameters and their corresponding mechanical parameters in 30 structural types.

In the analysis process, the correlation between the design parameters and the mechanical parameters was established through 560 sample sets. The vertical displacement of the structure and the stress of the cable are important indexes to measure the safety performance of the structure. According to the analysis process in Section 4.2, various mechanical parameters are classified first, as shown in Table 5.

Table 4. Correspondence between different design parameters and mechanical parameters.

Size of Upper Chord Beam (mm)	Size of Strut (mm)	Diameter of Lower Chord Cable (mm)	Number of Struts	Arc of Lower Chord Cable (°)	Initial Tension of the Lower Chord Cable (KN)	Vertical Displacement (mm)	Cable Stress (Mpa)
H700 × 250 × 20 × 36	150×8	50	7	8	600	−21.8589	292.9
H700 × 250 × 20 × 36	160×8	55	7	8	700	−20.7268	269
H700 × 250 × 20 × 36	170×8	60	7	8	800	−19.7968	247.4
H700 × 250 × 20 × 36	170×8	65	7	8	900	−19.1291	225.4
H700 × 250 × 20 × 36	150×8	70	7	9	1000	−16.0137	195.9
H700 × 250 × 16 × 20	160×8	50	7	9	600	−16.1379	159.8
H700 × 250 × 16 × 20	170×8	55	7	9	700	−17.3404	135.4
H700 × 250 × 16 × 20	150×8	60	7	10	800	−20.2685	129
H700 × 250 × 16 × 20	160×8	65	7	10	900	−30.8147	121
H700 × 250 × 16 × 20	170×8	70	7	10	1000	−41.0492	113
H700 × 250 × 20 × 36	150×8	50	9	8	600	−21.751	306.6
H700 × 250 × 20 × 36	160×8	55	9	8	700	−20.4623	283
H700 × 250 × 20 × 36	170×8	60	9	8	800	−19.3475	261.7
H700 × 250 × 20 × 36	170×8	65	9	8	900	−18.317	240.8
H700 × 250 × 20 × 36	150×8	70	9	9	1000	−14.7422	211.8
H700 × 250 × 16 × 20	160×8	50	9	9	600	−14.12	191.1
H700 × 250 × 16 × 20	170×8	55	9	9	700	−13.7923	172.6
H700 × 250 × 16 × 20	150×8	60	9	10	800	−13.7882	147.2
H700 × 250 × 16 × 20	160×8	65	9	10	900	−13.2104	132.5
H700 × 250 × 16 × 20	170×8	70	9	10	1000	−13.9175	119.8
H700 × 250 × 20 × 36	150×8	50	11	8	600	−21.9187	316.5
H700 × 250 × 20 × 36	160×8	55	11	8	700	−20.5432	293
H700 × 250 × 20 × 36	170×8	60	11	8	800	−19.333	271.7
H700 × 250 × 20 × 36	170×8	65	11	8	900	−18.1724	250.5
H700 × 250 × 20 × 36	150×8	70	11	9	1000	−13.6152	202
H700 × 250 × 16 × 20	160×8	50	11	9	600	−12.551	175
H700 × 250 × 16 × 20	170×8	55	11	9	700	−12.0984	161.7
H700 × 250 × 16 × 20	150×8	60	11	10	800	−11.3253	160.4
H700 × 250 × 16 × 20	160×8	65	11	10	900	−11.1626	147.3
H700 × 250 × 16 × 20	170×8	70	11	10	1000	−11.1514	136.1

Table 5. Classification of various mechanical parameters.

Parameter Type	Class	Corresponding Interval
Vertical displacement	A	[−10 mm, 0 mm)
	B	[−20 mm, −10 mm)
	C	[−30 mm, −20 mm)
	D	[−40 mm, −30 mm]
Cable stress	A	[100 Mpa, 175 Mpa)
	B	[175 Mpa, 250 Mpa)
	C	[250 Mpa, 325 Mpa)
	D	[325 Mpa, 400 Mpa]

In the process of the mechanical property analysis, vertical displacement and cable stress analysis models are established. Driven by random forests, the levels of the design parameters and of the mechanical parameters are taken as the input and output layers of the model, respectively. Through the analysis of the model, the key factors that affect the changes in the various mechanical parameters are found, as shown in Figure 7.

It can be seen from Figure 7 that the key factors that affect the vertical displacement and the cable stress in this structure are the number of struts and the radian of the lower chord cable, respectively. Therefore, when the safety performance of the structure exceeds the specification limit, the structural maintenance can be accurately carried out by adjusting the corresponding design parameters. In order to intuitively represent the influence of the key factors on the structural mechanical parameters, two types of design parameters, and their corresponding mechanical parameters, are extracted from the correlation mechanism of the design parameters and the mechanical parameters. The relationship between the key design parameters and the mechanical parameters is shown in Figure 8. In the string-supported beam structure, the greater the number of struts, the greater the initial tension of the lower chord cable, and the vertical displacement of the structure is relatively small. When the arc of the lower chord cable is smaller and the number of struts is greater, the corresponding cable stress is relatively large. Therefore, when the mechanical parameters of the structure exceed the limit, the safety maintenance measures can be accurately formulated.

The limit values of the structural mechanical parameters are shown in Table 6. During the test, the mechanical parameters that correspond to different design parameters are compared and analyzed by setting the load conditions. At the same time, the change in the mechanical parameters is investigated by changing the key factors that affect the structural safety. In the process of the experimental verification, it was proven that the number of struts and the arc of the lower chord cable have the greatest influence on the vertical displacement and the cable stress of the structure. The experimental verification of the structural-safety-influencing factor analysis is shown in Figure 9. The intelligent analysis method for the safety-influencing factors of prestressed steel structures based on digital twinning and random forest that was formed in this study provides a basis for the safety maintenance of the structure.

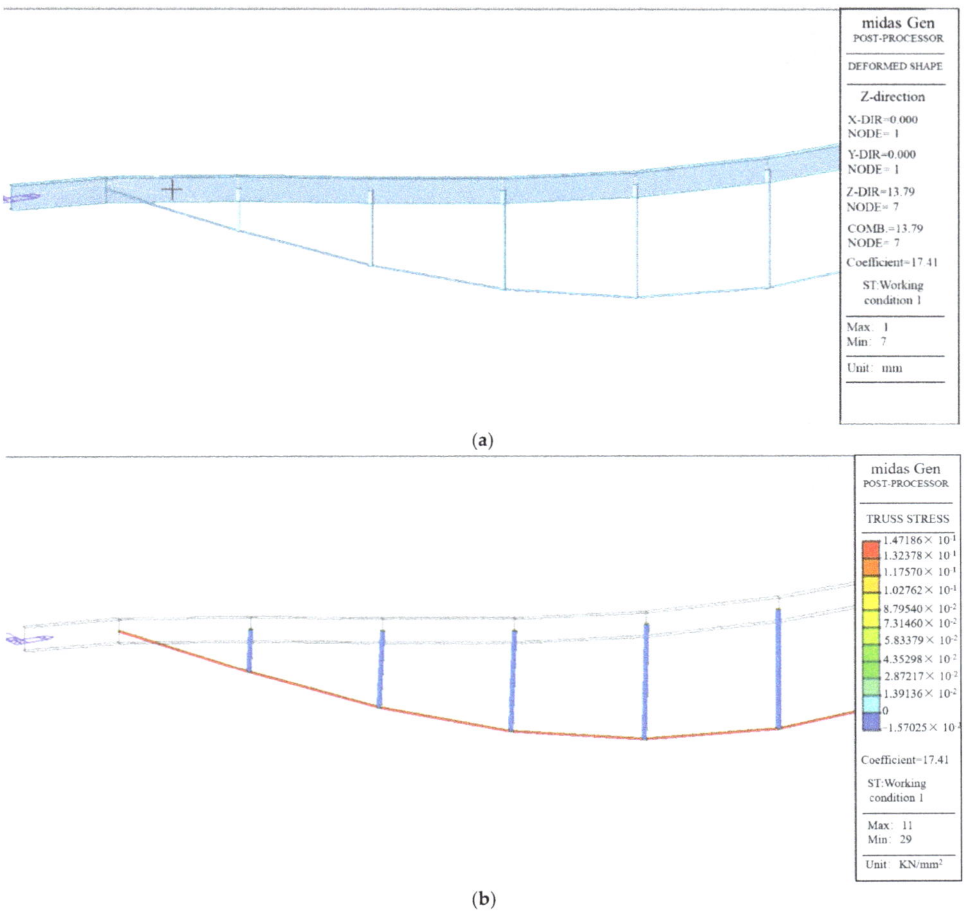

(a)

(b)

Figure 6. Mechanical parameters of twin model structure: (**a**) vertical displacement; and (**b**) cable stress.

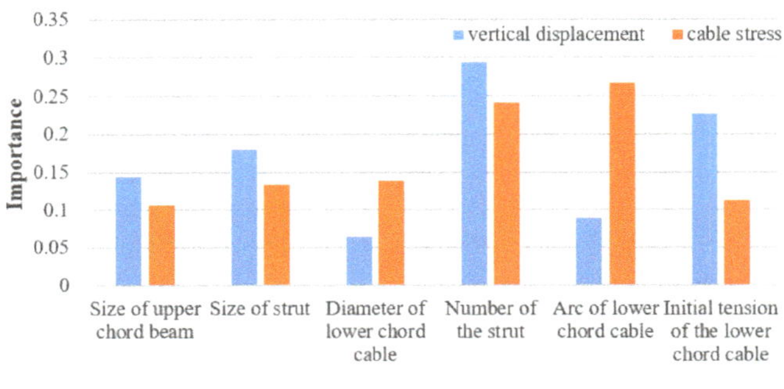

Figure 7. Influence of various design parameters on various mechanical parameters.

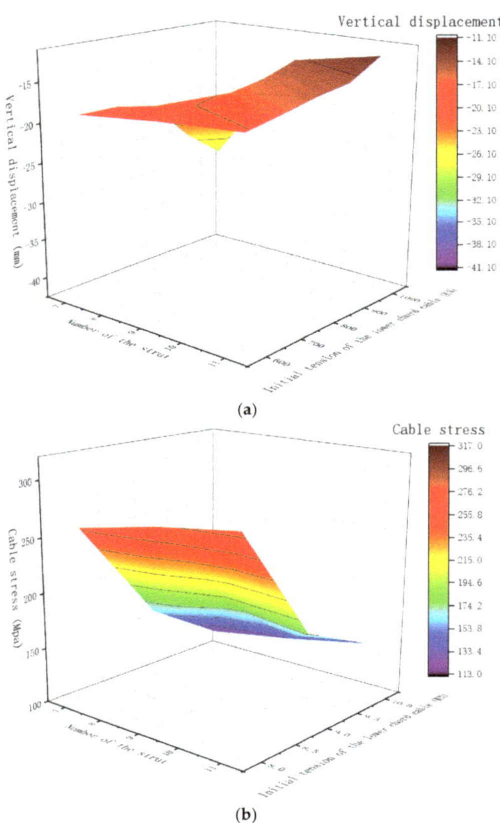

(a)

(b)

Figure 8. Relationship between key design parameters and mechanical parameters: (**a**) relationship between key design parameters and vertical displacement; (**b**) relationship between key design parameters and cable stress.

Table 6. Limits of mechanical parameters of structures.

Mechanical Parameter	Specific Limits
Perpendicular displacement	Less than 1/250 of structural span
Cable stress	Less than 1/2.5 of allowable stress

Figure 9. Experimental verification of structural safety influence factor analysis.

5.4. Results Analysis

How to realize an intelligent safety performance analysis of long-span spatial structures is an important research direction. The traditional analysis method only simulates the mechanical properties by the finite element method, which cannot accurately and efficiently obtain the key factors that affect the structural safety performance. In this study, and on the basis of the previous research and the application of DTs and RF, the intelligent analysis of the safety-influencing factors of prestressed steel structures based on DTs and RF is proposed. Taking the chord-supported beam roof of a convention and exhibition center as the research object, the effectiveness of this method is verified. Firstly, the twin model of the structure is established on the basis of the weighted average method. By comparing the simulation value and the collected value of the internal force of the component, it is verified that the twin model has high fidelity. The analysis model that is established in this study, which is based on DTs, can effectively map the state of the test structure. On the basis of the twin model, the mechanical parameters of the structure are obtained by adjusting the different design parameters under load conditions. In the research process, the correlation mechanism between the design parameters and the mechanical parameters was established. The vertical displacement of the structure and the stress of the cable are important indexes for measuring the safety performance of the structure. Driven by RF, the key design factors that affect the two types of mechanical parameters are obtained. Therefore, the intelligent analysis method that is formed in this study can provide a reliable basis for the safety maintenance of the structure. At the same time, large-span spatial structures are mostly used for buildings of high importance, such as large public buildings. In the construction and operation of the structure, it is necessary to strictly ensure safety and reliability. Therefore, the analysis of the structural safety factors is very necessary. The establishment of a twin model can accurately reflect the state of the structure. This method can predict the mechanical properties of the structure by setting the load conditions. At the same time, RF is an important tool for data classification and analysis. Therefore, the combination of the two improves the intelligence level of the structural analysis. The combination of DTs and RF can accurately obtain the key factors, on the one hand, and can accurately reflect the safety state of the structure, on the other hand. This research method can assist with the formulation of structural safety maintenance measures.

6. Discussion and Conclusions

In this study, an intelligent analysis method for the safety-influencing factors of prestressed steel structures based on DTs and RF is proposed. This method effectively improves the maintenance accuracy and efficiency of the structural safety state. The dynamic simulation of the structural safety state is realized by DTs, and the design parameters and mechanical parameters of the structure are obtained. The importance analysis of the influencing factors of the structural safety is realized by RF. The intelligent analysis of the influencing factors of the structural safety performance formed by the integration of RF and RF was applied to the beam string structure, and the effectiveness of this research method is verified. By analyzing the key factors that affect the structural mechanical parameters, a reliable basis is provided for the safety maintenance of the structure. The main conclusions are as follows:

(1) In the process of the safety intelligent analysis of prestressed steel structures, the establishment of a high-precision twin model is the first step. The high-fidelity mapping of the finite element model to the real structure can be realized by the weighted average method. The resulting high-fidelity twin model can directly extract the structural design parameters and their corresponding mechanical parameters;

(2) The twin model obtains the structural parameters. Driven by RF, an intelligent analysis model of the structural-safety-influencing factors can be formed. The model is divided into five modules, with full consideration to the virtual–real interaction and the spatiotemporal evolution of the structural-safety-state changes;

(3) Driven by the intelligent analysis model, the correlation mechanism between the design parameters and the mechanical parameters of the structure is formed. The design parameters are the influencing factors of the structural safety. The safety performance of the structure is reflected according to the mechanical parameters. The key factors that affect the safety performance of the structure are obtained as the basis for the formulation of safety maintenance measures.

In the process of the experimental verification of a beam string structure, it was proven that the number of struts and the arc of the lower chord cable have the greatest influence on the vertical displacement and the cable stress of the structure. Therefore, when the mechanical parameters of the structure exceed the limit, the key factors can be corrected to improve the safety performance of the structure. The research method provides a basis for the safety, high precision and high efficiency maintenance of the structure. In the future, the intelligence level of the structural safety maintenance will be improved through the consideration of more realistic conditions. Under various load conditions, the relationship between the design parameters and the mechanical parameters can be quantitatively determined by correcting the key factors and by formulating efficient safety maintenance measures.

Author Contributions: Conceptualization, H.Z.; methodology, H.Z.; software, H.Z.; validation, H.Z. and Y.W.; writing—original draft preparation, H.Z.; writing—review and editing, H.Z.; project administration, H.Z.; funding acquisition, H.Z. All authors have read and agreed to the published version of the manuscript.

Funding: This research received no external funding.

Institutional Review Board Statement: Not applicable.

Informed Consent Statement: Not applicable.

Data Availability Statement: The data presented in this study are available upon request from the corresponding author. The data are not publicly available because of confidentiality.

Acknowledgments: The authors would like to thank the South China University of Technology, Guangzhou, China, for their support throughout the research project.

Conflicts of Interest: The authors declare no conflict of interest. The funders had no role in the study's design; in the collection, analyses or interpretation of data; in the writing of the manuscript or in the decision to publish the results.

References

1. Lu, J.; Xue, S.D.; Li, X.Y.; Liu, R.J. Key construction technology of annular crossed cable-truss structure. *J. Tianjin Univ. (Sci. Technol.)* **2021**, *54*, 101–110. (In Chinese)
2. Guo, J.M.; Zhou, D. Pretension simulation and experiment of a negative Gaussian curvature cable dome. *Eng. Struct.* **2016**, *127*, 737–747. [CrossRef]
3. Deng, H.; Zhang, M.R.; Liu, H.C.; Dong, S.; Zhang, Z.; Chen, L. Numerical analysis of the pretension deviations of novel crescent-shaped tensile canopy structural system. *Eng. Struct.* **2016**, *119*, 24–33. [CrossRef]
4. Lu, J.; Xue, S.D.; Li, X.Y.; Liu, R.J. Study on membrane roof schemes of annular crossed cable-truss structure. *Int. J. Space Struct.* **2019**, *34*, 85–96. [CrossRef]
5. Xue, S.D.; Tian, X.S.; Liu, Y.; Li, X.Y.; Liu, R.J. Mechanical behavior of single-layer saddle-shape crossed cable net without inner-ring. *J. Build. Struct.* **2021**, *42*, 30–38. (In Chinese)
6. Liu, R.J.; Zou, Y.; Xue, S.D.; Li, X.Y.; Wang, C. Influence on static performance of loop-free suspen-dome after removal of cables. *J. Build. Struct.* **2020**, *41*, 1–9. (In Chinese)
7. Krishnan, S. Structural design and behavior of prestressed cable domes. *Eng. Struct.* **2020**, *209*, 110294. [CrossRef]
8. Chen, Z.H.; Ma, Q.; Yan, X.Y.; Lou, S.Y.; Chen, R.H.; Si, B. Research on Influence of Construction Error and Controlling Techniques of Compound Cable Dome. *J. Hunan Univ. (Nat. Sci.)* **2018**, *45*, 47–56. (In Chinese)
9. Ge, J.Q.; Liu, B.N.; Wang, S.; Zhang, G.J.; Zhang, M.S.; Huang, J.Y.; Liu, X.G. Study on design of prestressed tensegrity cable structures. *J. Build. Struct.* **2019**, *40*, 73–80. (In Chinese)
10. Thai, H.T.; Kim, S.E. Nonlinear static and dynamic analysis of cable structures. *Finite Elem. Anal. Des.* **2011**, *47*, 237–246. [CrossRef]
11. Guo, Y.L.; Zhang, X.Q. Experimental study on the influences of cable length errors in Geiger cable dome designed with unadjustable cable length. *China Civil. Eng. J.* **2018**, *51*, 52–68. (In Chinese)
12. Zhang, A.L.; Sun, C.; Jiang, Z.Q. Calculation method of prestress distribution for levy cable dome with double struts considering self-weight. *Eng. Mech.* **2017**, *34*, 211–218. (In Chinese)
13. Wang, Y.; Guo, Z.; Luo, B.; Shi, K. Study on the determination method for the equivalent pre-tension in cables of spatial prestressed steel structure. *China Civ. Eng. J.* **2013**, *46*, 53–61. (In Chinese)
14. Arezki, S.; Kamel, L.; Amar, K. Effects of temperature changes on the behavior of a cable truss system. *J. Constr. Steel Res.* **2017**, *129*, 111–118.
15. Basta, A.; Serror, M.H.; Marzouk, M. A BIM-based framework for quantitative assessment of steel structure deconstructability. *Autom. Constr.* **2020**, *111*, 103064. [CrossRef]
16. Liu, Z.; Shi, G.; Zhang, A.; Huang, C. Intelligent Tensioning Method for Prestressed Cables Based on Digital Twins and Artificial Intelligence. *Sensors* **2020**, *20*, 7006. [CrossRef]
17. Yang, Y.; Zhang, Y.; Tan, X. Review on Vibration-Based Structural Health Monitoring Techniques and Technical Codes. *Symmetry* **2021**, *13*, 1998. [CrossRef]
18. Grieves, M. *Virtually Perfect: Driving Innovative and Lean Products through Product Lifecycle Management*; Space Coast Press: CocoaBeach, FL, USA, 2011.
19. (16) (PDF) A Framework for Prefabricated Component Hoisting Management Systems Based on Digital Twin Technology. Available online: https://www.researchgate.net/publication/358944762_A_Framework_for_Prefabricated_Component_Hoisting_Management_Systems_Based_on_Digital_Twin_Technology (accessed on 7 March 2022).
20. Shirowzhan, S.; Tan, W.; Sepasgozar, S.M.E. Digital Twin and Cyber GIS for Improving Connectivity and Measuring the Impact of Infrastructure Construction Planning in Smart Cities. *Int. J. Geo-Inf.* **2020**, *9*, 240. [CrossRef]
21. Daniali, S.M.; Barykin, S.E.; Kapustina, I.V.; Khortabi, F.M.; Sergeev, S.M.; Kalinina, O.V.; Mikhaylov, A.; Veynberg, R.; Zasova, L.; Senjyu, T. Predicting Volatility Index According to Technical Index and Economic Indicators on the Basis of Deep Learning Algorithm. *Sustainability* **2021**, *13*, 14011. [CrossRef]
22. Le Cun, Y.; Bengio, Y.; Hinton, G. Deep learning. *Nature* **2015**, *521*, 436–444. [CrossRef]
23. Liu, Z.S.; Zhang, A.S.; Wang, W.S.; Wang, J.J. Dynamic Fire Evacuation Guidance Method for Winter Olympic Venues Based on Digital Twin-Driven Model. *J. Tongji Univ. (Nat. Sci.)* **2020**, *48*, 962–971. (In Chinese)
24. Lu, Q.; Parlikad, A.K.; Woodall, P.; Xie, X.; Liang, Z.L.; Konstantinou, E.; Heaton, J.; Schooling, J.M. Developing a dynamic digital twin at building and city levels: A case study of the West Cambridge campus. *J. Manag. Eng.* **2019**, *36*, 05020004. [CrossRef]
25. Acharya, D.; Khoshelham, K.; Winter, S. BIM-PoseNet: Indoor camera localisation using a 3D indoor model and deep learning from synthetic images. *ISPRS J. Photogramm. Remote Sens.* **2019**, *150*, 245–258. [CrossRef]
26. Bhuiyan, M.A.; Dinçer, H.; Yüksel, S.; Mikhaylov, A.; Danish, M.S.S.; Pinter, G.; Uyeh, D.D.; Stepanova, D. Economic indicators and bioenergy supply in developed economies: QROF-DEMATEL and random forest models. *Energy Rep.* **2022**, *8*, 561–570. [CrossRef]
27. Liu, Y.R.; Wang, L.; Li, M. Kalman filter-random forest-based method of dynamic load identification for structures with interval uncertainties. *Struct. Control Health Monit.* **2022**, *29*, e2935. [CrossRef]
28. Soleimani, F. Analytical seismic performance and sensitivity evaluation of bridges based on random decision forest framework. *Structures* **2021**, *32*, 329–341. [CrossRef]
29. Zeng, B.; Zhou, Z.; Zhang, Q.F.; Xu, Q.; Meng, S.P. Analytical and experimental research on damage identification of cable-stayed arch-truss based on data fusion. *China Civil. Eng. J.* **2020**, *53*, 28–37; 86. (In Chinese)

30. Bassier, M.; Genechten, B.V.; Vergauwen, M. Classification of Sensor Independent Point Cloud Data of Building Objects using Random Forests. *J. Build. Eng.* **2019**, *21*, 468–477. [CrossRef]
31. Miguel, A.; Paulo, C.; António, A.C. BIMSL: A generic approach to the integration of building information models with real-time sensor data—ScienceDirect. *Autom. Constr.* **2017**, *84*, 304–314.
32. Wang, B.; Gao, L.; Juan, Z. Travel Mode Detection Using GPS Data and Socioeconomic Attributes Based on a Random Forest Classifier. *IEEE Trans. Intell. Transp. Syst.* **2018**, *19*, 1547–1558. [CrossRef]
33. Tan, J.; Xie, X.; Zuo, J.; Xing, X.; Liu, B.; Xia, Q.; Zhang, Y. Coupling Random Forest and Inverse Distance Weighting to Generate Climate Surfaces of Precipitation and Temperature with Multiple-Covariates. *J. Hydrol.* **2021**, *598*, 126270. [CrossRef]

Article

Experimental Investigation of Fracture Performances of SBHS500, SM570 and SM490 Steel Specimens with Notches

Yan Liu [1], Shuto Ikeda [1], Yanyan Liu [2], Lan Kang [3,4,*] and Hanbin Ge [1,*]

[1] Department of Civil Engineering, Meijo University, Nagoya 468-8502, Japan; liuyan@meijo-u.ac.jp (Y.L.); 213433003@ccmailg.meijo-u.ac.jp (S.I.)
[2] National Engineering Research Center of Biomaterials, Nanjing Forestry University, No.159 Longpan Road, Nanjing 210037, China; liuyanyan@njfu.edu.cn
[3] School of Civil Engineering and Transportation, South China University of Technology, Guangzhou 510641, China
[4] State Key Laboratory of Subtropical Building Science, South China University of Technology, Guangzhou 510641, China
*   Correspondence: ctlkang@scut.edu.cn (L.K.); gehanbin@meijo-u.ac.jp (H.G.)

Abstract: High-strength steels (HSSs) with nominal yield stress not less than 460 MPa have been increasingly employed in bridge structures. Compared with SM490 normal-strength steel (NSS), HSSs, including SBHS500 and SM570, have higher strength but lower ductility, and brittle fracture can easily occur in the HSSs members with notches. Therefore, 48 tension specimens with U-notch or V-notch made of SBHS500, SM570 and SM490 structural steels are carried out. The influences of notch depth, U-notch radius, V-notch degree and chemical composition on the mechanical and fracture performances of the steel specimens are investigated. It is concluded from experimental results that SBHS500 and SM570 HSSs with higher yield stress have a relatively higher elastic stress concentration factor, crack initiation appears earlier, and brittle fracture is more likely to occur. Compared to SM570 HSS, SBHS500 HSS has better weldability.

Keywords: fracture performance; SBHS500; SM570; SM490; high strength steel; notch

---

Citation: Liu, Y.; Ikeda, S.; Liu, Y.; Kang, L.; Ge, H. Experimental Investigation of Fracture Performances of SBHS500, SM570 and SM490 Steel Specimens with Notches. *Metals* **2022**, *12*, 672. https://doi.org/10.3390/met12040672

Academic Editor: Maciej Motyka

Received: 6 March 2022
Accepted: 7 April 2022
Published: 14 April 2022

Publisher's Note: MDPI stays neutral with regard to jurisdictional claims in published maps and institutional affiliations.

1. Introduction

In engineering structures and components, notches are difficult to avoid, and some notch-like geometries are necessary for structural design [1–3]. Notches produce stress and strain concentration, then the stress state with high stress triaxiality, and these make it easy for brittle fracture to occur [4–8]. U-notch and V-notch are the two main notch types [9,10]. High-strength steel (HSS) usually refers to steels with nominal yield stress not less than 460 MPa [11,12]. High yield stress might lead to poor plastic deformation capacity. Accordingly, it is necessary to investigate the crack initiation, propagation and final failure performances of notched HSS specimens, which is important to ensure the safety of HSS structures with notches.

Great efforts have been put into fracture behavior of HSSs without notches in previous studies [13–16]. These include Q460 [11], Q550 [16], Q690 [4,15,16], Q890 [16], ASTM A572 [17], DP980 [18], ASTM A36 [19], ASTM A572 [19] and ASTM A992 [19] HSSs. Additionally, for the Japanese bridge steel SM490 [20], the ductile fracture mechanism of SM490 base metal, weld and heat affect zone was investigated for welded SM490 specimens in the reference [20], and a three-stage and two-parameter ductile fracture model was proposed. Tension tests on U-notch and V-notch specimens with different detailed geometries were carried out to investigate the ductile fracture behavior of SM490 steel [6]. An improved three-stage and two-parameter ductile fracture model was proposed to accurately predict the ductile fracture behavior of specimens with notches by considering the effect of stress triaxiality in the softening stage.

In the past, SM490 steel was widely used in steel bridges in Japan. However, HSSs, including SBHS500 and SM570, are employed in Japanese bridges in recent years to reduce steel consumption by 10% [21,22] compared with SM490 steel because of their high yield stress. In the previous study, the post-fire mechanical properties of SBHS500 HSS were experimentally investigated [23]. Although HSS bridges have a significant advantage in improving the bearing capacity of structures, the application of HSSs (including SBHS500 and SM570) in seismic structures is limited, owing to their relatively poor ductility compared with that of normal strength steel (NSS). HSS bridges have several HSS members with notches. This shortcoming of poor ductility limits the application of SBHS500 and SM570 HSSs in seismic areas, especially for the HSS members with notches.

In this study, the fracture performances of notched steels are investigated for three types of structural steels used in Japanese bridges, including SM490, SM570 and SBHS500, in which SM570 and SBHS500 are HSSs standardized by the Japanese Industrial Standards (JIS) Committee in the code JIS G 3140 [24], and SM490 is NSS standardized by JIS G 3106 [25]. Tension tests on U-notch and V-notch specimens with different detailed geometries are carried out to study fracture performances of the steel specimens at different stress triaxialities [6]. Crack initiating performances and crack initiation regions are identified. Fracture surfaces of tested specimens are deeply analyzed. Moreover, effects of notch depth, U-notch radius, V-notch degree and chemical composition on the yield stress and ultimate stress of HSSs are investigated experimentally.

2. Experimental Program and Tested Specimens

Uniaxial tension tests were carried out to obtain the ductile fracture behavior of SBHS500, SM570, and SM490 structural steels, using single-side notched flat bar specimens made of these three types of steels, which consist of two series of notched specimens with V-notch (abbreviated to VBS) and U-notch (abbreviated to UBS). In this experimental investigation, uniaxial tension tests on flat bar specimens with various notches were employed to investigate the dependence of ductile fracture behavior on stress triaxiality. The geometric dimensions of VBS and UBS specimens tested are illustrated in Figure 1. Different stress triaxialities are provided by changing the notch degree in VBS specimens and the notch radius in UBS specimens [6]. For the VBS specimens, all of the specimens have the same 3 mm notch depth, and four different notch degrees (30°, 60°, 90° and 120°) are employed. Regarding the UBS specimens, all of the specimens have the same 5 mm notch depth, and four different notch radii (1 mm, 2 mm, 3 mm and 5 mm) are used. Two specimens are allocated in each set and a total of 48 specimens were prepared for the experimental tests. The actual flat bar thickness of VBS and UBS specimens are listed in Table 1, and the photo of specimens is shown in Figure 2. Surface crack initiation behaviors for all types of specimens were observed and recorded continuously by one high-speed camera (Canon Inc., Tokyo, Japan). Though there are solutions, such as damage inspection by magnetic dye penetrant or fractographic measurement, ductile crack initiation is defined as the point when crack length extends to 1–2 mm according to visual or video camera observation [20].

Table 1. Actual flat bar thickness of VBS and UBS specimens (Unit: mm) *.

	VBS-V30-1	VBS-V30-2	VBS-V60-1	VBS-V60-2	VBS-V90-1	VBS-V90-2	VBS-V120-1	VBS-V120-2
SBHS500	12.18	12.20	12.20	12.13	12.15	12.15	12.13	12.20
SM570	12.13	12.30	12.33	12.13	12.13	12.15	12.15	12.13
SM490	12.05	12.00	12.05	12.08	12.00	12.00	12.03	12.00
	UBS-R1-1	**UBS-R1-2**	**UBS-R2-1**	**UBS-R2-2**	**UBS-R3-1**	**UBS-R3-2**	**UBS-R5-1**	**UBS-R5-2**
SBHS500	12.40	12.08	12.20	12.15	12.18	12.20	12.18	12.10
SM570	12.23	12.13	12.20	12.15	12.15	12.23	12.23	12.18
SM490	12.03	12.03	12.00	12.03	12.05	12.00	12.03	12.00

* The value in this table is measured by a vernier caliper (±0.005 mm).

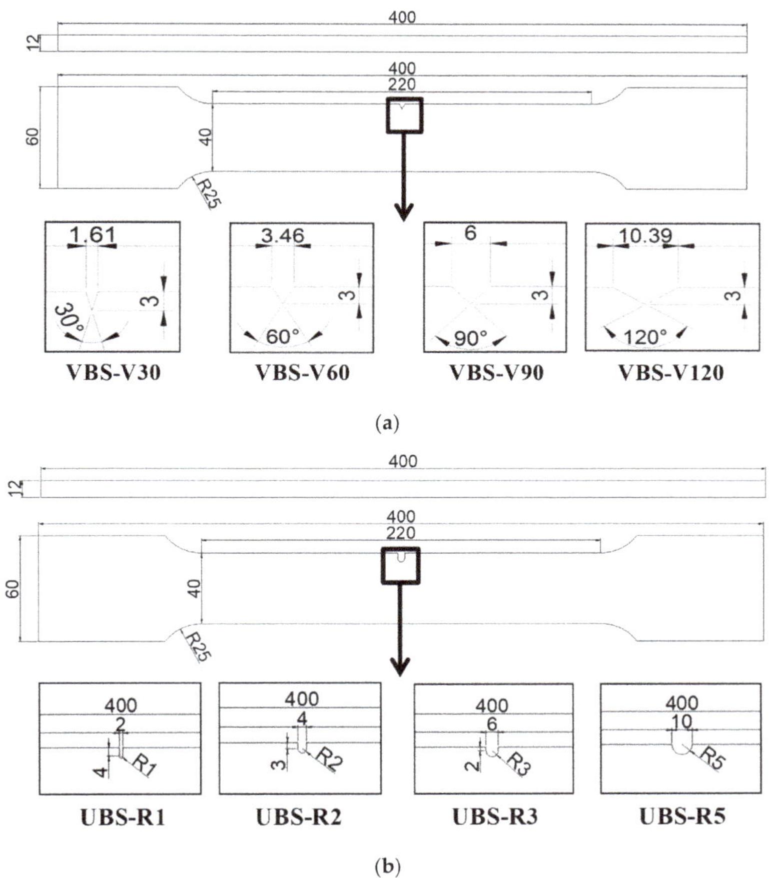

Figure 1. Geometric dimensions of VBS and UBS specimens tested (Unit: mm): (**a**) VBS (V-notch) specimens; (**b**) UBS (U-notch) specimens.

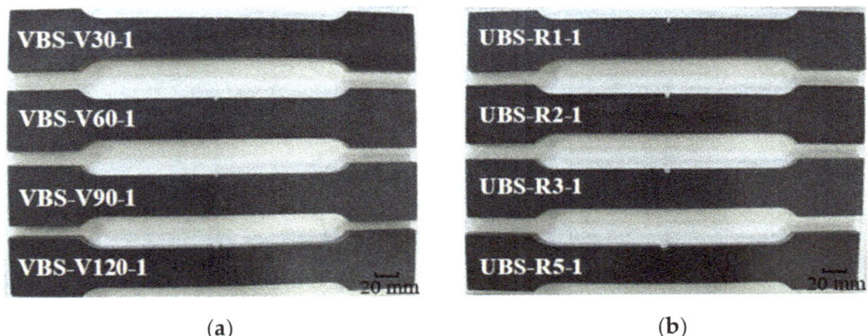

Figure 2. Photo of specimens: (**a**) VBS specimens; (**b**) UBS specimens.

All specimens were tested using a 500 kN MTS material testing machine (MTS, Minnesota, MN, USA) under displacement control, as shown in Figure 3a. The specimens were loaded at a rate of 0.02 to 0.05 mm/s (nominal corresponding strain rate 1×10^{-4}/s to

2.5×10^{-4}/s). The observed point displacements were measured using a contact Ω extensometer (TML, Tokyo, Japan) in a gauge length of 200 mm, as shown in Figure 3b. During the testing, the load P and observed point displacement were measured and recorded by a data logger (TDS-530, TML, Tokyo, Japan).

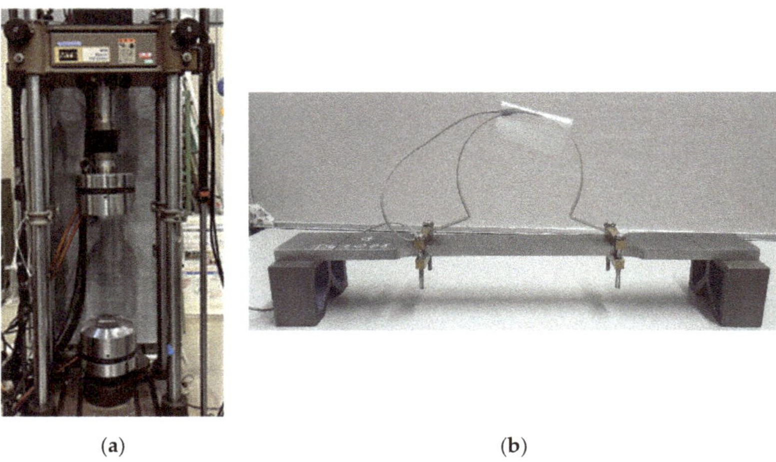

<table>
<tr><td>(a)</td><td>(b)</td></tr>
</table>

Figure 3. The test equipment: (**a**) MTS Material testing machine; (**b**) Ω extensometer.

3. Experimental Results and Discussions

3.1. Experimental Results

For comparison, Figure 4 shows the engineering stress-engineering strain curves obtained from base metal specimens (non-notched flat bar specimens) made of SBHS500, SM570, and SM490 structural steels, respectively, in which three tensile coupons of each material (named as -C1, -C2, and -C3) are tested in this study. The engineering stress is defined as the load divided by section area, and the engineering strain is defined as the displacement from an extensometer divided by 200 mm. It can be observed that the three tensile coupons' engineering stress-engineering strain curves of each material, as shown in Figure 4, are very close. The mechanical properties of SBHS500, SM570, and SM490 structural steels are listed in Table 2, in which the values are the average of results obtained from three tensile coupons. For the SM570 and SM490 steels with obvious yield platform, the yield stress is regarded as the stress when the material reaches yielding. Because the SBHS500 HSS has no yield platform, 0.2% proof stress is regarded as the yield stress, where 0.2% proof stress is the engineering stress corresponding to the residual plastic strain level of 0.2%. It is to be noted that the SBHS500 and SM570 HSSs have similar mechanical properties, including elastic modulus, yield stress, ultimate stress, and elongation, although SBHS500 HSS has no obvious yield platform, which exists in SM570 HSS. However, SM490 NSS has relatively low yield stress and relatively large elongation.

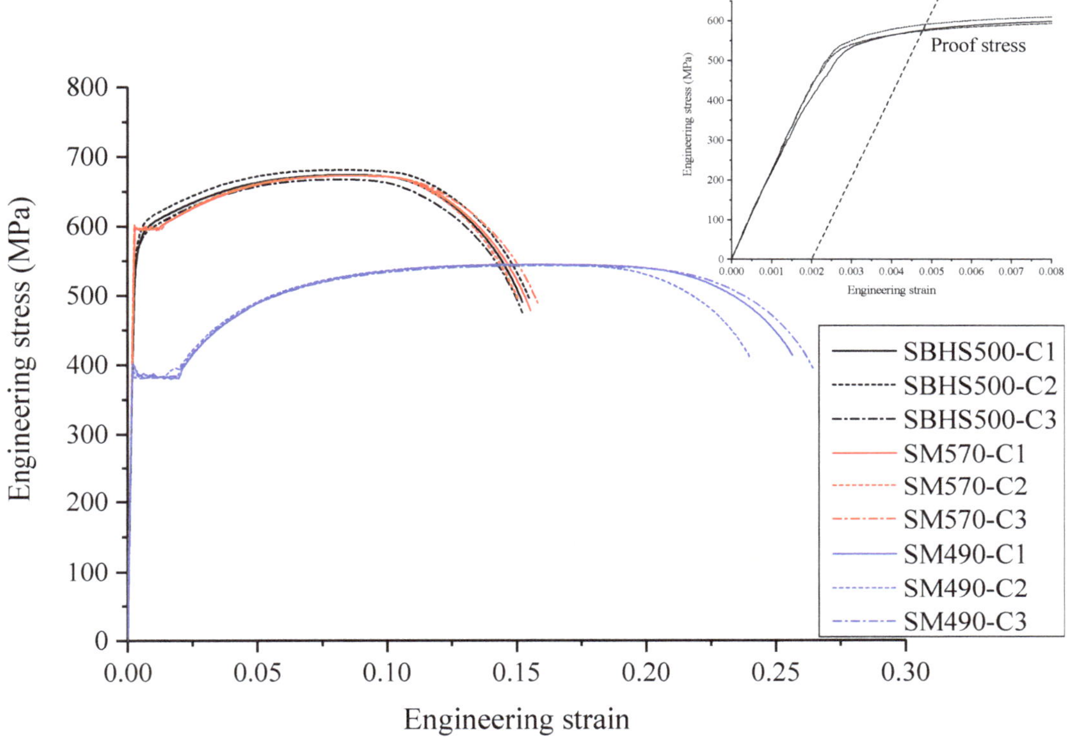

Figure 4. Engineering stress–engineering strain curves of SBHS500, SM570, and SM490 structural steels.

Table 2. Mechanical properties of SBHS500, SM570 and SM490 structural steels.

	E (GPa) [1]	v [2]	σ_y (MPa) [3]	ε_y [4]	σ_u (MPa) [5]	δ (%) [6]
SBHS500	208 ± 1.137	0.270 ± 0.020	578 ± 12.16	0.0020 ± 0.0000	675 ± 15.34	16.7 ± 0.01
SM570	216 ± 1.549	0.278 ± 0.009	597 ± 1.809	0.0028 ± 0.0013	674 ± 0.153	15.5 ± 0.04
SM490	213 ± 1.488	0.273 ± 0.009	388 ± 0.150	0.0018 ± 0.0057	546 ± 2.603	26.0 ± 1.30

[1] E = Young's modulus, [2] v = Poisson's ratio, [3] σ_y = yield stress, [4] ε_y = yield strain, [5] σ_u = ultimate stress, [6] δ = elongation.

The experimental load–displacement curves of VBS and UBS specimens made of SBHS500, SM570 and SM490 structural steels are shown in Figure 5. It is evident from Figure 5 that the load–displacement curves of the two specimens of the same type and material are in agreement with each other, including elastic and plastic parts, although softened and fractured parts of the two specimens of the same type and material have a difference of less than 10%. Regarding UBS specimens, with the increase in U-notch radius, ductile crack initiation occurs later. Figure 6 illustrates the strain-displacement curves of VBS-V90 specimens made of SBHS500, SM570 and SM490 structural steels, where the strain and displacement are directly obtained from a large strain gauge (YFLA-2, TML, Tokyo, Japan) in Figure 7a and extensometer, respectively. The maximum strain that can be measured by this large strain gauge is 20%. It can be observed from Figure 6 that the notched specimens of SBHS500 HSS have a relatively larger strain rate compared to those of SM570 and SM490 steels, so the notched specimens of SBHS500 HSS have a relatively faster crack propagation speed.

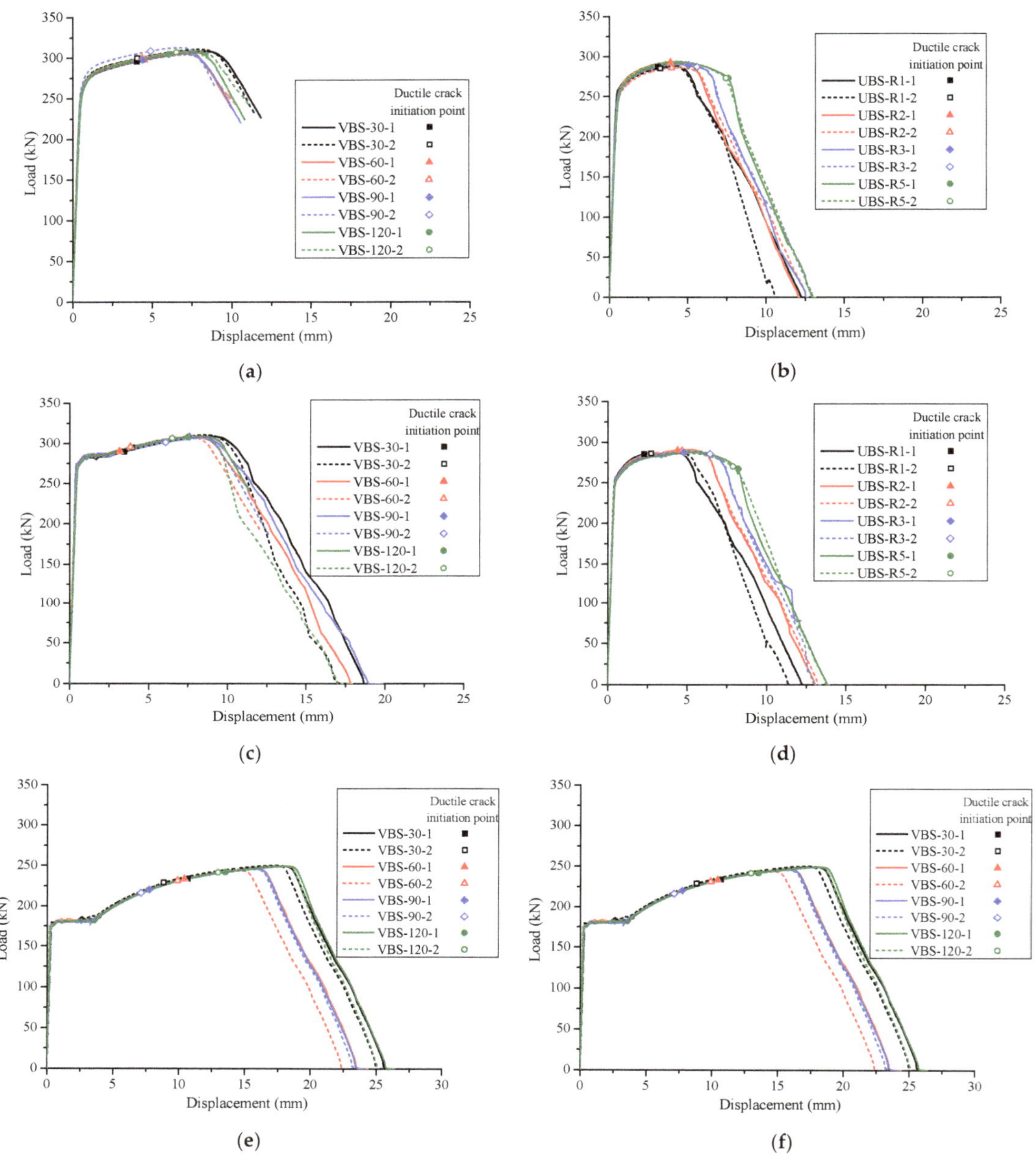

Figure 5. Load–displacement curves of VBS and UBS specimens made of various HSS materials: (**a**) VBS specimens made of SBHS500; (**b**) UBS specimens made of SBHS500; (**c**) VBS specimens made of SM570; (**d**) UBS specimens made of SM570; (**e**) VBS specimens made of SM490; (**f**) UBS specimens made of SM490.

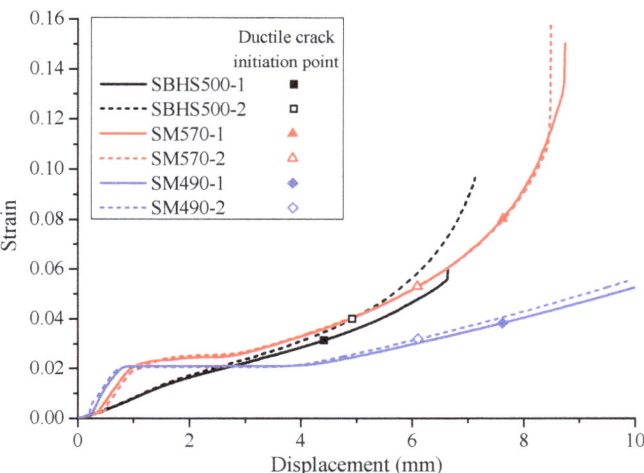

Figure 6. Strain–displacement curves of various materials (VBS-V90 specimens).

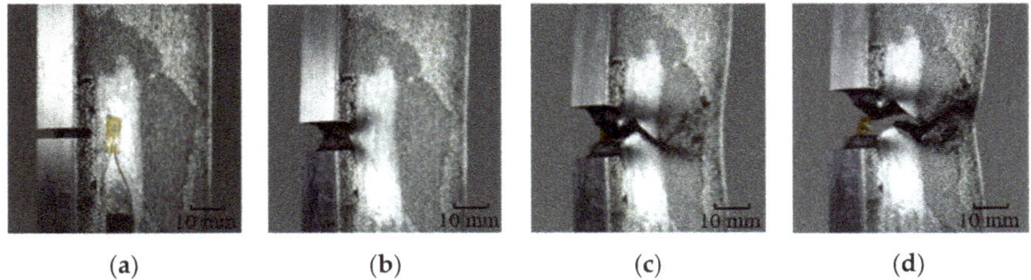

(a)   (b)   (c)   (d)

Figure 7. Crack initiation, propagation and final fracture of Specimen UBS-R1-1: (**a**) Before crack initiation; (**b**) crack initiation; (**c**) crack propagation; (**d**) final fracture.

The strain gauge was arranged at the position 5 mm away from the notch tip, as shown in Figure 7a. When the displacement is about 1 mm, and the specimen is in the elastic range, SM570 and SM490 specimens have the same strain increase rate; however, the SBHS500 specimen has a relatively smaller strain increase rate. When the displacement is 4 mm, the strain of the SM490 NSS specimen is about 2%, and that of SM570 and SBHS500 HSS specimens is about 3%. It can be observed and obtained that the plastic deformation capacity of SM570 and SBHS500 HSS specimens is relatively worse than that of SM490 specimens, and the inelastic strain of SM570 and SBHS500 HSS specimens concentrates at the notch region of specimens. On the contrary, the SM490 NSS specimens with better plastic deformation capacity have a greater gauged displacement. Therefore, a smaller strain at the notch region can be observed in the SM490 NSS specimens compared to SM570 and SBHS500 HSS specimens.

Average elongation values of VBS and UBS specimens for ductile fracture tests are listed in Table 3. Regarding the VBS specimens, it is evident that the elongation of the specimen has no obvious relationship with the notch degree. For the UBS specimens, the obvious relationship between the elongation and the notch radius is not observed except for the specimens with a notch radius of 1 mm. However, the elongation of UBS specimens is 2% greater than that of VBS specimens on average. The phenomenon's reason is that the stress triaxiality and inelastic strain concentration degree of VBS specimens are higher than those of UBS specimens. It can be concluded from Table 3 and Figure 8 that the elongation of SBHS500 HSS specimens is the smallest of all specimens made of three steel materials in this study, and brittle fracture occurs in the SBHS500 HSS specimens. Additionally, the elongations of SM570 and SM490 specimens change more greatly along with the notch shape change; however, the varied range of elongation of SBHS500 HSS specimens along with the notch shape is fewer than that of SM570 and SM490 specimens.

Table 3. Average elongation value of VBS and UBS specimens for ductile fracture tests *.

	SBHS500	SM570	SM490		SBHS500	SM570	SM490
VBS-V30	5.8 ± 0.01%	9.1 ± 0.07%	12.9 ± 0.45%	UBS-R1	5.9 ± 0.30%	5.9 ± 0.19%	9.6 ± 0.01%
VBS-V60	6.7 ± 0.07%	7.5 ± 0.69%	11.7 ± 0.12%	UBS-R2	6.1 ± 0.00%	6.5 ± 0.04%	10.5 ± 0.04%
VBS-V90	4.9 ± 0.00%	8.1 ± 1.51%	12.1 ± 0.15%	UBS-R3	6.4 ± 0.09%	6.4 ± 0.30%	10.4 ± 0.00%
VBS-V120	6.9 ± 0.02%	7.3 ± 0.42%	12.9 ± 0.13%	UBS-R5	6.6 ± 0.03%	6.9 ± 0.01%	10.5 ± 0.02%

* The value in this table is the average value of two specimens.

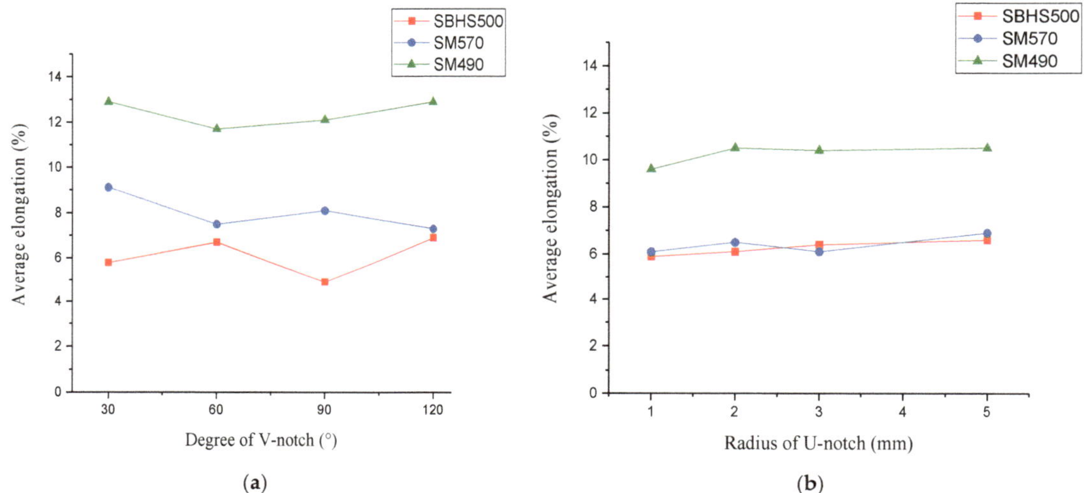

Figure 8. Average elongation value of VBS and UBS specimens for ductile fracture tests: (**a**) Average elongation–V-notch radius relationship; (**b**) average elongation–U-notch radius relationship.

3.2. Crack Initiation

"Linear notch mechanics" proposed by Nishitani [26,27] is regarded as an engineering method to evaluate the mechanical properties of members with the notch. The elastic stress concentration factor is approximately defined as follows:

$$K = 1 + 2\sqrt{\frac{d}{R}} \tag{1}$$

where K is the elastic stress concentration factor, d is the notch depth, and R is the notch radius. For the UBS specimens, when the notch radius is equal to 1 mm and 2 mm, the crack initiation occurs before the ultimate loading point; when the displacement is equal to 3 mm and 5 mm, the crack initiation occurs after the ultimate loading point.

The parameters of UBS specimens calculated and obtained from tests are listed in Table 4. It can be observed from Table 4 that the $\bar{\varepsilon}_{crack}$ value of UBS specimens decreases with the increase in the notch radius. Compared to the SM490 NSS specimens with better plastic deformation capacity, the SBSH500 and SM570 HSS specimens with worse plastic deformation capacity have smaller strain values at crack initiation ($\bar{\varepsilon}_{crack}$).

Table 4. Parameters of UBS specimens calculated and obtained from tests.

		UBS-R1	UBS-R2	UBS-R3	UBS-R5
	d (mm) [2]	5	5	5	5
	R (mm) [3]	1	2	3	5
	K [1]	5.472	4.162	3.582	3.000
SBHS500	$\bar{\varepsilon}_{crack}$ [4]	0.0159 ± 0.0004	0.0198 ± 0.0002	0.0268 ± 0.0012	0.0377 ± 0.0004
	$\bar{\varepsilon}_{crack}/\varepsilon_y$ [5]	7.950	9.900	13.400	18.850
SM570	$\bar{\varepsilon}_{crack}$	0.0125 ± 0.0010	0.0228 ± 0.0007	0.0282 ± 0.0041	0.0403 ± 0.0008
	$\bar{\varepsilon}_{crack}/\varepsilon_y$	4.471	8.155	10.073	14.395
SM490	$\bar{\varepsilon}_{crack}$	0.0398 ± 0.0009	0.0573 ± 0.0018	0.0610 ± 0.0012	0.0705 ± 0.0016
	$\bar{\varepsilon}_{crack}/\varepsilon_y$	22.111	31.833	33.889	39.167

[1] K = stress concentration factor, [2] d = notch depth, [3] R = notch radius, [4] $\bar{\varepsilon}_{crack}$ = average value of strain at crack initiation for two specimens, [5] ε_y = yield strain of HSS.

The $\bar{\varepsilon}_{crack}/\varepsilon_y$-$1/K$ relationship of the UBS specimens is shown in Figure 9. A linear relationship is observed between the value of $\bar{\varepsilon}_{crack}/\varepsilon_y$ and $1/K$. With the increase of $1/K$, the $\bar{\varepsilon}_{crack}/\varepsilon_y$ of three series of the specimens increases. The coefficient of determination of these three curves is more than 0.9. Regarding the SBHS500 and SM570 HSSs specimens with a high yield stress, with the decrease in notch radius, the elastic stress concentration factor increases and the strain at crack initiation decreases. This fact might lead to a brittle fracture occurring before the plastic deformation is fully employed.

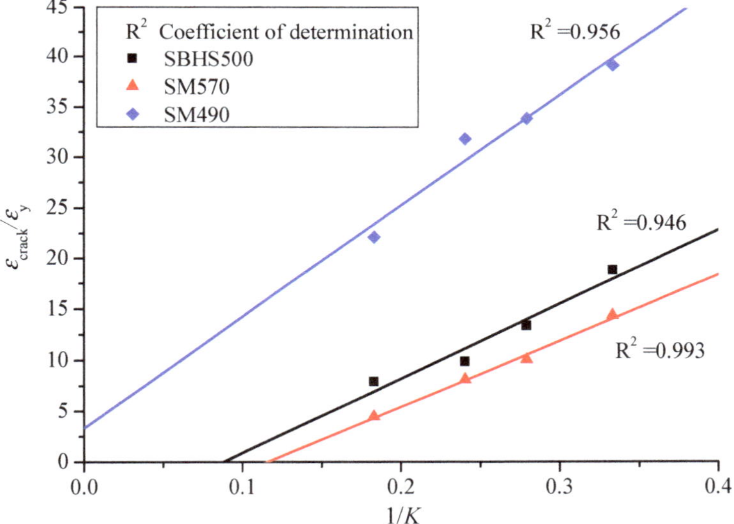

Figure 9. εcrack/εy-1/K relationship of UBS specimens.

3.3. Fracture Surface

It can be observed from tests that for VBS and UBS specimens, as shown in Figure 10, cracking is induced by the growth and coalescence of numerous nucleated micro-voids from the surface of the notch root region (Zone A), then propagation in a herringbone pattern (Zone B), leading to final shear mode failure (Zone C) [20]. Zone A and Zone B are coarse surfaces; Zone C is very smooth and along a local shear band oriented at an angle of about 45° in relation to the tensile axis [12].

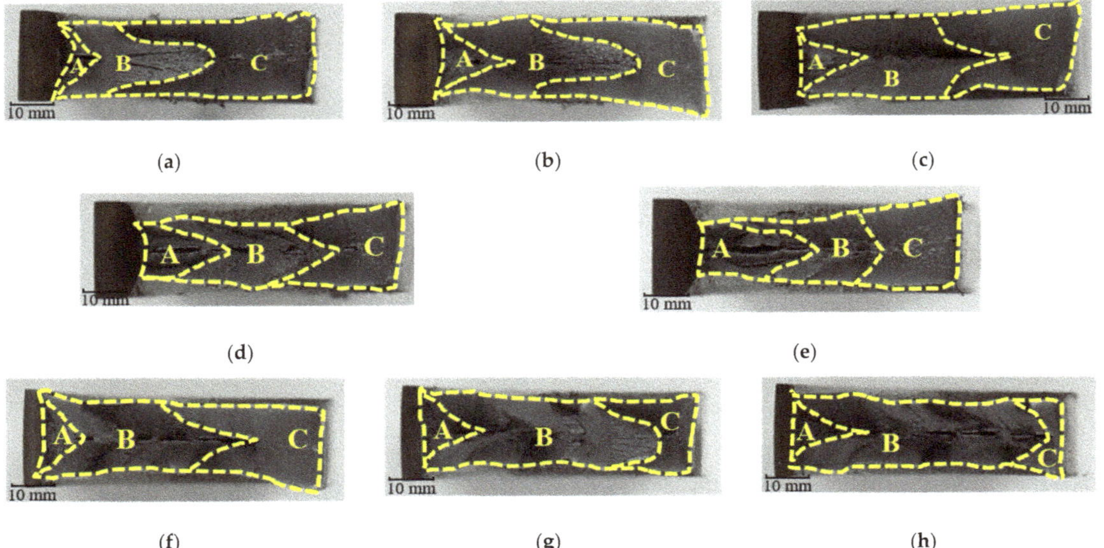

Figure 10. Fracture surface of specimens: (**a**) UBS-R1 (SBHS500); (**b**) UBS-R1 (SM570); (**c**) UBS-R1 (SM490); (**d**) UBS-R3 (SBHS500); (**e**) UBS-R5 (SBHS500); (**f**) VBS-V30 (SM570); (**g**) VBS-V60 (SM570); (**h**) VBS-V90 (SM570). Zone A: crack initiation region; Zone B: crack propagation region; Zone C: shear zone.

It can be concluded from Figure 10a–c that the specimens made of SM490 with the best plastic deformation capacity have the relatively larger crack initiation region and crack propagation region; however, the specimens made of SBHS500 with the worst plastic deformation capacity have the relatively smaller crack initiation region and crack propagation region. It can be observed in Figure 10a,d,e that the crack initiation region greatly increases with the increase in the U-notch radius. From Figure 10f–h, it can be observed that the crack initiation region slightly increases with the increase in V-notch degree. Moreover, there is obvious delamination in the failure surface of SBHS500 and SM570 HSS specimens.

3.4. Effect of Notch Depth

To investigate the effect of notch depth on fracture behavior, Figure 11 illustrates the VBS test results of SM490 specimens with a notch depth of 3 mm in this study and those of SM490 specimens with a notch depth of 6 mm in the reference [6]. Figure 12 shows the UBS test results of SM490 specimens with a notch depth of 5 mm in this study and those of SM490 specimens with a notch depth of 6 mm in the reference [6].

In this section, the yield and ultimate stresses are defined as the 0.2% proof stress and the maximum stress during tests, respectively. When there is no obvious yield platform, the yield stress can be defined as 0.2% proof stress [28]. In the VBS tests, as listed in Table 5, the yield stress of VBS specimens with a notch depth of 3 mm in this study is 400 MPa; however, that of VBS specimens with a notch depth of 6 mm in the previous study is 450 MPa. The ultimate stress of VBS specimens with a notch depth of 3 mm and 6 mm is 550 MPa and 480 MPa, respectively. In other words, with the increase in notch depth, the yield stress increases, but the ultimate stress decreases. The same phenomenon occurs in the UBS specimens, as shown in Figure 12. The yield stress of UBS specimens with a notch depth of 6 mm is greater than that with a notch depth of 5 mm; however, the ultimate stress is the opposite. The decent rate of a load of specimens in this study is not related to notch shape and notch depth. The greater the notch depth, the earlier the crack initiation. A greater notch depth might result in brittle fracture because the plastic deformation capacity cannot be fully employed. Therefore, a detailed notch geometry should be obtained in practical engineering because a greater notch depth might lead to brittle fracture of members.

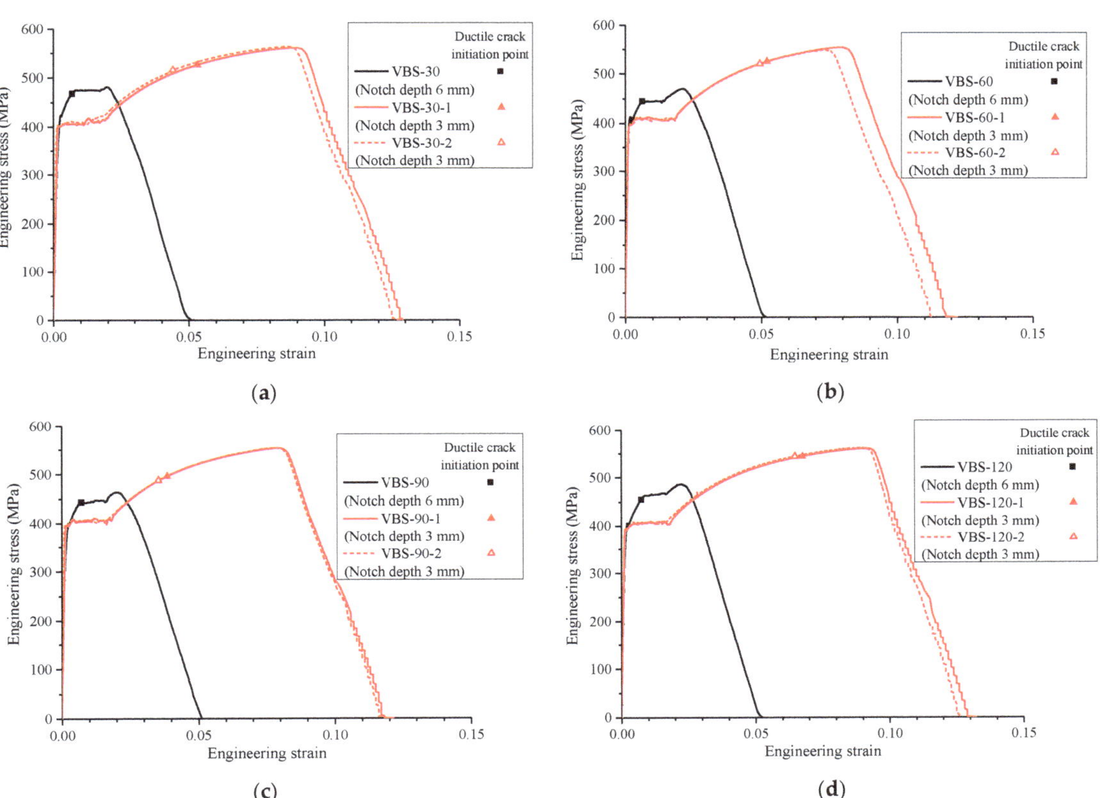

Figure 11. Effect of notch depth (VBS specimens made of SM490): (**a**) VBS-V30 specimens; (**b**) VBS-V60 specimens; (**c**) VBS-V90 specimens; (**d**) VBS-V120 specimens.

Table 5. Effect of notch depth (UBS specimens made of SM490).

Depth	σ_y (MPa)	σ_u (MPa)
3 mm	400	550
6 mm	450	480

Because the tested specimens in this study are single-groove, the HSS material in the notch region firstly reaches to yielding, as marked by the black circle in Figure 13a, and then the full cross-section enters yielding. During this process, the neutral axis moves to the side without the notch because the elastic modulus will decrease once the material enters plastic. In this study, the yielding stress of specimens tested is regarded as the point that a full cross-section enters plasticity because the yielding point of the notch region cannot be determined easily.

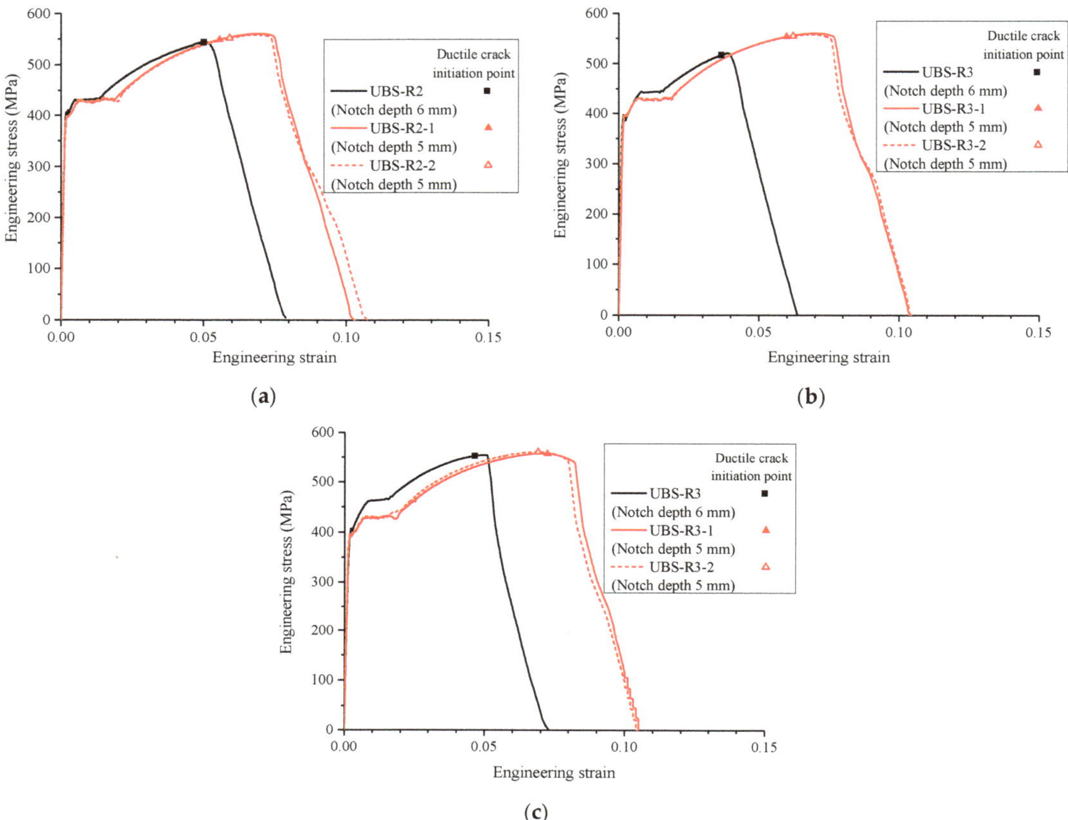

Figure 12. Effect of notch depth (UBS specimens made of SM490): (**a**) UBS-R2 specimens; (**b**) UBS-R3 specimens; (**c**) UBS-R5 specimens.

3.5. Relationships between Detailed Notch Geometry with Yield and Ultimate Stresses

The relationships between detailed notch geometry (U-notch radius and V-notch degree) with yield and ultimate stresses are shown in Figure 14. The yield and ultimate stresses in Figure 14 are defined as the yield load and ultimate load divided by the notched minimum sectional area, respectively. It can be observed that the yield stress has no obvious relationship with the detailed notch geometry, including both the U-notch radius and V-notch degree. The yield stress of VBS and UBS specimens is about 1.1 times greater than that of base metal specimens. Similarly, the ultimate stress has no obvious relationship with the detailed notch geometry. Accordingly, the effect of notch depth is higher than the effects of notch degree or notch radius.

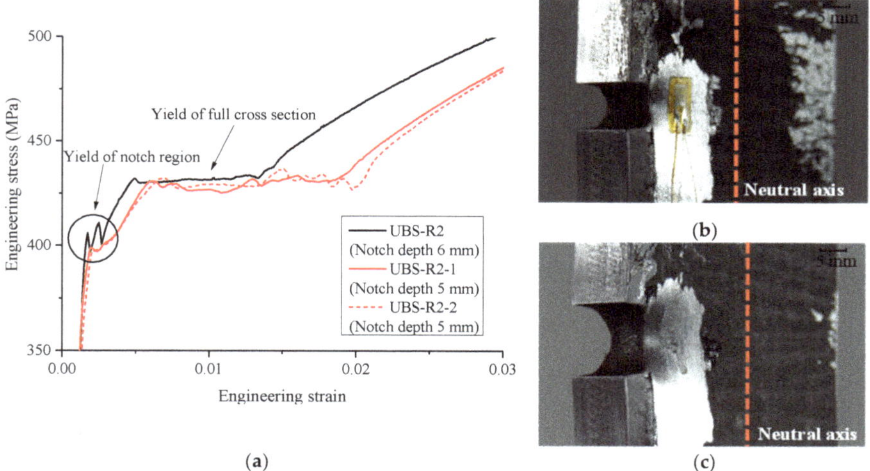

Figure 13. Effect of neutral axis change on yield stress: (**a**) UBS-R2 specimens; (**b**) before neutral axis change; (**c**) after neutral axis change.

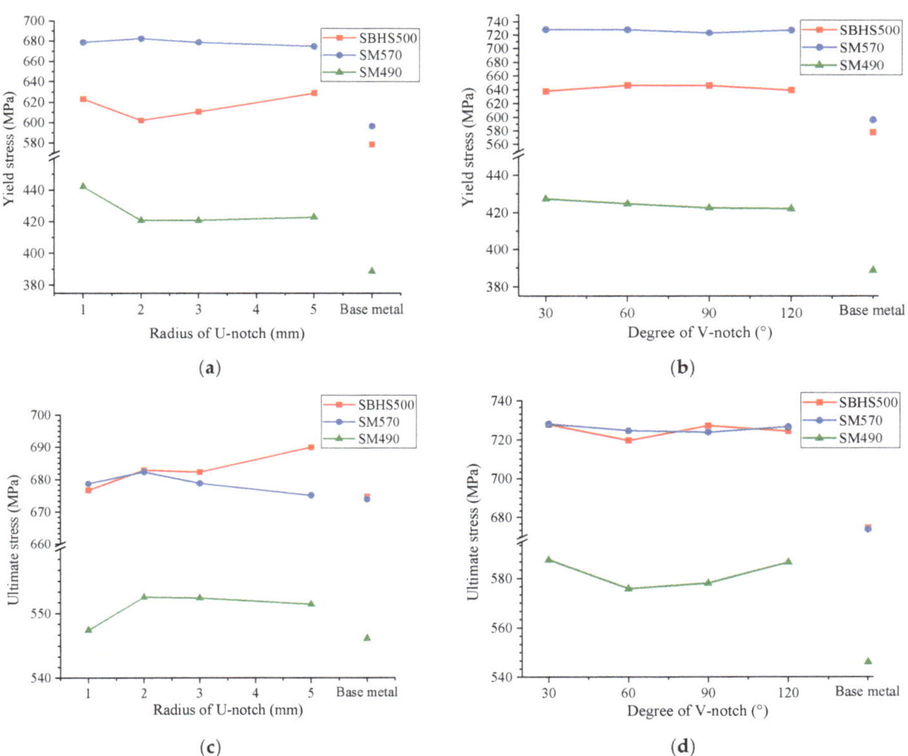

Figure 14. Effect of U-notch radius and V-notch degree: (**a**) Yield stress–U-notch radius relationship; (**b**) yield stress–V-notch degree relationship; (**c**) ultimate stress–U-notch radius relationship; (**d**) ultimate stress–V-notch degree relationship.

3.6. Weldability

It is remarked that from the metallurgical point of view, both the grain sizes and microstructural arrays of distinctive materials and alloys have important roles in the resulting material's properties (e.g., mechanical and corrosion behaviors), as previously reported [29–36]. In this study, the weldability of these steels is studied based on the chemical composition.

Table 6 lists the chemical compositions of HSSs in this study, including SM570 and SBHS500, from HSS manufacturers. The chemical composition of SM490 NSS is not provided by the manufacturer. As shown in Table 6, C, Si, Mn, P, and S are the main five compositions, in which C, Si and Mn play an important role in the control of mechanical and impact properties, and P and S are impurities. Except for these five compositions, other compositions can be added to HSS materials to improve impact properties or to refine particle size. Generally, increasing C content can lead to an increase in yield and ultimate stresses. In this study, SBHS500 and SM570 HSSs have almost the same ultimate stress; however, the yield stress of SM570 is greater than that of SBHS500 because the C content of SM570 HSS is 0.02% higher than that of SBHS500. Adding the composition of Cr can result in improving the corrosion, oxidation, and abrasion resistance; adding the Cr content has no effect on the mechanical properties of HSS. The carbon equivalent value C_{eq} and the welding crack sensitivity index P_{CM}, which is a weldability index, are defined as follows:

$$C_{eq} = C + \frac{Mn}{6} + \frac{Si}{24} + \frac{Ni}{40} + \frac{Cr}{5} + \frac{Mo}{4} \tag{2}$$

$$P_{CM} = C + \frac{Mn}{20} + \frac{Si}{30} + \frac{Cu}{20} + \frac{Ni}{60} + \frac{Cr}{20} + \frac{Mo}{15} + \frac{V}{10} + 5B \tag{3}$$

Table 6. Chemical composition of SM570 and SBHS500 (%).

	B	N	C	Si	Mn	P	S	Cu	Ni	Cr	Mo	Nb	V
SM570	0.0012	0.0160	0.1200	0.2100	1.6000	0.0110	0.0020	0.010	0.020	0.010	0.000	0.020	0.060
SBHS500	0.0010	0.0153	0.1000	0.2200	1.5300	0.0090	0.0020	0.010	0.010	0.140	0.000	0.023	0.060

Table 7 lists the calculated values of C_{eq} and P_{CM} in Equations (2) and (3). It is illustrated that the C_{eq} value of SBHS500 is equal to that of SM570; however, the P_{CM} value of SBHS500 is 0.02% less than that of SM570. Therefore, the preheating of welding of SBHS500 specimens or members can be carried out at a relatively low temperature, the crack due to heating during welding is not easy to occur, and SBHS500 specimens or members have better weldability.

Table 7. Calculated values of P_{CM} and C_{eq}.

	P_{CM}	C_{eq}
SM570	0.220	0.402
SBHS500	0.203	0.397

4. Conclusions

The ductile fracture performances of two types of HSSs, including SBHS500 and SM570, and an NSS SM490 are investigated by a series of V-notch and U-notch fracture tests. Four various V-notch degrees and four U-notch radii are employed to create different stress triaxialities. A total of 48 specimens are tested. The following conclusions can be obtained:

(1) For the three structural steels, with the decreases of the degree of V-notch or radius of U-notch, both the ductile crack initiation point and drop in the load–displacement responses of VBS and UBS specimens appear earlier;

(2) The elongation of specimens made of SM490 and SM570 vary greatly along with the notch shape change; however, the elongation changes of specimens made of SBHS500 along with the notch shape change is relatively small;

(3) NSS SM490 is the structural steel with the best plastic deformation capacity in the three structural steels of this study. SBHS500 and SM570 HSSs have the worse plastic deformation capacity because of their higher yield stress;

(4) SBHS500 and SM570 HSSs with higher yield stress have a relatively higher elastic stress concentration factor, the crack initiation appears earlier, and the brittle fracture is more likely to occur;

(5) Compared to SM570 HSS, SBHS500 HSS has a lower P_{CM} and better weldability.

Author Contributions: Y.L. (Yan Liu) performed the experiment; S.I. and Y.L. (Yanyan Liu) performed the data analyses; L.K. wrote the manuscript; H.G. proposed the idea and provided the funding. All authors have read and agreed to the published version of the manuscript.

Funding: This research was funded by [the National Natural Science Foundation of China] grant number [52178286], [the Natural Science Foundation of Guangdong Province] grant number [2020A1515011070] and [the Guangdong Provincial Key Laboratory of Modern Civil Engineering Technology] grant number [2021B1212040003].

Informed Consent Statement: Informed consent was obtained from all subjects involved in the study.

Data Availability Statement: Not applicable.

Acknowledgments: The study is also supported in part by grant from the Advanced Research Center for Natural Disaster Risk Reduction, Meijo University. All the sources of support are gratefully acknowledged.

Conflicts of Interest: The authors declare no conflict of interest.

References

1. Valentini, R.; Tedesco, M.M.; Corsinovi, S.; Bacchi, L. Delayed Fracture in Automotive Advanced High Strength Steel: A New Investigation Approach. *Steel Res. Int.* **2019**, *90*, 1900136. [CrossRef]
2. Fujie, W.; Taguchi, M.; Kang, L.; Ge, H.; Xu, B. Ductile crack initiation evaluation in stiffened steel bridge piers under cyclic loading. *Steel Compos. Struct.* **2020**, *36*, 463–480.
3. Kang, L.; Wang, Y.; Liu, X.; Uy, B. Investigation of residual stresses of hybrid normal and high strength steel (HNHSS) welded box sections. *Steel Compos. Struct.* **2019**, *33*, 489–507.
4. Hai, L.-T.; Li, G.-Q.; Wang, Y.-B.; Wang, Y.-Z. Experimental and numerical investigation on Q690 high strength steel beam-columns under cyclic lateral loading about weak axis. *Eng. Struct.* **2021**, *236*, 112107. [CrossRef]
5. Liu, Y.; Jia, L.-J.; Ge, H.; Kato, T.; Ikai, T. Ductile-fatigue transition fracture mode of welded T-joints under quasi-static cyclic large plastic strain loading. *Eng. Fract. Mech.* **2017**, *176*, 38–60. [CrossRef]
6. Kang, L.; Ge, H.; Fang, X. An improved ductile fracture model for structural steels considering effect of high stress triaxiality. *Constr. Build. Mater.* **2016**, *115*, 634–650. [CrossRef]
7. Jia, L.-J.; Ge, H.; Shinohara, K.; Kato, H. Experimental and Numerical Study on Ductile Fracture of Structural Steels under Combined Shear and Tension. *J. Bridg. Eng.* **2016**, *21*, 04016008. [CrossRef]
8. Wang, W.-Z.; Jiang, L.-Y.; Sun, R.-Q.; Wang, X.-T. Tensile Tests on High Strength Steel Q345 Notched Plates with the Moderate Thickness. In Proceedings of the International Conference on Smart Materials and Intelligent Systems, Chongqing, China, 23–25 December 2011; p. 1450.
9. Xiang, P.; Qing, Z.; Jia, L.-J.; Wu, M.; Xie, J. Damage evaluation and ultra-low-cycle fatigue analysis of high-rise steel frame with mesoscopic fracture models. *Soil Dyn. Earthq. Eng.* **2020**, *139*, 106283. [CrossRef]
10. Li, W.; Liao, F.; Zhou, T.; Askes, H. Ductile fracture of Q460 steel: Effects of stress triaxiality and Lode angle. *J. Constr. Steel Res.* **2016**, *123*, 1–17. [CrossRef]
11. Kang, L.; Wu, B.; Liu, X.; Ge, H. Experimental study on post-fire mechanical performances of high strength steel Q460. *Adv. Struct. Eng.* **2021**, *24*, 2791–2808. [CrossRef]
12. Kang, L.; Suzuki, M.; Ge, H.; Wu, B. Experiment of ductile fracture performances of HSSS Q690 after a fire. *J. Constr. Steel Res.* **2018**, *146*, 109–121. [CrossRef]
13. Xin, H.; Veljkovic, M. Evaluation of high strength steels fracture based on uniaxial stress-strain curves. *Eng. Fail. Anal.* **2021**, *120*, 105025. [CrossRef]
14. Park, S.-J.; Cerik, B.C.; Choung, J. Comparative study on ductile fracture prediction of high-tensile strength marine structural steels. *Ships Offshore Struct.* **2020**, *15*, S208–S219. [CrossRef]

15. Huang, X.; Ge, J.; Zhao, J.; Zhao, W. Experimental and Fracture Model Study of Q690D Steel under Various Stress Conditions. *Int. J. Steel Struct.* **2021**, *21*, 561–575. [CrossRef]
16. Wang, Y.-Z.; Li, G.-Q.; Wang, Y.-B.; Lyu, Y.-F.; Li, H. Ductile fracture of high strength steel under multi-axial loading. *Eng. Struct.* **2020**, *210*, 110401. [CrossRef]
17. Sajid, H.U.; Kiran, R. Post-fire mechanical behavior of ASTM A572 steels subjected to high stress triaxialities. *Eng. Struct.* **2019**, *191*, 323–342. [CrossRef]
18. Kim, M.; Lee, H.; Hong, S. Experimental determination of the failure surface for DP980 high-strength metal sheets considering stress triaxiality and Lode angle. *Int. J. Adv. Manuf. Technol.* **2018**, *100*, 2775–2784. [CrossRef]
19. Sajid, H.U.; Kiran, R. Influence of high stress triaxiality on mechanical strength of ASTM A36, ASTM A572 and ASTM A992 steels. *Constr. Build. Mater.* **2018**, *176*, 129–134. [CrossRef]
20. Kang, L.; Ge, H.; Kato, T. Experimental and ductile fracture model study of single-groove welded joints under monotonic loading. *Eng. Struct.* **2015**, *85*, 36–51. [CrossRef]
21. Miki, C.; Ichikawa, A.; Kusunoki, T.; Kawabata, F. Proposal of New High Performance Steels for Bridges (BHS500, BHS700). *Doboku Gakkai Ronbunshu* **2003**, *2003*, 1–10. (In Japanese) [CrossRef]
22. Kang, L.; Suzuki, M.; Ge, H. A study on application of high strength steel SM570 in bridge piers with stiffened box section under cyclic loading. *Steel Compos. Struct.* **2018**, *26*, 583–594.
23. Hirohata, M.; Teraguchi, D.; Kitane, Y. Mechanical properties of steels for bridge high performance structure subjected to heating and cooling process simulating fire. *Steel Constr. Eng.* **2019**, *26*, 79–86. (In Japanese)
24. *JIS G 3140:2011*; Higher Yield Strength Steel Plates for Bridges. Japan Industrial Committee: Tokyo, Japan, 2011.
25. *JIS G 3106:2015*; Rolled Steels for Welded Structure. Japan Industrial Committee: Tokyo, Japan, 2015.
26. Nishitani, H. Measure of stress field in a notch corresponding to stress intensity factor in a crack. *Trans. JSME* **1983**, *49*, 1353–1359. (In Japanese) [CrossRef]
27. Nishitani, H. Linear notch mechanics as an extension of linear fracture mechanics. In *Computational and Experimental Fracture Mechanics*; Computational Mechanics Publications: Tokyo, Japan, 1994; Volume 16, pp. 187–211.
28. Rasmussen, K. Full-range stress-strain curves for stainless steel alloys. *J. Constr. Steel Res.* **2003**, *59*, 47–61. [CrossRef]
29. Petch, N. The cleavage strength of polycrystals. *J. Iron Steel Inst.* **1953**, *174*, 25–31.
30. Donelan, P. Modelling microstructural and mechanical properties of ferritic ductile cast iron. *Mater. Sci. Technol.* **2000**, *16*, 261–269. [CrossRef]
31. Osório, W.; Santos, A.C.; Quaresma, J.; Garcia, A. Mechanical properties as a function of thermal parameters and microstructure for Zn-Al castings. *J. Mater. Proc. Technol.* **2003**, *143*, 703–709. [CrossRef]
32. Lloyd, D.J.; Court, S.A. Influence of grain size on tensile properties of Al-Mg alloys. *Mater. Sci. Technol.* **2003**, *19*, 1349–1354. [CrossRef]
33. Osório, W.R.; Cheung, N.; Spinelli, J.E.; Goulart, P.R.; Garcia, A. The effects of a eutectic modifier on microstructure and surface corrosion behavior of Al-Si hypoeutectic alloys. *J. Solid State Electrochem.* **2007**, *11*, 1421–1427. [CrossRef]
34. Rosa, D.M.; Spinelli, J.E.; Osório, W.R.; Garcia, A. Effects of cell size and macrosegregation on the corrosion behavior of a dilute Pb–Sb alloy. *J. Power Sources* **2006**, *162*, 696–705. [CrossRef]
35. Kostryzhev, A.; Singh, N.; Chen, L.; Killmore, C.; Pereloma, E. Comparative effect of Mo and Cr on microstructure and mechanical properties in NbV-microalloyed bainitic steels. *Metals* **2018**, *8*, 134. [CrossRef]
36. Tesser, E.; Silva, C.; Artigas, A.; Monsalve, A. Effect of Carbon Content and Intercritical Annealing on Microstructure and Mechanical Tensile Properties in FeCMnSiCr TRIP-Assisted Steels. *Metals* **2021**, *11*, 1546. [CrossRef]

 *metals*

Article

Numerical and Experimental Research on Similarity Law of the Dynamic Responses of the Offshore Stiffened Plate Subjected to Low Velocity Impact Loading

Haibo Zhou [1], Yang Han [2], Yi Zhang [2], Wei Luo [3], Jingxi Liu [4,5] and Rong Yu [6,*]

[1] Wuhan Second Ship Design and Research Institute, Wuhan 430064, China; zhouhaibo198712@163.com
[2] First Engineering Co., Ltd., China Construction Third Bureau, Wuhan 430048, China;
    hanyangsuzhou@foxmail.com (Y.H.); tmdjump@gmail.com (Y.Z.)
[3] China Ship Development and Design Centre Co., Ltd., Wuhan 430064, China; superfighter1979@sina.com
[4] School of Naval Architecture and Ocean Engineering, Huazhong University of Science and Technology,
    Wuhan 430074, China; liu_jing_xi@hust.edu.cn
[5] Hubei Key Laboratory of Naval Architecture and Marine Hydrodynamics,
    Huazhong University of Science and Technology, Wuhan 430074, China
[6] School of Mechanical Technology, Wuxi Institute of Technology, Wuxi 214121, China
[*] Correspondence: yurong@wxit.edu.cn

Abstract: Similarity laws of scaled models of offshore platform deck structures under low velocity impact loading are proposed in the present research. The similarity laws of scaled models with different scaling factors are established in forms of dimensionless factors with consideration of flow stress differences of the materials. A dimensionless displacement is defined by dividing displacement by plate thickness and a dimensionless force is defined by dividing force by flow stress and plate thickness; then, a dimensionless force-displacement relationship is established. Dynamic responses of three geometrically similar stiffened structures with scaling factors of 1:4, 1:2, and 1:1 subjected to the dropping impact of a rigid triangular pyramidic impactor are investigated by an experimental test and a finite element analysis. Results show that dimensionless force-displacement curves of geometrically similar plates coincide with each other; meanwhile, the difference of maximum impact force for the three structures with various scaling factors is less than 5%, and the difference of maximum impact depth is less than 1%, which definitely show the effectiveness of the scaling laws based on dimensionless factors. The present research provides useful insight into the similarity laws of dynamic responses of deck structures subjected to falling object impact and would be used in the crashworthiness research and design process of the offshore structures.

Keywords: offshore structure; falling object collision; similarity law; scaled test; finite element analysis

Citation: Zhou, H.; Han, Y.; Zhang, Y.; Luo, W.; Liu, J.; Yu, R. Numerical and Experimental Research on Similarity Law of the Dynamic Responses of the Offshore Stiffened Plate Subjected to Low Velocity Impact Loading. *Metals* **2022**, *12*, 657. https://doi.org/10.3390/met12040657

Academic Editors: Zhihua Chen, Hanbin Ge, Siu-lai Chan and Cristiano Fragassa

Received: 28 February 2022
Accepted: 5 April 2022
Published: 12 April 2022

Publisher's Note: MDPI stays neutral with regard to jurisdictional claims in published maps and institutional affiliations.

1. Introduction

Low velocity impacts of ships and offshore structures, including collision and grounding, floating object collision and falling object impact, are prone to catastrophic disasters. Thus, researchers worldwide pay particular attention to investigations of ship structure collisions, as plenty of useful conclusions have been obtained by analytical investigations, numerical research and experimental tests [1,2]. Among these investigations, important factors such as impact load, stress distribution, damage initiation and propagation would be obtained by experimental research; furthermore, experimental test results provide benchmark studies for numerical simulations and analytical research. Thus, experimental methods play an irreplaceable role in collision research.

A small-scaled model test has been mainly adopted in ship collision research for the purpose of reducing difficulty and economic consumption [3,4]. One of the most important factors is to establish the similarity laws between the scaled model and the prototype. Jones [5] proposed the similarity law to transfer the scaling model impact test results into the prototype impact results. Calle [6] proposed a scaling law with consideration of material yield, material strain hardening, damage initiation and strain rate effect.

The above-mentioned research provides support for the extension of scale model test results to full-scaled prototypes for ship collisions. However, the research mentioned above mainly focuses on collisions of ship side structures, and few investigations have been proposed on research of deck structures under the impact of dropping objects; thus, similarity laws of structure models with different scaling factors should be considered. The present research conducts tests on similarity laws of scaled models of an offshore platform deck structure under low velocity impact of dropping objects. The first step is to propose the similarity laws between a small-scaled model and a full-scaled prototype, and to introduce a couple of dimensionless factors; the following step is to establish three geometrically similar stiffened plate structures with different scaling factors; then, experimental and numerical simulation research on low velocity impact has been completed and the last step is to analyze the experimental and numerical results with the dimensionless factors. Finally, the similarity laws are verified.

2. Similarity Laws

Generally, a dynamic impact test is conducted on a geometrically similar small-scaled model to obtain dynamic responses with consideration of economical consumption. In this series of research, a prototype means a model which is the same as the stiffened plate of a real offshore structure; they have the same length, width, the same shell thickness, the same stiffeners and the same stiffener spaces. Meanwhile, the scaled model is geometrically similar to the prototype, whereas they are of the same shape but a different size. Furthermore, the scaled models and the prototype are usually manufactured with the same material and the same fabrication method.

Thus, the first priority is to establish scaling principles between scaled model and prototype. For a full-scaled prototype model and the corresponding geometrically similar small-scaled model, the relationships between the characteristics of the scaled model and the prototype are established by using the scaling factor [5]. The scaling factor β between the geometrically similar small-scaled model and the full-scaled prototype is defined as:

$$\beta = L_s / L_p \tag{1}$$

where L_S means length of the scaled structure and L_p means length of the full-scaled prototype. As well, in a low velocity impact scenario the impact displacement, shell thickness, impact force and energy absorption follow the following scaling principle:

$$\Delta_s : \Delta_p = \beta : 1 \tag{2}$$

$$t_s : t_p = \beta : 1 \tag{3}$$

$$F_s : F_p = \beta^2 : 1 \tag{4}$$

$$E_s : E_p = \beta^3 : 1 \tag{5}$$

These similarity laws were established under the assumption that the small-scaled model and the full-scaled prototype were fabricated by the same material with the same values of mass density (ρ), elastic module (E), Poisson's ratio (μ), yield strength (σ_s) and ultimate strength (σ_b). However, experimental test results show that yield strength and ultimate strength of the plates with different thicknesses are slightly different in many cases. Thus, the influence of the differences of yield strength should be considered in the scaling laws. Previous experimental and theoretical studies show that the impact force is proportional to the flow stress of the materials [4,7]:

$$F \propto \sigma_0 \tag{6}$$

where $(\sigma_0) = (\sigma_s + \sigma_b)/2$ means the flow stress of the materials, and which equals to the average value of yield stress and ultimate strength. Thus, the influence of the differences of flow stress should also be considered in the scaling law; an effective way is to establish a couple of dimensionless factors. A dimensionless displacement is defined by dividing displacement by shell thickness and a dimensionless impact force is defined by dividing force by flow stress and shell thickness:

$$\left\{ \begin{array}{l} \overline{\Delta}_s = \frac{\Delta_s}{t_s}, \ \ \overline{\Delta}_p = \frac{\Delta_p}{t_p} \\ \overline{F}_s = \frac{F_s}{(\sigma_0)_s \cdot t_s^2}, \ \ \overline{F}_p = \frac{F_p}{(\sigma_0)_p \cdot t_p^2} \end{array} \right. \tag{7}$$

In the same way, the dimensionless energy absorption is also defined as:

$$\left\{ \begin{array}{l} \overline{E}_s = \frac{E_s}{(\sigma_0)_s \cdot t_s^2 \cdot t_s} = \frac{E_s}{(\sigma_0)_s \cdot t_s^3} \\ \overline{E}_p = \frac{E_p}{(\sigma_0)_p \cdot t_p^2 \cdot t_p} = \frac{E_p}{(\sigma_0)_p \cdot t_p^3} \end{array} \right. \tag{8}$$

In Equations (1)–(8), the symbols with subscript "s" represent the physical quantity of the small-scaled model, whereas the symbols with subscript "p" represent the physical quantity of the full-scaled prototype; the same as below, as shown in Table 1. Theoretically, the dimensionless impact force and energy absorption of the scaled model should be equal to those of the full-scaled prototype under the same dimensionless displacement. In another way, the dimensionless force-displacement curves and dimensionless energy absorption curves of models with different scaling factors should be the same shape.

Table 1. Meaning of the symbols of small-scaled model and full-scaled prototype.

	Small-Scaled Model	Full-Scaled Prototype
Length	L_s	L_p
Shell thickness	t_s	t_p
Displacement	Δ_s	Δ_p
Impact force	F_s	F_p
Energy absorption	E_s	E_p
Flow stress of material	$(\sigma_0)_s$	$(\sigma_0)_p$
Dimensionless displacement	$\overline{\Delta}_s$	$\overline{\Delta}_p$
Dimensionless impact force	$\overline{F}_s$	$\overline{F}_p$
Dimensionless energy absorption	$\overline{E}_s$	$\overline{E}_p$

3. Materials and Methods

Three types of geometrically similar stiffened plate models with different scaling factors were designed, an impact tower for drop weight impact was established, and an FEA (finite element analysis) of the dynamic impact scenarios was conducted for further investigations.

3.1. Specimens

Rectangular stiffened steel plates, which were derived from the deck structure of an offshore platform, were conducted in the present research. Three similar specimens, including two small-scaled models with a 1:4 and a 1:2 scaling factor, and one full-scaled prototype, were proposed. As shown in Figures 1 and 2, three geometrically similar specimens were stiffened by three uniformly distributed L-shaped stiffeners. The thicknesses of the plates were 3.9 mm, 7.8 mm and 15.6 mm, respectively.

Since the sharp edge of the dropping container would cause deck damage in case of falling, a rigid triangular pyramidic impactor derived from a container was conducted with the length of the edge "L" equal to 200 mm, as shown in Figure 3. The impactor was fabricated with abrasive steel, and a heat treatment process ensured its surface hardness.

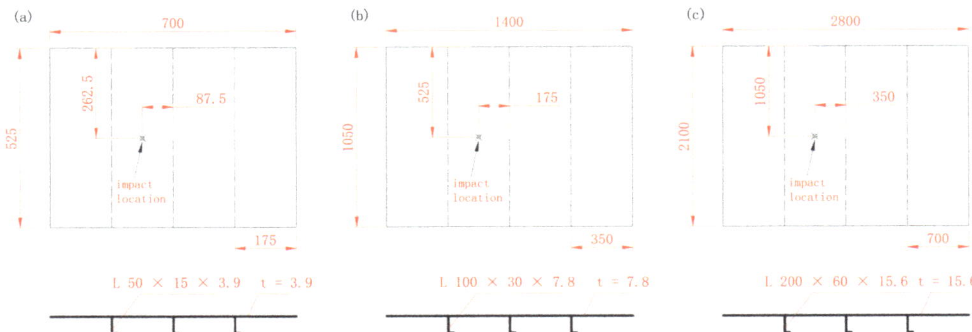

Figure 1. Geometric configurations of the geometrically similar stiffened plates (unit: mm): (**a**) 1:4 scaled; (**b**) 1:2 scaled; (**c**) full-scaled ("L" means that the stiffeners are angle steel with "L" shaped cross-section, "t" means thickness of the plate).

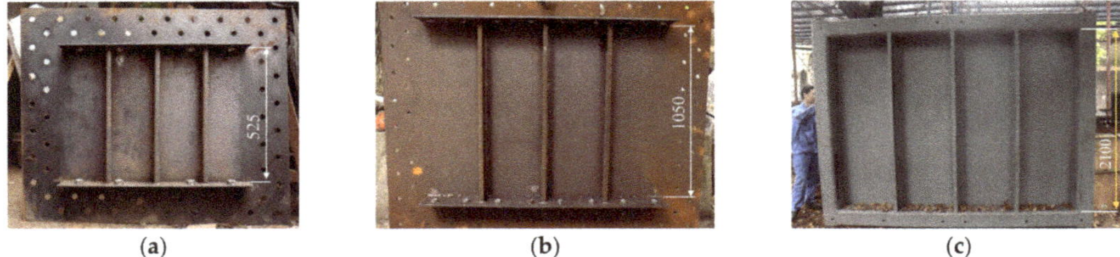

Figure 2. Photographs of the specimens (unit: mm): (**a**) 1:4 scaled model; (**b**) 1:2 scaled model; (**c**) full-scaled prototype.

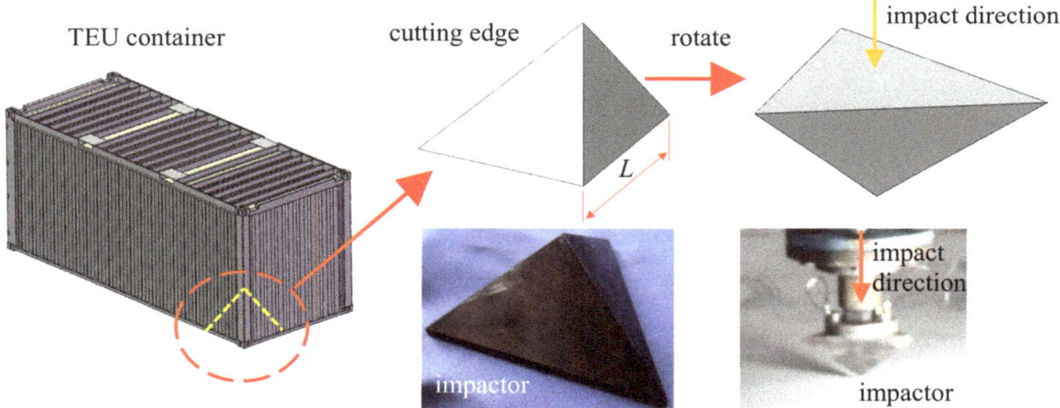

Figure 3. TEU (Twenty feet Equivalent Unit) container and the triangular pyramidic impactor.

3.2. Experimental Set-Up

The low velocity impact was carried out by the weight drop impact test tower. As shown in Figure 4, the test specimen was located at the bottom of the tower. In order to prevent the specimen from bounding or sliding away, the specimen was clamped by a dozen of bolts located around the stiffened plate during the impact process. The impactor was guided by two vertical rails and consisted of three parts: an impactor, a force transducer and the mass. During the impact process, the initial impact energy was achieved by adjusting the drop height and the mass property of the impactor; contact force versus time history could be achieved by the piezo-electric force transducer connected to the impactor under a sampling frequency of 40 kHz. The vertical displacement of the impactor was tested by the laser displacement transducer, and furthermore, the initial impact velocity was calculated by the differential of displacement. Then, energy absorption during the impact procedure could be calculated by trapezoidal numerical integration of the force–displacement curve. The piezo-electric force transducer, with a code number of L1100–909543, was fabricated by Xiyuan Electronic Technology Co., Ltd, Yangzhou, China. The laser displacement transducer, with a code number of HG–C 1400, was fabricated by Panasonic Co., Ltd, Kasugai, Japan.

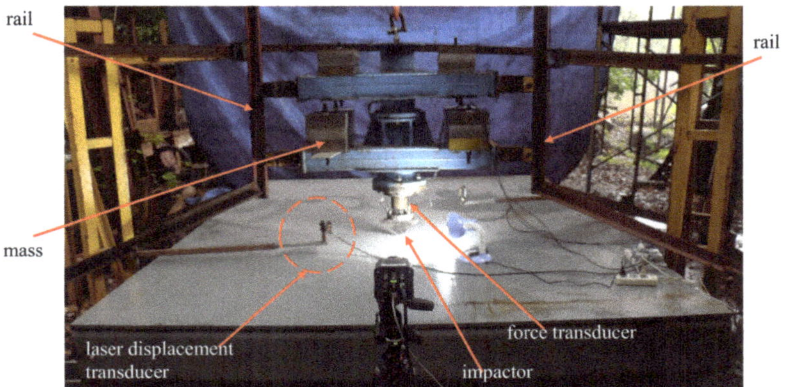

Figure 4. Weight drop impact test tower.

3.3. Finite Element Method

For the numerical simulation of low velocity impact, the commercial package ABAQUS/Explicit with a non-linear explicit algorithm was implemented. As shown in Figure 5, the numerical model consisted of two parts, the impactor and the specimen. The triangular pyramidic impactor was modeled as a rigid body, and a concentrated mass point was defined to simulate the mass of the impactor. The specimens with different scaling factors were modeled by shell model. Marinatos [8] suggested that the aspect ratio of the elements was kept as close as possible to 1:1, and $l_e/t \leq 1$ (where l_e means the element size, and t means the shell thickness) led to better reproduction of the experimental results; thus, S4R elements (4-node doubly curved shell reduced integration elements) with sizes of 4 mm, 8 mm and 16 mm were used to discretize the specimens in the 1:4 scaled model, the 1:2 scaled model and the full-scaled prototype, respectively. In the numerical models, an automatic surface to surface contact strategy provided by ABAQUS/Explicit (Version 6.11, created by Dassault System, permission from Huazhong University of Science and Techonolgy, Wuhan, China) was selected to simulate the contact problem. The friction coefficient was set to 0.3 for the tangential contact surfaces, which was proven to be efficient to match with the experimental results in previous studies [9,10]. Boundaries of the stiffened plate were modeled as fully clamped, whereas the freedom of the pyramidic impactor was restricted except in the vertical impact direction. Furthermore, a predefined field built into the software was adopted to define the initial impact velocity of the impactor.

The built-in ductile damage model built into the ABAQUS software was employed to characterize the mild steel material. Tensile tests were conducted under the guidance of GB/T 228-2002, a metallic materials–tensile test at ambient temperature. The uniaxial tensile tests were implemented by a universal tension test machine with a code number of WAW–600 E, fabricated by Chuance Test Machine Co., Ltd, Jinan, China. A dog-bone specimen shown in Figure 6 was cut down from the plate by wire electrical discharge, and uniaxial tensile curves of the specimens were carried out by the universal tension test machine under a tension velocity of 3 mm/min.

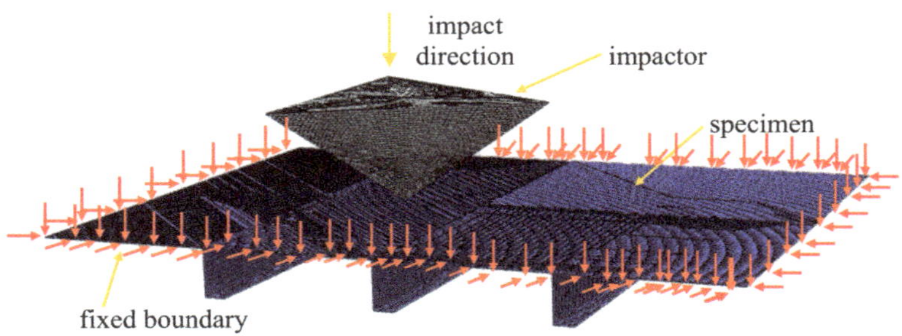

Figure 5. The finite element model of low velocity impact (the red arrows mean that the freedoms of the boundary in three directions have been fixed).

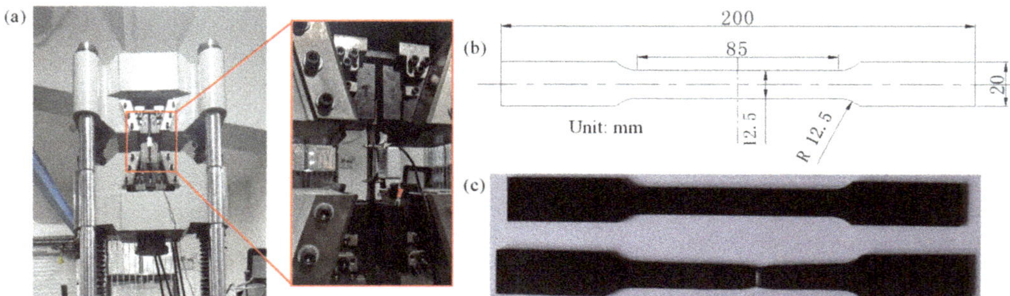

Figure 6. Quasi-static tensile test of the steel plates: (**a**) tensile test machine; (**b**) geometric configurations of the tensile specimen; (**c**) photograph of the tensile test specimen.

Based on the tensile test results, a combined material relationship was adopted to express the true stress-strain curves. The true stress-strain relationship before necking was proposed as follows:

$$\varepsilon_t = ln(1 + \varepsilon_n), \ \sigma_t = \sigma_n(1 + \varepsilon_n) \tag{9}$$

where ε_n and σ_n represent the nominal strain and nominal stress, respectively, and ε_t and σ_t represent the true strain and true stress, respectively. Since fracture strain defined in the finite element software was larger than that obtained by the tensile test, the true stress-strain relationship $\sigma_t - \varepsilon_t$ beyond necking was expressed by using upper power law and lower power law [11]:

$$\sigma_t = \sigma_t^0 \left[0.5 \times \left(1 + \varepsilon_t - \varepsilon_t^0 \right) + 0.5 \times \left(\varepsilon_t / \varepsilon_t^0 \right)^{\varepsilon_t^0} \right] \tag{10}$$

where σ_t^0 and ε_t^0 represent the true stress and true strain at the necking point [10]. Nominal stress-strain curves and true stress strain curves of the plates with different thicknesses are shown in Figure 7. Then, the material properties of the steel were defined in the ABAQUS software. The elastic module of the steel is defined as 210 GPa, a density of 7800 kg/m^3 and a Poisson's ratio of 0.3, and plastic behaviors of the steel were defined by the curves in Figure 7.

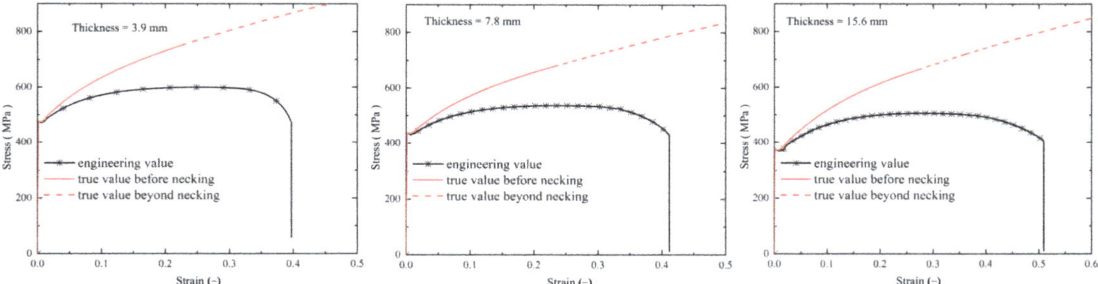

Figure 7. Stress-strain curves of the plates.

A ductile failure criterion built into the ABAQUS software, which simulated the damage initiation and propagation of the ductile mild steel with acceptable accuracy [12,13], was adopted to simulate failure behavior. In this criterion, damage initiation and evolution were determined by an indicator, which was a function of the equivalent plastic strain and stress triaxiality. Once damage initiation occurred at a certain point, the material stiffness degradation occurred at that point correspondingly. When the stiffness degradation reached a critical value, the material point was assumed to fail, and the corresponding element was deleted. In the low velocity impact scenarios, such as ship collision and falling object impact, the impact velocity was far lower than the wave speed in the mild steel, and the strain rate effect slightly affected the dynamic responses. Thus, the strain rate effect was neglected in the numerical simulation of the impact process [10,14].

4. Results and Discussion

In terms of the impact force response and the deformation mode, the numerical simulation results match well with the experimental results. In both the experimental test and the numerical simulation, the initial impact energy was adjusted by adjusting the mass property and the impact velocity; details were listed in Table 2. As shown in Figures 8 and 9, the plates were torn by the sharp rigid impactor and the deformed shape was similar to the shape of the impactor. Due to the sharp corner of the rigid impactor, firstly, the plates underwent a premature fracture, then were torn with three cracks. Due to the strengthening effect of the stiffener, the damage area was located between two stiffeners next to the impact point, and the stiffeners did not suffer obvious plastic deformation. Figure 10 presents the impact force-displacement curves, and the numerical simulation results match well with the experimental test results. As the impactor fell, the impact force raised rapidly until the premature fracturing of the plate; then, the force raised with the tearing of the plate; at the last step, the impact rebounded due to the elastic energy of the stiffened plate.

Table 2. Impact velocity and energy in impact test.

Scaling Factor	Mass (kg)	Impact Velocity (m/s)	Initial Impact Energy (J)
1:4	362.2	5.00	4528
1:2	502.2	8.75	19,224
1:1	795.0	11.30	50,757

The similarity rules were verified by comparing dimensionless force-displacement relationships of models with different scaling factors. Yield stress, ultimate stress and flow stress of the plates with different thicknesses were listed in Table 3; thus, the curves shown in Figure 10 were converted to non-dimensional forms by the Equation (7), and the non-dimensional curves were shown in Figure 11. Since agreement between the FEA results and the impact test results were achieved in Figure 10, only the simulation results were listed in Figure 11 for the purpose of clear and concise expression. It was shown that the three curves coincide with each other perfectly; the maximum impact displacement and the residual plastic deformation significantly decreased with an increase in the scaling factor. The main reason for the differences was that the 1:2 scaled model and full-scaled prototype underwent a relatively "low energy impact" compared with the 1:4 scaled model. For the 1:4 scaled model, 1:2 scaled model and full-scaled prototype, the initial dimensionless impact energy, which were calculated by Equation (8), were 124.1, 73.7 and 26.4, respectively. The main reason for adopting a relatively "low energy impact" for the 1:2 scaled model and the prototype was that the impact tower cannot withstand such a high energy impact test. For instance, the initial energy should be set to 2.38 MJ for the full-scaled prototype under a dimensionless initial energy of 124.1.

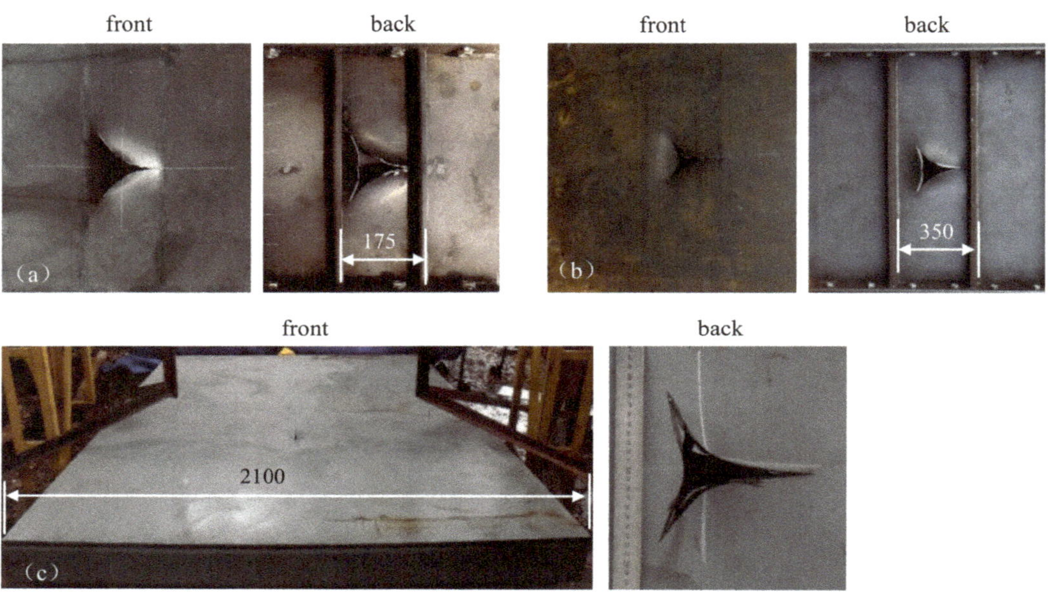

Figure 8. Damage mode of the three specimens in the impact test (unit: mm): (**a**) 1:4 scaled; (**b**) 1:2 scaled; (**c**) full-scaled.

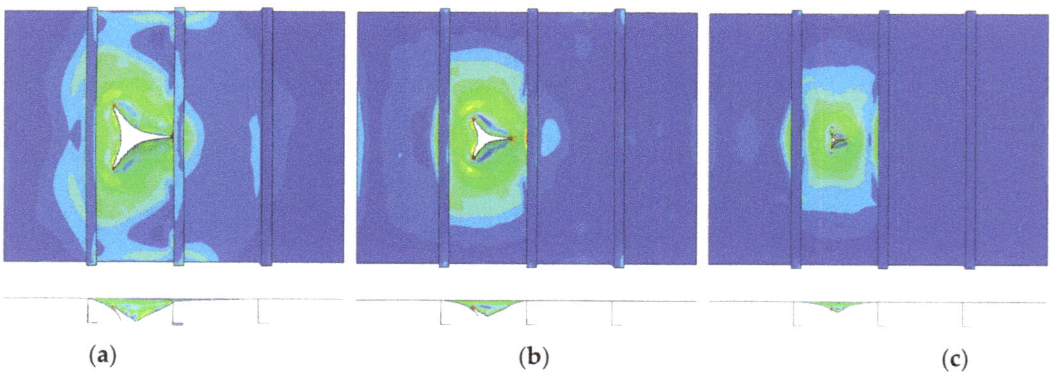

Figure 9. Damage mode of the three specimens in the numerical simulation: (**a**) 1:4 scaled; (**b**) 1:2 scaled; (**c**) full-scaled.

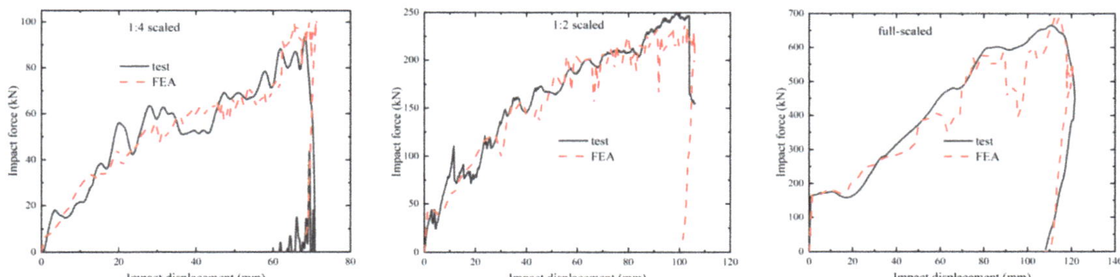

Figure 10. Comparison of impact force-displacement relationships.

Table 3. Necessary stress of the materials gained by a tensile test (unit: MPa).

Shell Thickness	Yield Strength	Ultimate Strength	Flow Stress
3.9 mm	473	758	615
7.8 mm	433	667	550
15.6 mm	376	635	506

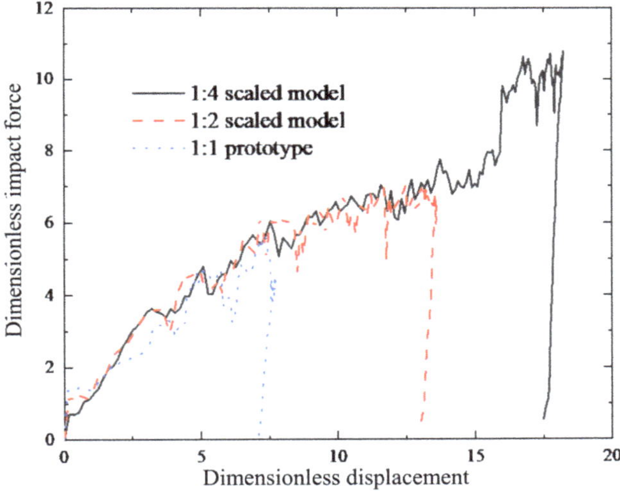

Figure 11. Comparison of the dimensionless force-displacement curves.

For further investigation, high energy impact scenarios with the same dimensionless initial energy were carried out by the FEA, and agreement was achieved. In the numerical models, the dimensionless initial impact energy was set to 124.1 under the same impact velocity of 5 m/s, which means that the mass property of the impactor was adjusted to fulfill the demands of the initial impact energy. More details in the finite element model were listed in Table 4. The calculated results of dimensionless force-displacement curves were shown in Figure 12; there were slight fluctuations in the curves mainly due to elastic waves in the structure and the element deletion strategy adopted in the finite element model. To avoid unexpected distortion of the elements, the failed elements were deleted during the simulation process. Thus, a slight drop–down of the impact force occurred as soon as an element was deleted. Except for the slight fluctuations, the three curves almost coincided with each other perfectly, which verified the correctness and effectiveness of the similarity laws.

Table 4. Mass property, impact velocity and energy in finite element analysis.

Scaling Factor	Mass (kg)	Impact Velocity (m/s)	Initial Impact Energy (J)	Initial Dimensionless Energy
1:4	362.2	5.00	4528	124.1
1:2	2591.4	5.00	32,393	124.1
1:1	19,072.3	5.00	2,384,038	124.1

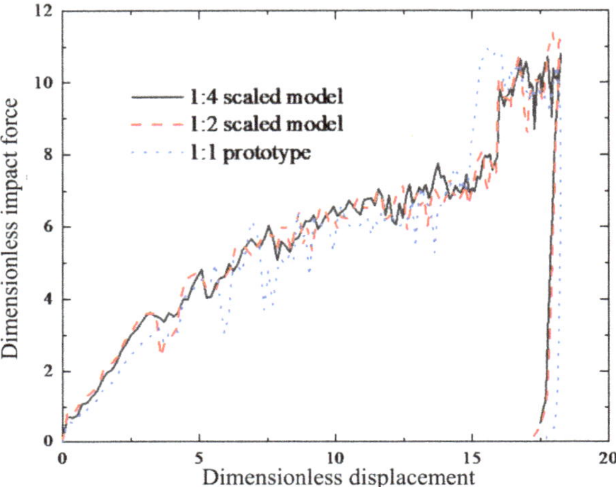

Figure 12. Comparison of dimensionless force-displacement curves under the same dimensionless impact energy.

5. Conclusions

Similarity laws of scaled models of offshore platform deck structures under low velocity impact of dropping objects were conducted in the present research. The similarity laws of scaled models with different scaling factors were proposed. The dynamic responses of three geometrically similar stiffened structures with scaling factors of 1:4, 1:2, and 1:1 under the dropping impact of a triangular pyramidic impactor were investigated by experimental and numerical methods, and the useful conclusions are listed below:

(1) Similarity laws between scaled models and prototypes of stiffened structures under low velocity impact were established, in forms of dimensionless factors including dimensionless force and displacement, with consideration of flow stress of the different plates.

(2) Finite element results and experimental tests show that the dimensionless force–displacement curves of different models match well, which show the effectiveness of the similarity law.

(3) Stiffened plates of an offshore platform deck would suffer a premature fracture under the impact of a sharp triangular pyramidic impactor; the structure could still withstand impact energy absorption after fracture initiation due to tearing of the plate.

The present research provides useful insight into the similarity laws of dynamic responses of deck structures subjected to falling object impact, and were used in the crashworthiness research of offshore structures.

Author Contributions: Conceptualization, H.Z., J.L.; methodology, Y.H., Y.Z., W.L., J.L.; investigation, R.Y., H.Z.; writing—original draft, R.Y., H.Z.; writing—review and editing, R.Y.; funding acquisition, J.L., R.Y. All authors have read and agreed to the published version of the manuscript.

Metals **2022**, 12, 657

Funding: This research was funded by the Doctoral Research Start-up Fund of Wuxi Institute of Technology, Grant No. BT2022-03, and National Natural Science Foundation of China, Grant No. 52071150.

Institutional Review Board Statement: Not applicable.

Informed Consent Statement: Not applicable.

Data Availability Statement: The data presented in this study are available upon request from the corresponding author.

Conflicts of Interest: The authors declare no conflict of interest.

References

1. Jonas, W.R.; Amdahl, J.; Chen, B.Q. MARSTRUCT benchmark study on nonlinear FE simulation of an experiment of an indenter impact with a ship side-shell structure. *Mar. Struct.* **2018**, *59*, 142–157.
2. Zhang, S.R.; Villavicencio, R.; Pedersen, P.T. Ship collision damages: Case studies. In *Developments in the Collision and Grounding of Ships and Offshore Structures*; Soares, G., Ed.; Taylor and Francis Group: London, UK, 2020; pp. 17–23.
3. Gruben, G.; Solvernes, S.; Berstad, T. Low-velocity impact behaviour and failure of stiffened steel plates. *Mar. Struct.* **2017**, *54*, 73–91. [CrossRef]
4. Zhang, M.; Sun, Q.B.; Liu, J.X. A study of the rupture behavior of a ship side plate laterally punched by a full-shape bulbous bow indenter. *Ocean Eng.* **2019**, *182*, 48–60. [CrossRef]
5. Jones, N. *Structural Impact*; Cambridge University Press: Cambridge, UK, 2011.
6. Calle, A.G.; Oshiro, R.E.; Alves, M. Ship collision and grounding: Scaled experiments and numerical analysis. *Int. J. Impact Eng.* **2017**, *103*, 195–210. [CrossRef]
7. Zhang, S.M. Plate tearing and bottom damage in ship grounding. *Mar. Struct.* **2002**, *15*, 101–117. [CrossRef]
8. Marinatos, J.N.; Samuelides, M.S. Towards a unified methodology for the simulation of rupture in collision and grounding of ships. *Mar. Struct.* **2015**, *42*, 1–32. [CrossRef]
9. Liu, B.; Villavicencio, R.; Zhang, S. A simple criterion to evaluate the rupture of materials in ship collision simulations. *Mar. Struct.* **2017**, *54*, 92–111. [CrossRef]
10. Cheng, Y.; Liu, K.; Li, Y. Experimental and numerical simulation of dynamic response of U-type corrugated sandwich panels under low-velocity impact. *Ocean Eng.* **2022**, *245*, 110492. [CrossRef]
11. Yun, L. Uniaxial True Stress-Strain after Necking. *AMP J. Technol.* **1996**, *5*, 37–48.
12. Khodadadian, A.; Noii, N.; Parvizi, M.; Abbaszadeh, M.; Wick, T.; Heitzinger, C. A Bayesian estimation method for variational phase-field fracture problems. *Comput. Mech.* **2020**, *66*, 827–849. [CrossRef] [PubMed]
13. Noii, N.; Khodadadian, A.; Ulloa, J. Bayesian inversion for unified ductile phase-field fracture. *Comput. Mech.* **2021**, *68*, 943–980. [CrossRef]
14. Radford, D.D.; Mcshane, G.J.; Deshpande, V.S. Dynamic Compressive Response of Stainless-Steel Square Honeycombs. *J. Appl. Mech.* **2007**, *74*, 658–667. [CrossRef]

Article

Prestressed Steel Material-Allocation Path and Construction Using Intelligent Digital Twins

Zhansheng Liu [1,2], Guoliang Shi [1,2], Jie Qin [3,*], Xiangyu Wang [4,5] and Junbo Sun [6]

1   Faculty of Architecture, Civil and Transportation Engineering, Beijing University of Technology, Beijing 100124, China; liuzhansheng@bjut.edu.cn (Z.L.); shiguoliang@emails.bjut.edu.cn (G.S.)
2   The Key Laboratory of Urban Security and Disaster Engineering of the Ministry of Education, Beijing University of Technology, Beijing 100124, China
3   Architectural Engineering College, North China Institute of Science and Technology, Langfang 065201, China
4   School of Civil Engineering and Architecture, East China Jiao Tong University, Nanchang 330013, China; xiangyu.wang@curtin.edu.au
5   Australasian Joint Research Centre for Building Information Modelling, Curtin University, Perth, WA 6102, Australia
6   School of Design and the Built Environment, Curtin University, Perth, WA 6102, Australia; tunneltc@gmail.com
*   Correspondence: qjcumt@263.net

Abstract: This study is aimed at the fact that material allocation and construction progress cannot be intelligently controlled in the construction of prestressed steel structures. An intelligent planning method of a material-allocation path for prestressed steel-structure construction, based on digital twins (DTs), is proposed. Firstly, the characteristics of material allocation in the process of structural construction are analyzed, and a five-dimensional integrated DT framework for intelligent path-planning is built. Driven by the DT framework, the progress and environmental information of the construction site are collected in real time. At the same time, the field working conditions are dynamically simulated in the virtual model, so as to realize the interactive mapping between physical space and virtual space. In each construction process, by integrating the progress of each process at each construction location, and the storage and allocation of materials, a multidimensional model for the intelligent planning of material allocation is formed. The information fusion of virtual and real space is carried out using an entropy method to analyze the construction buffer time and material allocation time at each location of the construction site. On this basis, combined with the Dijkstra algorithm, the transportation time associated with the path is calculated according to the field distribution of each location. A feasibility analysis is carried out in the virtual model and imported into the field dynamic-marking system. Combined with radio frequency technology to guide material allocation on site, the intelligent planning of the material-allocation path is realized. In this study, taking the construction of the National Speed Skating Pavilion of the 2022 Beijing Winter Olympics as an example, the DT technology and Dijkstra algorithm are applied to the intelligent planning of the material-allocation path. It is fully verified that the intelligent method can effectively coordinate the relationship between schedule control and material allocation.

Keywords: digital twin; Dijkstra algorithm; prestressed steel structure; material allocation; intelligent planning; construction management

Citation: Liu, Z.; Shi, G.; Qin, J.; Wang, X.; Sun, J. Prestressed Steel Material-Allocation Path and Construction Using Intelligent Digital Twins. *Metals* **2022**, *12*, 631. https://doi.org/10.3390/met12040631

Academic Editors: Zhihua Chen, Hanbin Ge and Siu-lai Chan

Received: 1 March 2022
Accepted: 4 April 2022
Published: 6 April 2022

Publisher's Note: MDPI stays neutral with regard to jurisdictional claims in published maps and institutional affiliations.

1. Introduction

With the continuous improvement of construction technology, large-span spatial structures, which are composed of various forms of string beams, cable domes and cable-membrane structures, have been widely used [1]. To cope with the future labor shortage, it is necessary to achieve intelligent management of large stadium construction, so as to enable a less-manned or unmanned construction site. However, due to the large construction volume of prestressed steel structures, it is difficult to intelligently control the construction elements. Currently, the method of integrating multiple elements of construction, establishing a data coordination mechanism, and forming an intelligent construction management mode has become a research hotspot in engineering management [2,3]. This area has attracted the engagement of many experts and scholars in the construction industry. Through efficient management of the construction process, the energy consumption of a project is reduced [4].

Goh et al. [5] conducted a detailed simulation study on modular construction operations. By applying lean production theory, cycle time and process time can be reduced, and process efficiency and labor productivity can be improved. In view of the lack of a unified and transparent quality-information management system in the construction process, Sheng et al. [6] developed a framework based on blockchain to manage quality information. The proposed framework can disperse the management of quality information, so as to realize consistent and safe quality-information management. Moon et al. [7] adapted sensor-based smart insoles to monitor the frequent workload of building materials on construction sites, and concluded that foot pressure during walking could be used to estimate the weight of building materials currently owned by workers. On the basis of work breakdown structure (WBS) and Bayesian Network, You et al. [8] considered the time-sequence relationship and resource constraint conditions between each unit of WBS, and established the critical chain project management Bayesian network model (CCPMBN) with examples. Zhang et al. [9] studied and proposed the integration technology of Building Information Modeling (BIM) and 3D Geographic Information System (3DGIS). This study solved the problems of data sharing and mining utilization among different stages of design, construction, and management, and realized functions from 3DGIS visualization, roaming, and 3D space analysis, to BIM construction management, construction dynamic simulation, and construction schedule overview.

In the aforementioned studies, a series of analysis methods for construction management are proposed and applied in engineering. However, some shortcomings still exist: (1) It is impossible to achieve a virtual–real interaction on the construction site, and to visually guide the construction process from the perspective of virtual space; (2) there is no integration of construction progress, materials and other information elements for comprehensive management, resulting in insufficient intelligence of construction process management. With the development of new-generation information technology and the promotion of industrial information systems, the application of information technology to engineering construction has become a research hotspot [10]. The application of the digital twins (DTs) concept in engineering practice can significantly improve the accuracy and intelligence of structural performance analysis. At the same time, in the construction process, intelligent algorithms are integrated for data analysis and processing to improve the accuracy and efficiency of management [11]. DTs and intelligent algorithms have been widely used in engineering; the integration of the two provides a reference for the improvement of the intelligent level of the construction industry, especially the construction process.

DTs is a technology that makes full use of models, data, and intelligence, and integrates multiple disciplines. It is a digital way to establish a dynamic virtual model of physical entities with multi-dimensional, multi-temporal scales, as well as multi-disciplinary and multi-physical measures to simulate and characterize the properties, behaviors and rules of physical entities in the real environment. It has begun to be applied to intelligent manufacturing, intelligent factories, Smart cities and other fields [12]. Liu et al. [13] proposed a real-time data-driven online prediction method for the operation state of a DTs workshop. They realized the online prediction of the workshop based on continuous transient simulation by integrating real-time data. Dong et al. [14] proposed five key modeling and simulation technologies for DTs in aircraft structure for fatigue-life management, and realized the interactive mapping between physical space and virtual space. This study provided a reference for the systematic research and engineering application of DTs for aircraft structures. Lee [15] proposed a time machine method for DT design and implementation, using historical data in the whole process of integrating DTs with a network physics system. This provides a reference for the research and engineering application of a DT system for aircraft structure. In order to monitor the health status of proton-exchange-membrane fuel cells during operation. Meraghni et al. [16] applied DTs as an intelligent manufacturing technology to establish a set of overall residual service-life prediction systems, and achieved high prediction accuracy. Gopalakrishnan et al. [17] pointed out that manufacturing needs digital transformation, and created a model-based feature information network (MFIN) based on DTs, which realized the digital description of components or systems. Thus, compared to manufacturing, DTs is relatively less used in the construction industry. For the intelligent transformation and upgrading of the construction industry, intelligent algorithms could also be integrated, driven by DTs [18] to achieve closed-loop control of the construction process. In recent years, many scholars in the construction industry have begun to apply intelligent algorithms to solving problems in civil engineering. Intelligent algorithms can extract high-level features from the original data for perceptual decision making, and improve the objectivity and accuracy of information analysis. In order to solve the problem of structural health monitoring and find suitable structural damage identification features, Li et al. [19] used a convolutional neural network to extract structural features and identify damage, which proved the advantages of the convolutional neural network in automatic feature extraction. In order to accurately identify the damage to concrete structures, Xu et al. [20] used acoustic emission technology to monitor the four-point bending failure test of reinforced concrete beams. Moreover, they established a deep belief network (DBN) to train an acoustic emission signal sample set, to improve the accuracy of structural damage identification. Solhmirzaei et al. [21] proposed a data-driven machine learning (ML) framework for predicting failure modes and the shear capacity of ultra-high-performance concrete (UHPC) beams. This framework can identify the failure mode of UHPC beams and simplify the expression used to predict their shear capacity. Valipour et al. [22] used F-ANP to effectively obtain the coupling relationship between the highway PPP project risk factors, which improved the capability of project risk management. Liu et al. [23] used the Dijkstra algorithm to plan an evacuation path, and proposed a DT-driven dynamic guidance method for fire evacuation. This study realized the functions of real-time collection of environmental information, three-dimensional visualization of indoor layout, fire alarm, indoor personnel positioning and evacuation path planning. In addition, the planning algorithm provides new ideas and tools for data analysis and prediction [24]. Therefore, the combination of a planning algorithm and DTs in the construction process can significantly improve the accuracy and efficiency of engineering management.

Intelligent control of the building structure construction process has become a research hotspot in the construction industry. Combining DTs and intelligent algorithms, this paper puts forward the intelligent planning method of a construction-material-allocation path for a prestressed steel structure. In this study, firstly, the characteristics of material allocation for prestressed steel-structure construction are summarized. Based on this, a DT framework for the intelligent planning of a material-allocation path for a prestressed steel-structure construction is built, to realize the comprehensive control of material and progress. Driven by the twin framework, from the perspective of virtual–real interaction, a virtual–real interaction model for path intelligent planning is created, and twin data are formed using the entropy method. The Dijkstra algorithm is used to process the twin data, and the optimized path is imported into the dynamic identification system of the construction site. The 'one thing and one code' of the material are realized using radio frequency technology to guide the material allocation on the site. Based on the above theoretical method, this study takes emergency material allocation in the construction of the 2022 Beijing Winter Olympics National Speed Skating Hall as an example of practical application. The effectiveness of this method is preliminarily verified using practical application.

2. DT-Driven Intelligent Planning Framework for Material-Allocation Paths

In view of the characteristics of material allocation in the construction process of prestressed steel structures, how to effectively control the relationship between construction progress and material allocation has become an important scientific research topic and engineering practice in construction management [25,26]. Driven by the concept of DTs, through the integration of physical field construction and virtual model simulation, a framework for intelligent planning of a material-allocation path is built, so as to carry out construction management accurately and efficiently.

2.1. Characteristics of Material Allocation for Structural Construction

Prestressed steel structures [27] have been increasingly applied in large public buildings due to their advantages of large space, reasonable force, diversified structural forms, and fast construction speed. Therefore, in the construction of prestressed steel structures, some attributes are provided by the large construction volume:

(1)   Linkage: In the process of structural construction, many construction elements such as 'human, machine, material, method, and environment' are involved, and each element is integrated across fields and multi-services, forming a linkage construction system. A change in one construction element, will cause a response from the whole construction system.

(2)   Complexity: In the process of construction management, to realize the macro control of the whole process, it is necessary to sort out the relationship between progress, quality, cost, and safety, especially the coordination between material allocation and construction progress. In addition, the large construction volume of prestressed steel structures and the variety of materials required undoubtedly increase the complexity of construction management.

(3)   Diversity: Most of the construction of prestressed steel structures is located in complicated construction sites, accompanied by multiple allocation paths of construction materials, and faced with the problem of selecting the most appropriate allocation path efficiently and accurately.

In order to ensure that each construction link can have sufficient construction buffer time and avoid the phenomenon of running out of work, this study proposes the use of DT technology to reasonably plan the allocation path of materials. Thus, the coordination between construction progress and material allocation time is realized.

2.2. The DT Framework of Intelligent Path Planning

According to the characteristics of material allocation in the construction process, it is necessary to establish a DT model for the real-time optimization of schedule control and material allocation, to improve the accuracy and intelligence of construction management [28]. The purpose of DTs is to copy the real physical entity using visual virtual space modeling and simulate the dynamic behavior of the entity in the real environment. Through the virtual mapping of entities and their production processes, the performance of products is accurately evaluated, and the production accuracy and efficiency of product development and manufacturing are improved [29]. Driven by the integration of DTs and artificial intelligence, the multi-factor, multi-process and multi-service time-history parallel simulation and virtual–real integrated control of intelligent construction systems can be realized [30]. This study builds a DT framework for the intelligent planning of material-allocation paths for prestressed steel-structure construction based on the concept of DTs, which is shown in Figure 1.

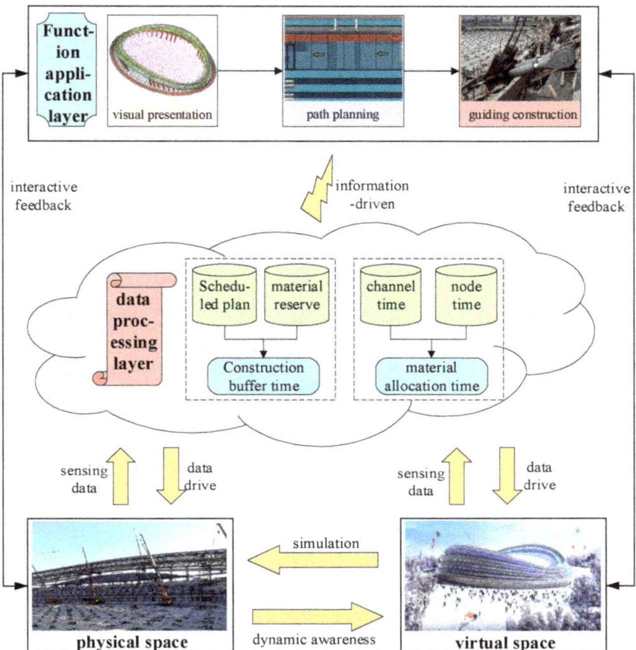

Figure 1. DT framework for intelligent planning of construction-material-allocation path.

The DT framework for the intelligent planning of a material-allocation path for prestressed steel-structure construction is composed of five dimensions, namely: physical space, virtual space, the data processing layer, the functional application layer, and the connection layer among all dimensions. Its mathematical language is expressed as Equation (1).

$$F_{DT} = (PS, \ VS, \ DL, \ FL, \ CL) \tag{1}$$

In Equation (1), F_{DT} represents the DT frame; PS represents the physical space; VS represents the virtual space; DL equals the data processing layer; FL means the function application layer; and CL represents the connection layer among all dimensions.

In the physical space, by capturing the schedule, material reserve and field distribution of each node in the field construction, it provides real working condition support for the simulation in the virtual space. In the virtual space, the site layout model of the construction site is established to truly map the working conditions and layout of the site. At the same time, construction behavior roaming is carried out in the virtual model to simulate the allocation of materials on the site. Additionally, the condition of the site is simulated from the construction state, thus realizing the interactive mapping between the virtual space and the physical space. In the data processing layer—driven by an intelligent algorithm—the buffer time of each construction position (the time when materials can support construction), the time spent on allocating materials in each path, and the feasibility of allocating paths are analyzed. Finally, the visual presentation of the construction site and the planning of the material-allocation path are carried out in the functional application layer, so as to guide the on-site construction. At the same time, the data-driven function is realized through the information extraction of the scene and the simulation analysis of the virtual model. The functional service layer interacts with the virtual and real space, thus connecting the various dimensions of the framework.

Driven by the DT framework for the intelligent planning of a material-allocation path for prestressed steel-structure construction, this study proposes an intelligent planning method. According to the construction site, a virtual model with high fidelity is established, which can simulate the field distribution of the site and the construction state of each position in the virtual space. In order to improve the construction efficiency, the Dijkstra algorithm is utilized to analyze the information extracted from the site and the information simulated from the virtual model. Guided by the dynamic marking system on site, this can intelligently judge the feasibility of allocating materials by each route, and finally, select the most reasonable allocation route. The intelligent planning method of a material-allocation path driven by DTs is shown in Figure 2.

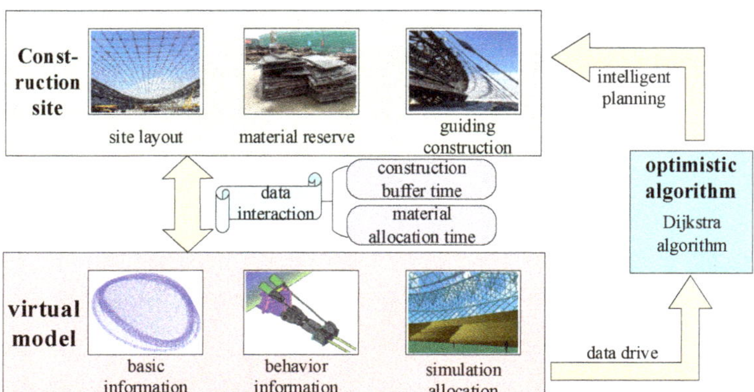

Figure 2. Intelligent planning method of material-allocation path driven by DTs.

3. Creation of Virtual–Real Interaction Model for Intelligent Path Planning

According to the intelligent planning method of a material-allocation path driven by DTs, it is necessary to collect the physical information on site in real time, and carry out the dynamic simulation of site construction in the virtual model. Thus, the interactive mapping between virtual space and physical space is achieved. The virtual–real interaction model for intelligent path planning is built to support the data processing of the Dijkstra algorithm.

3.1. Real-Time Collection of Physical Information

In the planning of a material-allocation path, the collection of physical information mainly includes the construction progress of each node, the material reserve situation, the overall layout of the construction site, and the transportation speed of personnel and equipment on site. The mathematical expression of physical information collected in real time is Equation (2).

$$PI = (CP, \ MR, \ SL, \ TS, \ MT) \tag{2}$$

In Equation (2), PI represents physical information; CP means the construction progress of each node on the site; MR represents the material reserve of each node on site, and the buffer time of each node construction can be calculated according to the material reserve. Incoporating the control of the construction progress of each node, the relationship between construction progress and material reserve can be clarified, and the material allocation time on site can be reduced; SL is the overall layout of the construction site, which can analyze the transportation channel and the blockage of the site; TS equals the transport speed of personnel and equipment on site, and the time required on each path can be calculated by the layout and transport speed of the site; and MT indicates the types of materials required at each construction position, such as cable clamps, anchors, etc.

For the physical information on the scene, Internet of Things technologies, such as monitoring equipment and RFID, can be used for real-time acquisition, and the virtual–real interaction can be realized through dynamic simulation of the virtual model. In particular, with regard to material information collection, RFID technology is used to encode the material, and the basic information—such as the material type and construction location of the material—will facilitate the real-time capture and accurate distribution of information in the allocation process. At the same time, the construction buffer time and material allocation time of each node can be calculated from the collection of field information.

3.2. Dynamic Simulation of Virtual Model

The construction of the virtual model mainly includes the layout of the construction site and the relevant operation information of the construction process, which form the geometric model and behavior model for the construction process. According to the actual construction process on the site, each dimension model is correlated and integrated to realize the deep, multi-angle and comprehensive simulation of the construction site. Through the simulation analysis in the virtual model, the path can be displayed intuitively in the construction site, thus improving the efficiency of construction and the accuracy of management. Therefore, the information in the virtual model (VI) is divided into two categories—basic information (BI) and behavior information (BI^*)—which are simulated in the geometric model and the behavior model, respectively. The specific mathematical language is expressed as Equations (3)–(5).

$$VI = (BI,\ BI^*) \tag{3}$$

$$BI = (SL^*,\ CL,\ CC) \tag{4}$$

$$BI^* = (CT,\ NT,\ CP^*,\ MR^*) \tag{5}$$

In Equations (4) and (5), SL^* represents the site layout in the building; CL represents the layout of the channel in the building; CC means the construction of each node on the channel; CT represents the time used for allocating materials in each channel; NT is the time required to pass through each node on the channel; CP^* indicates the progress of each construction node analyzed in the virtual model; and MR^* equals the material reserves of each construction node analyzed in the virtual model.

In the process of establishing the virtual model, firstly, the basic information on the construction site structure layout, channel layout, and the construction of each node on the channel are modeled at the geometric level. The geometric model is established using a BIM modeling software such as Revit. By establishing a geometric model with high fidelity, the geometric characteristics of the construction process can be truly mapped. Additionally, the actual working conditions of the construction site can be intuitively displayed, which provides field information support for the analysis of the subsequent behavior model [31]. The geometric model can also intuitively show the allocation path after intelligent planning. In the behavior model, combined with the basic information of the scene provided by the geometric model, the time required for the allocation of materials in each channel and the time required for each node in the channel can be analyzed by roaming. In the behavior model, the relationship between the progress of each construction node and the material reserve can also be calculated through the construction simulation. Moreover, the buffer time of construction can be calculated, which provides constraints for the planning of material allocation time and path. The behavior information of the construction process is identified in the geometric model, and the construction of the behavior model provides simulation data support for the path planning in the geometric model. The information dynamic simulation in the virtual model is shown in Figure 3.

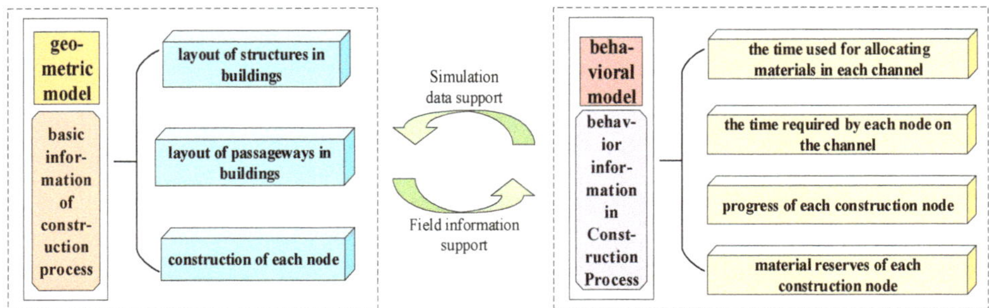

Figure 3. Information dynamic simulation in the virtual model.

3.3. Information Interaction between Physical Space and Virtual Space

Through the real-time collection of physical information and the dynamic simulation of information in a virtual model, twin data are formed. In this process, it is necessary to establish a virtual–real interaction mechanism to support the data processing of the Dijkstra algorithm and realize the intelligent planning of the material-allocation path. The virtual–real interaction mechanism is shown in Figure 4.

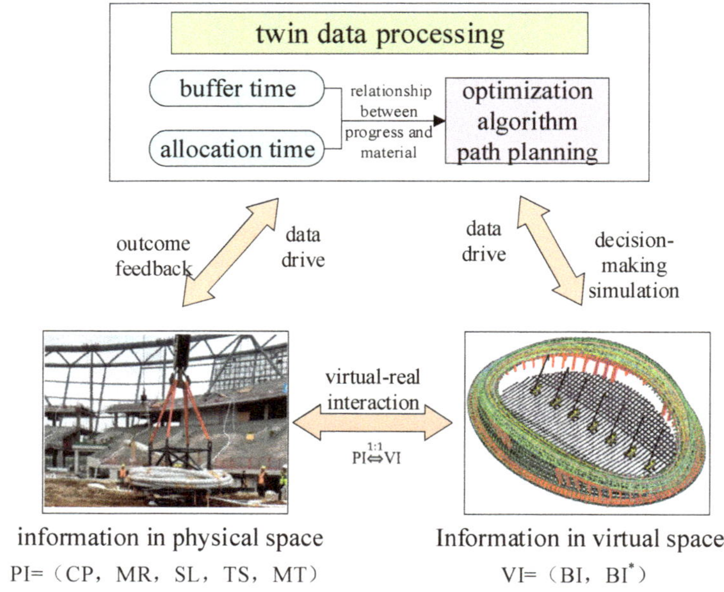

Figure 4. Virtual–real interaction mechanism.

Based on the analysis of the construction site, a one-to-one mapping correlation between virtual space and physical space is realized, namely, $PI \stackrel{1:1}{\Rightarrow} VI$. In the process of the virtual–real interaction, the most important achievement is that of information fusion. In view of the key information such as field layout, schedule, and material reserve in the construction process, sensors and monitors are arranged on the site. Through a high-speed, high-stability and low-delay data transmission protocol (such as HTTP, SMTP, SNMP, FTP, etc.), as well as a wired or wireless mode (such as Zigbee, Bluetooth, WIFI, etc.), a hardware and software guarantee for data transmission is realized. In this study, the data transmission protocol is HTTP, and data are transmitted through Bluetooth and WIFI, which enable the display of construction information in the virtual model, as well as the interactive feedback of virtual and real space. On the basis of virtual–real interaction, the buffer time (BT) of each node's material reserve supports the construction progress; moreover, the time used to allocate materials in each channel (CT), and the time required for each node through the channel (NT), are calculated by the intelligent algorithm in the twin data processing layer. Due to the inaccuracy of field information collection and virtual space simulation, the error of time will also be analyzed. From the practice on site, it is found that the fusion of the time gained using the entropy method can better reflect the construction process. Therefore, the entropy method [32] is used to fuse the information collected in physical space and the information simulated in virtual space, so as to ensure the effectiveness of the data. The original data on physical space and virtual space are expressed as Equation (6).

$$A = \begin{pmatrix} x_{11} & x_{12} \\ \vdots & \vdots \\ x_{m1} & x_{m2} \end{pmatrix} \tag{6}$$

In Equation (6), the data are judged in two ways—physical space and virtual space—and there are m analysis objects.

Firstly, the original data are standardized by Equation (7).

$$X_{ij} = \frac{x_{ij} - \min(x_j)}{\max(x_j) - \min(x_j)} (i = 1, 2, \cdots, m; j = 1, 2) \tag{7}$$

The proportion of the *i*th record under *j*th indicator is calculated by Equation (8).

$$P_{ij} = \frac{X_{ij}}{\sum_{i=1}^{m} X_{ij}} \tag{8}$$

The entropy of the *j*th indicator is calculated by Equation (9).

$$e_j = -k * \sum_{i=1}^{m} P_{ij} * \log(P_{ij}), k = \frac{1}{\ln(m)} \tag{9}$$

The difference coefficient of the *j*th indicator is calculated by Equation (10).

$$g_j = 1 - e_j \tag{10}$$

The weight of the *j*th indicator is calculated by Equation (11).

$$W_j = \frac{g_j}{\sum_{i=1}^{m} g_j} \tag{11}$$

The final fusion result is calculated by Equation (12).

$$X_i = \sum_{j=1}^{2} x_{ij} W_j \tag{12}$$

On one hand, the resulting twin data can directly guide the scene. On the other hand, they can be imported into the virtual model for the feasibility simulation analysis of decision-making, and ultimately provide data support for the intelligent planning of a material-allocation path based on the Dijkstra algorithm.

4. Intelligent Planning of Material-Allocation Path Based on Dijkstra Algorithm

By building a virtual–real interaction model for intelligent path planning, twin data are formed. In this study, the Dijkstra algorithm is used to process the twin data. In the analysis process, the algorithm is improved according to the characteristics of construction and intelligent improvement requirements. The obtained material-allocation path is then returned to the virtual model for feasibility simulation, and finally, the results of the path planning are imported into the dynamic marking system of the site. Combined with the real-time information capture function of RFID technology, the site construction is guided to form a complete intelligent planning process for the material-allocation path.

4.1. Improvement of Dijkstra Algorithm

In this study, the Dijkstra algorithm is selected as the planning algorithm for the shortest path optimization of material allocation, which can calculate the shortest path of any two points in a given planar topology [33]. Its basic principles are:

(1) There are two initialized sets, namely S and U. The set S contains only the source point v, namely S = {v}, and the shortest path of v is 0. The set U contains other nodes except node v, namely U = {S − v}.

(2) Select a nearest node u from the set U to join the node u in the set S, then the selected distance is the shortest path length from v to u.

(3) Taking node u as the new intermediate point, the shortest path length of each node j in the set U is modified. If the shortest path length (passing node u) from source point v to node j (j∈U) is shorter than the original shortest path (not passing node u), the shortest path length of node j is modified.

(4) Repeat steps (2), (3) until all nodes are contained in the set S.

The object of material-allocation path planning in this study is the entire construction site. Its topological structure is composed of multi-layer planar topology, and the field distribution of the path and the nodes connecting each path should be considered. Therefore, the traditional Dijkstra algorithm cannot meet the requirements of intelligent planning of the material-allocation path in the construction process. Thus, this study improves the Dijkstra algorithm as follows:

(1) The layout of each layer is extended to three-dimensional space. Combining with the construction characteristics of sports venues, the nodes from various planes are comprehensively considered. Assuming that there are n planes, the overall analysis object is f = (f1, f2, f3, . . . , fn). The projection of each plane is used to realize the comprehensive analysis of the construction site.

(2) In the construction process, the nodes between the paths may be in a state of construction blockage or material stacking. If the allocation path needs to pass through these nodes, the time to pass through these nodes needs to be considered, where the node is in a blockage state, and the time to pass through this node is set to be infinite.

(3) It is necessary to comprehensively consider the relationship between the buffer time and allocation time of each node. Equation (13) ensures that each node does not produce the phenomenon of nesting, and finally, evaluates the feasibility of the material deployment path by Equation (14).

$$BT_i > CT_i + NT_i \tag{13}$$

$$FI = \frac{\sum_{i=1}^{N}(CT_i + NT_i)}{\sum_{i=1}^{N} BT_i} \tag{14}$$

In Equations (13) and (14), BT_i represents the buffer time of each node; $CT_i + NT_i$ indicates the time of material delivery to each node; and FI represents the feasibility analysis index of the material-allocation path; the smaller the value is, the higher the feasibility is.

(4) Since the construction process is dynamic, in the path analysis, the material allocation time should be corrected in real time according to the changes in factors, such as field distribution information, in the virtual model.

4.2. Intelligent Material-Allocation Path Planning Process

Driven by the DT frame, the Dijkstra algorithm is integrated to intelligently plan the material-allocation path of prestressed steel-structure construction. Firstly, the real-time collection of physical information is performed, and the dynamic simulation of the construction process is carried out in the virtual model, thus forming the virtual–real interaction mechanism and extracting the twin data. The Dijkstra algorithm is used to analyze the twin data, and the relationship between the buffer time and allocation time of each node is comprehensively considered to plan the path. The feasibility of material allocation is evaluated, and the most reasonable way is selected. Driven by DTs, the material deployment path is imported into the virtual model for simulation analysis. The material-allocation path of the construction process is imported into the dynamic marking system of the construction site. Since the construction process is dynamic, the construction is guided by changing the guiding route of the construction site. At the same time, the allocation of materials is collected in real time using radio frequency technology to ensure the accuracy of the construction. The process of realizing the intelligent planning of a material-allocation path in the construction process is shown in Figure 5.

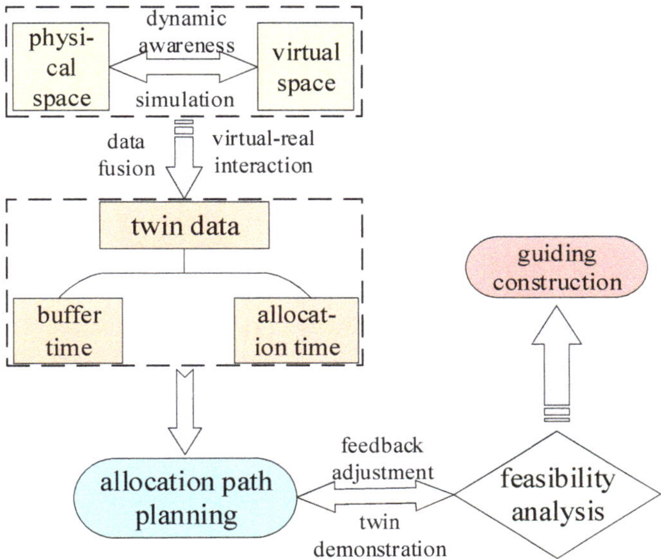

Figure 5. Process of intelligent planning of material-allocation path.

5. Case Study

Based on the analysis of theoretical methods, this study applies the intelligent planning method of a material-allocation path for prestressed steel-structure construction, based on DTs, to the emergency allocation of materials in the construction process of the 2022 Beijing Winter Olympics National Speed Skating Hall; the aim is to improve the efficiency and intelligence of its construction [34].

5.1. Engineering Overview

With a total construction area of 97,000 square meters, the National Speed Skating Pavilion, is located on the west side of Beijing Chaoyang District Olympic Forest Park and to the south side of the National Tennis Center Diamond Stadium. During the 2022 Beijing Winter Olympics, the National Speed Skating Pavilion will undertake speed skating competitions and training. The main structure of the speed skating hall is a cast-in-situ reinforced concrete structure, and the roof is a large-span saddle-shaped cable network structure. The structural span is 124 m × 198 m, supported by the outer steel ring truss, and the curtain wall cable is set outside the ring truss. The steel ring truss adopts the structural form of a three-dimensional truss, and the grid spacing is 4 m. The maximum specification of the chord in the truss is P1600 mm × 60 mm, and the joints are connected in the form of coherent welding. The ring truss and steel-reinforced concrete column are connected by finished spherical hinge support. The external curtain wall support structure adopts a steel cable and vertical wave steel keel. The construction site of the speed skating rink is shown in Figure 6.

Figure 6. Construction site of speed skating hall.

5.2. Virtual–Real Interaction Modeling

In the construction of the structure, because it is difficult for a tower crane to transport material to the connection part of the roof and the lower structure, the process often needs manual participation [35]. In order to improve the efficiency and intelligence, the construction progress and material reserves of important nodes are accurately controlled. In order to improve the accuracy of material allocation, the materials required by each node are collected on the spot and virtually modeled. The real-time interaction is carried out by the sensor equipment. The virtual–real interaction of materials is shown in Figure 7. The basic information such as the type of material, and the behavior information such as the construction position, are determined by the virtual model. The RFID tags are arranged in the real components to read the information in real time, and the management mode of 'one thing and one code' is realized.

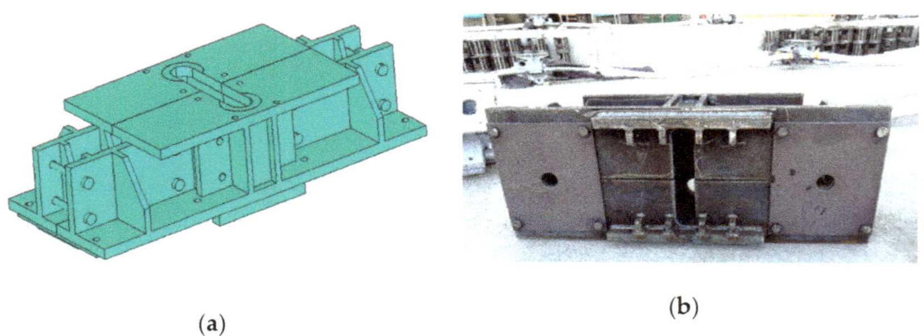

(a)   (b)

Figure 7. Virtual–real interaction of materials. (**a**) Virtual model; (**b**) Real component.

The layout diagram of the construction process analyzed in this study is shown in Figure 8, in which materials are stored at node A, and there is a shortage of materials at nodes D, E, and G. The nodes (A, B, C, D, E, F, G) in the figure represent the positions where the material passes. According to the construction schedule plan, material reserves and field distribution, the construction buffer time of each node is calculated in physical space and virtual space, respectively. Finally, the buffer time and allocation time of each node are fused using the entropy method; thus twin data are formed, which can effectively reduce the error caused by the inaccuracy of field information collection and virtual space simulation, and provide data support for the intelligent planning of the material-allocation path. The twin data of the buffer time and allocation time of each node are shown in Table 1. In the table, a single node indicates that there is a nesting phenomenon of the node, and analyzes its construction buffer time. The nodes are connected to represent the allocation path of the material, and the time includes the time passing through the starting node. Therefore, on the same path, different starting nodes will lead to different allocation times.

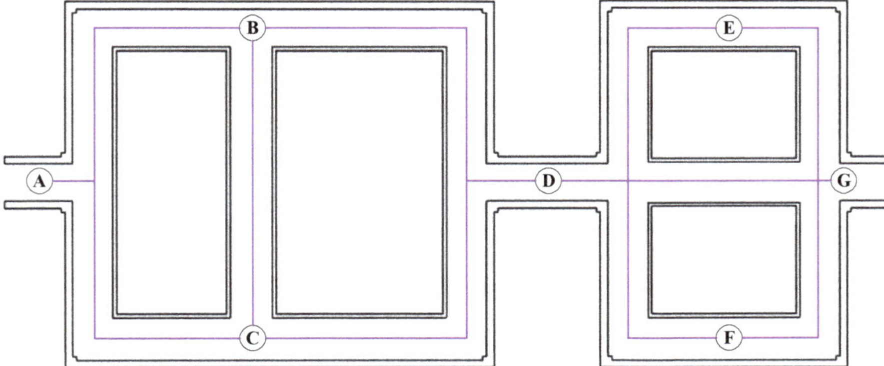

Figure 8. Field layout of construction process.

Table 1. Twin data of buffer time and allocation time of each node.

Location	Time of Physical Space Analysis (min)	Time of Virtual Space Analysis (min)	Fusion Time (min)
D	21.5	23.9	22.7
E	15.2	16.7	16.0
G	30.2	32.8	31.5
A→B	6.5	7.2	6.9
A→C	2.1	2.5	2.3
B→C	2.8	3.1	3.0
C→B	3.2	3.9	3.6
B→D	5.6	6.3	6.0
C→D	7.5	6.8	7.1
D→B	7.4	8.5	8.0
D→C	9.3	10.3	9.8
D→E	4.2	3.7	3.9
D→F	4.3	3.9	4.1
D→G	4.6	5.1	4.9
E→D	7.4	8.2	7.8
E→G	6.5	5.7	6.1
F→D	6.3	5.6	5.9
F→G	5.2	4.9	5.0
G→D	7.6	6.3	6.9
G→E	4.3	3.8	4.0
G→F	7.8	6.5	7.1

5.3. Path Planning

Together with the twin data formed by the virtual–real interaction, the Dijkstra algorithm is used to process the connection of each node, and the time required for the material-allocation path between each node is input. Considering the construction buffer time of each node and the passing time of each route, the planned material-allocation route and time are shown in Table 2 by combining Figure 8 with Table 1.

Table 2. Material-allocation path and time used.

Serial Number	Path	Time Used (min)
1	A→C→D	9.4
2	A→C→D→E	13.3
3	A→C→D→G	14.3

According to Table 1, it is necessary to adjust the path in Table 2. Firstly, the materials are allocated to node E, then returned from node E to node D. Finally, they are transferred from node D to node G, so as to ensure the normal construction of each node on site and the efficiency of the construction process. The formed material-allocation path is simulated in the virtual model, and its effectiveness is analyzed. The formed path is input to the dynamic marking system of the construction site to guide the on-site material allocation. The RFID information of the material is extracted in real time to ensure the accuracy of the allocation. The path planning for material allocation is shown in Figure 9.

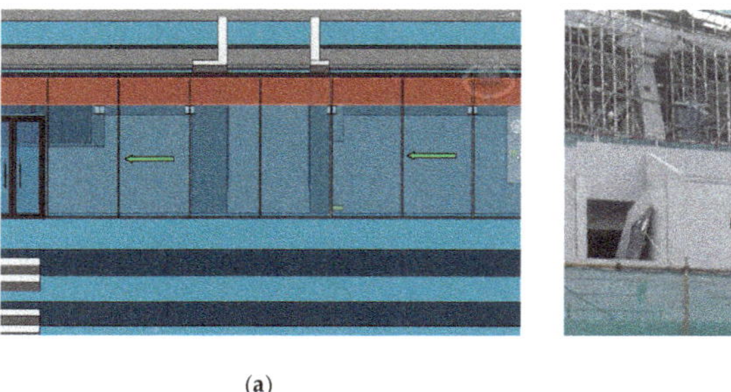

(a)   (b)

Figure 9. Path planning of material allocation. (**a**) Path simulation in virtual model; (**b**) Material allocation on site.

5.4. Effect Analysis

In the framework of the intelligent planning of a construction-material-allocation path driven by DTs, through the real-time collection of physical information on the construction site and the simulation analysis in the virtual model, the virtual–real interaction mechanism is formed. By integrating the Dijkstra algorithm to process twin data, the path of material allocation can be analyzed efficiently and accurately, and the construction of the site can be guided by the dynamic sign system. The material method formed in this study effectively saves time in the construction process and provides ideas for the realization of intelligent construction [36]. In the construction process, a twin model of site layout is established. The node of the construction position is digitized, and different construction channels are used as the distribution path of materials. The material distribution time and the material storage of the construction node are comprehensively considered. The Dijkstra algorithm is used to plan the optimal path. At the same time, the material that needs to be delivered is calibrated. This process requires RFID tags to ensure the consistency of materials. Finally, the dynamic marking system is arranged on the construction site to guide the distribution of materials. By indicating the optimized path on the spot, the delivery time is effectively shortened. For the whole process, material distribution in the same scenario is also carried out using the traditional guidance method. By comparing the time required by the traditional scheme and by the construction scheme proposed in this study, it is shown that the overall time saving of the latter is 21%. The scenario in this study involves the simultaneous delivery of multiple materials. In the research process, the optimization method focuses on time savings. In the future, cost factors will be considered to optimize the construction process and the layout of the working surface. The optimal construction scheme will be obtained by using the fewest dynamic markers. This will not only reduce the construction time, but will also reduce the construction cost.

6. Discussion and Conclusions

DT technology is the key technology for realizing the intelligent construction of structures. A DT framework for the intelligent planning of a material-allocation path for prestressed steel-structure construction—incorporated with the construction characteristics of prestressed steel structures—is built, and is driven by DTs. Moreover, an intelligent planning method of a material-allocation path driven by DTs is formed. According to the real-time collection of physical information on site and the dynamic simulation in the virtual model, a virtual–real interaction model for intelligent path planning is established and twin data are formed. The Dijkstra algorithm is used to process the twin data, and the path of material allocation is obtained. RFID and other Internet of Things technologies are integrated to guide the construction on site. In this study, the main conclusions are as follows:

(1)　Through the real-time collection of physical information and the dynamic simulation of virtual space, and by combining an entropy method, data fusion is carried out to form twin data, thus providing reliable data support for the intelligent planning of the material-allocation path in the construction process.

(2)　Based on the establishment of the virtual–real interaction model, the Dijkstra algorithm is used to process the twin data, and the construction buffer time and material allocation time of each node in the construction process are comprehensively considered; this contributes to a complete intelligent planning process of the material-allocation path, which provides technical support for improving the intelligence level of the construction process.

(3)　Through the fusion of the DT concept and the Dijkstra algorithm, the interactive mapping between virtual space and real construction can be realized. Moreover, the feasibility analysis of path planning and the integration of Internet of Things technology for on-site construction guidance can achieve closed-loop control of the construction process; this also offers a reference for virtual space to guide real construction.

The intelligent planning method of the material-allocation path proposed by this research was applied in the construction of the National Speed Skating Hall of the Beijing Winter Olympics in 2022, and effectively improved the construction intelligence. The examples of its application in engineering showed time savings of 21% for the construction of related processes. Driven by the intelligent planning method of the prestressed steel-structure construction-material-allocation path based on DTs, the intelligent control of structure construction can be carried out through the twin data fusion intelligent algorithm, formed by virtual and real interaction. Moreover, the cost of real structure construction management can be reduced by combining Internet of Things technologies. Future work will focus on the integration of all management elements in each link of the construction process, and on carrying out an intelligent analysis of the whole construction process. On the basis of this study, the coordination of the construction process and work surface will be studied. The efficient management of the whole construction process will be realized by considering more realistic and comprehensive factors.

Author Contributions: Conceptualization, Z.L.; methodology, Z.L.; software, Z.L.; validation, Z.L., G.S., J.Q., X.W. and J.S.; writing—original draft preparation, G.S.; writing—review and editing, Z.L.; project administration, Z.L.; funding acquisition, Z.L. All authors have read and agreed to the published version of the manuscript.

Funding: The research was funded by Intelligent prediction and control of construction safety risk of prestressed steel structures based on digital twin (grant number 5217082614).

Institutional Review Board Statement: Not applicable.

Informed Consent Statement: Not applicable.

Data Availability Statement: The data presented in this study are available upon request from the corresponding author. The data are not publicly available to retain confidentiality.

Acknowledgments: The authors would like to thank Beijing University of Technology, Beijing, China, for their support throughout the research project.

Conflicts of Interest: The authors declare no conflict of interest. The funders had no role in the study's design; in the collection, analysis, or interpretation of data; in the writing of the manuscript; or in the decision to publish the results.

References

1. Krishnan, S. Structural design and behavior of prestressed cable domes. *Eng. Struct.* **2020**, *209*, 110294. [CrossRef]
2. Moghaddam, M.R.; Khanzadi, M.; Kaveh, A. Multi-objective Billiards-Inspired Optimization Algorithm for Construction Management Problems. *Iran. J. Sci. Technol. Trans. Civ. Eng.* **2020**, *45*, 2177–2200. [CrossRef]
3. Wu, L.; Ji, W.; Feng, B.; Hermann, U.; AbouRizk, S. Intelligent data-driven approach for enhancing preliminary resource planning in industrial construction. *Autom. Constr.* **2021**, *130*, 103846. [CrossRef]
4. An, J.; Mikhaylov, A. Russian energy projects in South Africa. *J. Energy S. Afr.* **2020**, *31*, 58–64. [CrossRef]
5. Goh, M.; Goh, Y.M. Lean production theory-based simulation of modular construction processes. *Autom. Constr.* **2019**, *101*, 227–244. [CrossRef]
6. Sheng, D.; Ding, L.; Zhong, B.; Love, P.E.; Luo, H.; Chen, J. Construction quality information management with blockchains. *Autom. Constr.* **2020**, *120*, 103373. [CrossRef]
7. Moon, J.; Lee, A.; Min, S.D.; Hong, M. An Internet of Things sensor–based construction workload measurement system for construction process management. *Int. J. Distrib. Sens. Netw.* **2020**, *16*, 155014772093576. [CrossRef]
8. You, J.X.; Qin, Y. Critical Chain, Project Schedule Management Based on Bayesian Network Model. *J. Tongji Univ. Nat. Sci.* **2015**, *43*, 1606–1612. (In Chinese)
9. Zhang, W.S.; Wu, Q.; Qi, P.L. Integration Technology of BIM and 3DGIS and Its Application in Railway Bridge Construction. *China Railw. Sci.* **2019**, *40*, 45–51. (In Chinese)
10. Liu, Z.; Shi, G.; Jiang, A.; Li, W. Intelligent Discrimination Method Based on Digital Twins for Analyzing Sensitivity of Mechanical Parameters of Prestressed Cables. *Appl. Sci.* **2021**, *11*, 1485. [CrossRef]
11. Morkovkin, D.E.; Gibadullin, A.A.; Kolosova, E.V.; Semkina, N.S.; Fasehzoda, I.S. Modern transformation of the production base in the conditions of Industry 4.0: Problems and prospects. *J. Phys. Conf. Ser.* **2020**, *1515*, 032014. [CrossRef]
12. Tao, F.; Liu, W.; Zhang, M.; Hu, T.; Qi, Q.; Zhang, H.; Sui, F.; Wang, T.; Xu, H.; Huang, Z. Five-dimension digital twin model and its ten applications. *Comput. Integr. Manuf. Syst.* **2019**, *25*, 1–18. (In Chinese)
13. Liu, J.; Zhuang, C.; Liu, J.; Miao, T.; Wang, J. Online prediction technology of workshop operating status based on digital twin. *Comput. Integr. Manuf. Syst.* **2021**, *27*, 467–477. (In Chinese)
14. Dong, L.T.; Zhou, X.; Zhao, F.B.; He, S.X.; Lu, Z.Y.; Feng, J.M. Key technologies for modeling and simulation of airframe digital twin. *Acta Aeronaut. Astronaut. Sin.* **2021**, *42*, 113–141. (In Chinese)
15. Lee, J. Integration of Digital Twin and Deep Learning in Cyber-Physical Systems: Towards Smart Manufacturing. *Manuf. Lett.* **2020**, *38*, 901–910. [CrossRef]
16. Meraghni, S.; Terrissa, L.S.; Yue, M.; Ma, J.; Jemei, S.; Zerhouni, N. A data-driven digital-twin prognostics method for proton exchange membrane fuel cell remaining useful life prediction. *Int. J. Hydrogen Energy* **2021**, *46*, 2555–2564. [CrossRef]
17. Gopalakrishnan, S.; Hartman, N.W.; Sangid, M.D. Model-Based Feature Information Network (MFIN): A Digital Twin Framework to Integrate Location-Specific Material Behavior Within Component Design. *Manuf. Perform. Anal.* **2020**, *9*, 394–409. [CrossRef]
18. Lecun, Y.; Bengio, Y.; Hinton, G. Deep learning. *Nature* **2015**, *521*, 436. [CrossRef]
19. Li, X.S.; Ma, H.W.; Lin, Y.Z. Structural damage identification based on convolution neural network. *J. Vib. Shock* **2019**, *38*, 159–167. (In Chinese)
20. Xu, X.L.; Zhang, Y.; Li, X.H.; Li, Z.J.; Zhang, J.D. Study on damage identification method of reinforced concrete beam based on acoustic emission and deep belief nets. *J. Build. Struct.* **2018**, *39*, 400–407. (In Chinese)
21. Solhmirzaei, R.; Salehi, H.; Kodur, V.; Naser, M.Z. Machine learning framework for predicting failure mode and shear capacity of ultra high performance concrete beams. *Eng. Struct.* **2020**, *224*, 111221. [CrossRef]
22. Valipour, A.; Yahaya, N.; Md Noor, N.; Kildienė, S.; Sarvari, H.; Mardani, A. A fuzzy analytic network process method for risk prioritization in freeway PPP projects: An Iranian case study. *J. Civil Eng. Manag.* **2015**, *21*, 933–947. [CrossRef]
23. Liu, Z.; Zhang, A.S.; Wang, W.S.; Wang, J. Dynamic Fire Evacuation Guidance Method for Winter Olympic Venues Based on Digital Twin-Driven Model. *J. Tongji Univ. Nat. Sci.* **2020**, *48*, 962–971. (In Chinese)
24. An, J.; Mikhaylov, A.; Jung, S.U. Linear Programming approach for robust network revenue management in the airline industry. *J. Air Transp. Manag.* **2021**, *91*, 101979. [CrossRef]
25. Golkhoo, F.; Moselhi, O. Optimized Material Management in Construction using Multi-Layer Perceptron. *Can. J. Civ. Eng.* **2019**, *46*, 10. [CrossRef]
26. Ma, Z.L.; Yang, Z.T. Vehicle Routing Method for Redispatching Parts of Precast Components During Precast Production. *J. Tongji Univ. Nat. Sci.* **2017**, *45*, 1446–1453. (In Chinese)
27. Dong, S.L.; Tu, Y. Structural system innovation of cable dome structures. *J. Build. Struct.* **2018**, *39*, 85–92. (In Chinese)

28. Deng, C. Civil Engineering Management Level Improvement Strategy Based on Information Construction. *IOP Conf. Ser. Mater. Sci. Eng.* **2020**, *780*, 032011.
29. Wang, P.; Luo, M. A digital twin-based big data virtual and real fusion learning reference framework supported by industrial internet towards smart manufacturing—ScienceDirect. *J. Manuf. Syst.* **2021**, *58*, 16–32. [CrossRef]
30. Liu, Z.; Shi, G.; Zhang, A.; Huang, C. Intelligent Tensioning Method for Prestressed Cables Based on Digital Twins and Artificial Intelligence. *Sensors* **2020**, *20*, 7006. [CrossRef]
31. Wu, J.; Sadraddin, H.L.; Ren, R.; Zhang, J.; Shao, X. Invariant Signatures of Architecture, Engineering, and Construction Objects to Support BIM Interoperability between Architectural Design and Structural Analysis. *J. Constr. Eng. Manag.* **2021**, *147*, 04020148. [CrossRef]
32. Salehi, F.; Keyvanpour, M.R.; Sharifi, A. SMKFC-ER: Semi-supervised Multiple Kernel Fuzzy Clustering based on Entropy and Relative entropy. *Inf. Sci.* **2020**, *547*, 667–688. [CrossRef]
33. Mirahadi, F.; Mccabe, B.Y. EvacuSafe: A real-time model for building evacuation based on Dijkstra's algorithm. *J. Build. Eng.* **2020**, *34*, 101687. [CrossRef]
34. Gao, X.; Pishdad-Bozorgi, P.; Shelden, D.R.; Tang, S. Internet of Things Enabled Data Acquisition Framework for Smart Building Applications. *J. Constr. Eng. Manag.* **2021**, *147*, 04020169. [CrossRef]
35. Chan, A.P.C.; Chiang, Y.H.; Wong, F.K.W.; Liang, S.; Abidoye, F.A. Work–Life Balance for Construction Manual Workers. *J. Constr. Eng. Manag.* **2020**, *146*, 04020031. [CrossRef]
36. Wu, C.; Li, X.; Guo, Y.; Wang, J.; Ren, Z.; Wang, M.; Yang, Z. Natural language processing for smart construction: Current status and future directions. *Autom. Constr.* **2022**, *134*, 104059. [CrossRef]

Article

Numerical and Theoretical Investigation on the Load-Carrying Capacity of Bolted Ball-Cylinder Joints with High-Strength Steel at Elevated Temperatures

Jianian He [1], Baolong Wu [1], Nianduo Wu [2], Lexian Chen [1], Anyang Chen [3], Lijuan Li [1,*], Zhe Xiong [1] and Jiaxiang Lin [1]

1 School of Civil and Transportation Engineering, Guangdong University of Technology, Guangzhou 510006, China; jnhe@gdut.edu.cn (J.H.); 2112009068@mail2.gdut.edu.cn (B.W.); chenlexian2022@163.com (L.C.); gdgyxz263@gdut.edu.cn (Z.X.); jxiang.lin@gdut.edu.cn (J.L.)
2 China Construction Eight Engineering Division Corp., Ltd., Shanghai 200112, China; cscec8b@cscec.com
3 Shantou Polytechnic, Shantou 515071, China; aychen@stpt.edu.cn
* Correspondence: lilj@gdut.edu.cn

Abstract: Bolted ball-cylinder (BBC) joints are suitable for non-purlin space structures to effectively reduce structure height and save material costs. In this paper, we present a numerical and theoretical study for high-strength steel BBC joints at elevated temperatures. An finite element (FE) model was first developed, in which the effects of elevated temperatures were considered by introducing reduction factors for the material properties of steel, such as the yield stress and Young's modulus, to analyze the structural behavior of BBC joints subjected to compressive, tensile or bending loads. Based on parametric studies on 441 FE models, effects of the key parameters, including joint dimensions, material strength and temperatures, on the structural behavior of BBC joints are discussed. Then, theoretical analysis is conducted, and design methods are proposed to estimate the ultimate load-carrying capacity of BBC joints. Finally, we verified the accuracy of the design method by comparing the prediction with the FE results.

Keywords: bolted ball-cylinder joints; FE analysis; elevated temperature; parametric study; design methods

Citation: He, J.; Wu, B.; Wu, N.; Chen, L.; Chen, A.; Li, L.; Xiong, Z.; Lin, J. Numerical and Theoretical Investigation on the Load-Carrying Capacity of Bolted Ball-Cylinder Joints with High-Strength Steel at Elevated Temperatures. *Metals* **2022**, *12*, 597. https://doi.org/10.3390/met12040597

Academic Editors: António Bastos Pereira and Diego Celentano

Received: 22 February 2022
Accepted: 28 March 2022
Published: 30 March 2022

Publisher's Note: MDPI stays neutral with regard to jurisdictional claims in published maps and institutional affiliations.

1. Introduction

Space structure that contains axial members and joints is an efficient structural form to span larger areas without interior supports, such as stadiums and community halls [1–3]. Space structure has the advantages of light weight, modular construction, high rigidity and cost effectiveness. The mechanical behavior of joints in a space structure is of vital importance for its structural performance. Both welded and assembled joints are commonly used to connect the axial members. Welded joints such as welded tubular joints and welded spherical joints [4] are ideally assumed as rigid. Nevertheless, assembled joints generally exhibit a semi-rigid behavior. Up to now, various types of assembled joints, such as bolted-ball joints [5], socket joints [6] and plate joints [7], have been invented and investigated. Ahmadizadeh and Maalek [6] investigated the effect of socket joint flexibility on space structure behavior and found that node deformability resulted in significant changes in member forces, structure stiffness and load-carrying capacity. Ebadi-Jamkhaneh and Ahmadi [8] conducted an experimental and numerical study on MERO joints to investigate the effect of bolt tightness on the joint axial behavior. Han et al. [9] investigated the compressive behavior of a novel assembled hub joint in single-layer latticed domes and proposed theoretical models to predict the axial stiffness and strength. Xiong et al. [7,10] studied the behavior of single-layer reticulated shells with gusset joints and found that the semi-rigid behavior of joints has a significant influence on the buckling behavior of the single-layer shells. In general, the behavior of joints greatly affects the performance of a space structure.

In recent years, a new joint type named the bolted ball-cylinder (BBC) joint was proposed for non-purlin space structures [11]. In this structure, roof boards can be directly placed on the upper chords to effectively reduce the structural height and save costs. A bolted ball-cylinder joint consists of a solid semi-sphere and a circular tube. The chords, which are commonly in a rectangular tubular shape, are connected to the circular tube by bolts, and the diagonal members can be connected to the solid semi-sphere [12]. Zeng et al. [13] conducted experimental and numerical studies on the BBC joints subjected to tensile or compressive loads. They found that the strength of the circular tube mainly controlled the load-bearing capacity of the joints.

In recent decades, extensive high-performance and novel constructional materials have been applied to civil engineering [14–17], such as fiber-reinforced polymer, seawater sea sand concrete and high-strength aluminum. For steel structures, with the development of manufacturing techniques, using high-strength steel up to 1000 MPa is a trend in future construction for the efficient utilization of materials. Although several studies have been conducted for BBC joints, their structural behavior at elevated temperatures has not been investigated. It is known that design for fire resistance is a key requirement for steel structures [18]. Elevated temperatures could deteriorate the material properties such as the yield stress and Young's modulus of steel [19,20]. In current design standards (e.g., EC3 [21] and GB 51249 [22]), reduction factors are specified for the yield strength and Young's modulus of steel to account for the effect of elevated temperatures. Although no finite element (FE) studies have been conducted for BBC joints at elevated temperatures, some FE studies have been reported for steel joints at elevated temperatures, such as beam-to-column joints [23], T-joints [24,25] and tubular T-T joints [26].

In order to fill the knowledge gap of the structural behavior of BBC joints made of high-strength steel at elevated temperatures and promote the application of the novel BBC joints, we present here a numerical and theoretical study on BBC joints subjected to various loading scenarios, including compression, tension and bending. An FE model that accounts for the effect of temperatures was developed and verified. Based on FE analysis, parametric study was conducted for joints subjected to various temperatures and loading types to investigate the effects of key parameters on the structural behavior (i.e., strength, stiffness and load–displacement curves) of BBC joints. Finally, we propose design methods for the joints at elevated temperatures.

2. Finite Element Model

2.1. BBC Joint at Ambient Temperature

A bolted ball-cylinder (BBC) joint consists of a circular tube, a solid hemisphere, concave washers, bolts, end plates and tubular beams as shown in Figure 1, where D, t and H are the diameter, thickness and height of the circular tube, respectively; and b_1 and H_0 are the width and height of the tubular beam, respectively. A schematic diagram of each part in a BBC joint is shown in Figure 2. The FE model for a BBC joint at ambient temperature was simulated by the commercial FE software ABAQUS [27].

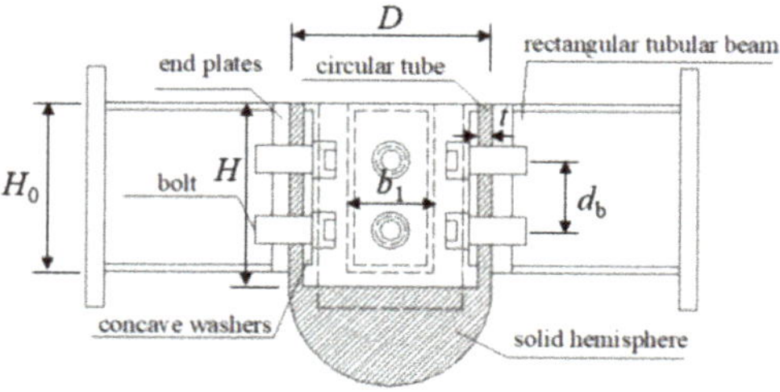

Figure 1. Bolted ball-cylinder joint.

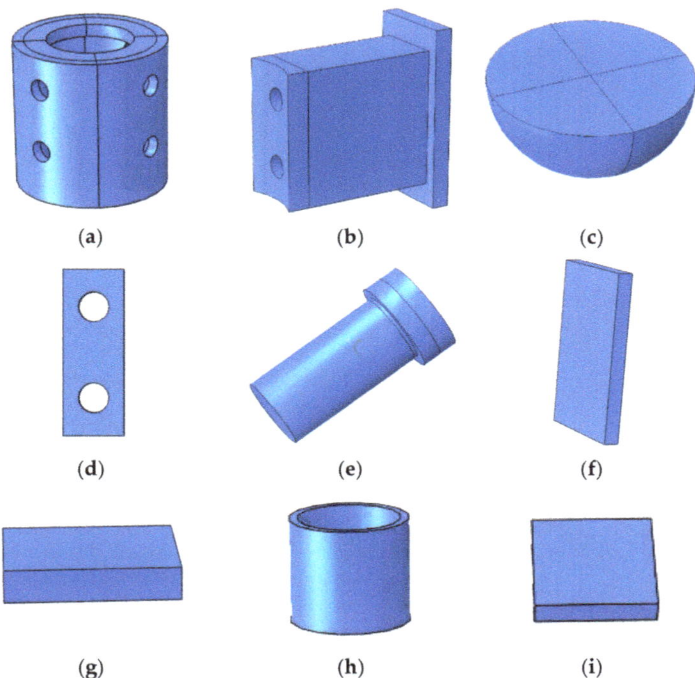

Figure 2. Schematic diagram of each part of BBC joint: (**a**) circular tube; (**b**) tubular beam with end plate; (**c**) solid hemisphere; (**d**) convex washer; (**e**) bolt; (**f**) loading plate; (**g**) supporting plate; (**h**) loading tube; (**i**) loading plate on tube.

Solid element C3D8I was used for the tubular beams and C3D8R was adopted for all the other components. All the parts were meshed by swept or structural meshing techniques to generate hexahedra-shaped elements (Figure 3). The element size for the model was selected as 5 mm, except the element size for the loading tube was 1.5 mm. This mesh size could achieve a balance between analysis accuracy and computing time. The classic metal plasticity model in ABAQUS (Dassault Systemes Simulia Corp. Johnston, RI, USA) was adopted for steel components. This material model implements the von Mises yielding criterion with an isotropic hardening feature. An elastic-perfect plastic model was used for bolts whereas the other components of the joint were modelled by a tri-linear model. The parameters for the material model were obtained from the tensile coupon test in [11], which was conducted in accordance with GB/T 228.1-2010 [28]. The measured yield strength (f_y), ultimate strength (f_u) and Young's modulus (E_S) are 215.7 MPa, 589.7 MPa and 209.5 GPa, respectively. Specially, the tri-linear model consists of a linear elastic portion up to f_y, a hardening portion up to f_u with hardening modulus of 0.05 E_s and a perfect plastic portion with constant stress of f_u. The high-strength bolt was 10.9 grade with nominal yield strength of 900 MPa and ultimate strength of 1000 MPa. For the supporting plate, loading tube and loading plate, the material is set as elastic with a very high modulus of 1000 times of steel Young's modulus.

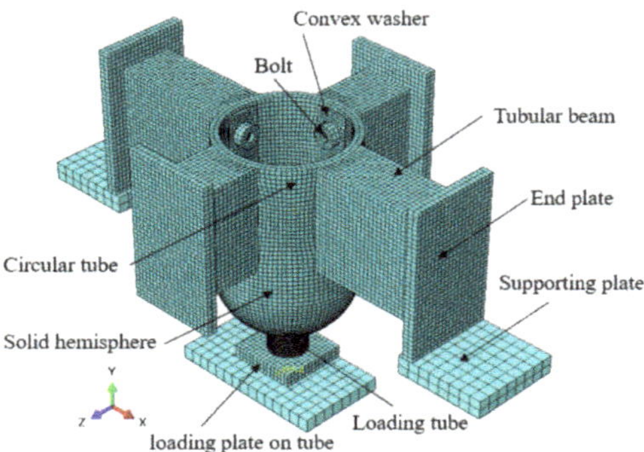

Figure 3. Finite element (FE) model of BBC joint under bending.

Contact was defined in the FE model to simulate the interactions between components (e.g., circular tube, washers and bolts) that prevent penetration but allow separation between them. Contact between the components was defined by "surface-to-surface" contact pairs. "Hard" contact was defined in the normal direction and the penalty friction with small sliding was applied for the tangential direction. The friction coefficient was set as 0.2 and the other parameters were set as default. Depending on the loading scenarios (i.e., compression, tension or bending), boundary conditions were applied on the supporting plates and displacements were applied on the loading plates.

Six analysis steps were added to the FE model for a smooth establishment of the pretension load in bolts and the contacts. The steps were as follows: apply temporary restraints and small force (i.e., 10 N) in bolts; remove the temporary restraints; set boundary conditions and a small displacement of 0.001 mm; modify the force in bolt as 5 kN; fix the bolt length; apply the actual displacement. The analysis was determined due to excessive deformation of the joint that led to a convergency difficulty. Nevertheless, the convergency problem occurred at a very late stage of the loading process and the ultimate capacity could be reached. After completing the analysis, load–displacement curves were obtained by extracting the reaction forces. It is necessary to mention that the preload in bolts is very low to simulate the snug tight condition of the bolts during experiments.

2.2. Verification of FE Model at Ambient Temperature

Three specimens from [11], which were JD5 under compression, JD8 under tension and JD10 under positive bending, were selected to verify the FE model. The dimensions of the specimens for JD5, JD8 and JD10 are D = 140, 180 and 160 mm; t = 10, 8 and 8 mm; and H = 140, 120 and 140 mm, respectively. Size of the tubular beams is $120 \times 60 \times 5$ mm for all the specimens. As shown in Figure 4, concentrical load was applied on the end plate of tubular beams for specimen JD5 and JD8, whereas the load was applied on the endplate of the loading tube. The layout of LVDTs (Linear Variable Displacement Transducers) for the specimens is shown in Figure 4. The recorded displacement for compressive and tensile specimens (i.e., JD5 and JD8) is taken as the relative displacement between point A and A′ (Figure 4). For specimen JD10 under positive bending, the recorded displacement is the average displacement at points B and B′ subtracted by the average displacement at points C and C′. From FE analysis, the simulated deformed shapes of BBC joints agree well with the experimental observations. The difference in the load–displacement curves obtained from FE analysis and the experimental results is generally less than 10%, as shown in Figure 5. Therefore, the FE model could accurately simulate the structural behavior of BBC joints at the ambient temperature.

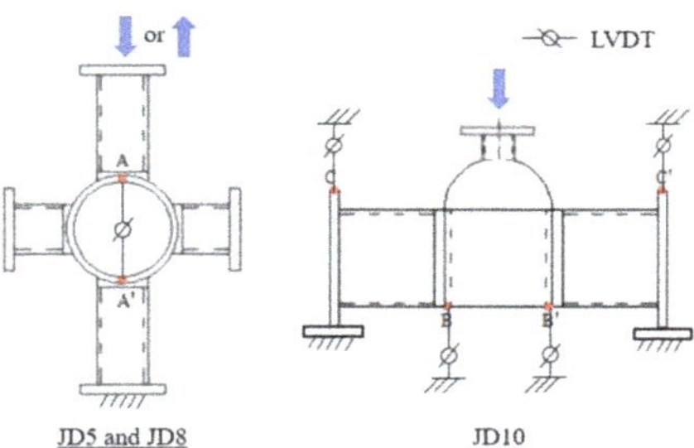

Figure 4. Loading conditions and layout of LVDTs for tested specimens.

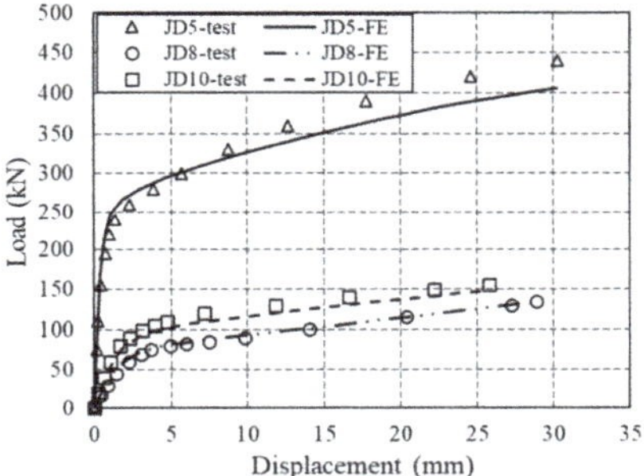

Figure 5. Verification of FE model.

2.3. FE Model Considering Engineering Practice

For experimental specimens in [11], the end plates of the tubular beams were fixed to the actuator and the beam was short. The additional restraints from loading plates led to an increase in the load-carrying capacity. However, in engineering projects, the tubular beam is long and the restraint from the beams is weaker than in the experiment. Based on FE analysis, the additional restraints caused a significant increase in the compressive capacity. Therefore, in parametric study, the loading plates were removed to better simulate the real behavior of the joints in structures.

In the design method for steel structures, yield strength is commonly used as the strength parameter. It is more reasonable to use the elastic-perfect plastic model in FE analysis for a convenient derivation of design methods. It was found that the ultimate capacity of BBC joints adopting elastic-perfect plastic model is lower than that adopting the tri-linear model. Therefore, the derived design method based on the elastic-perfect plastic model has safety margins to some extent. In the parametric study, the elastic-perfect plastic model was adopted for the FE model.

2.4. Constitutive Model for Steel at Elevated Temperature

It is known that the material properties of steel are degraded at elevated temperatures. In the "Code for fire safety of steel structures in buildings" [22], reduction factors η_T and χ_T are introduced to account for the effects of elevated temperatures on the yield strength and Young's modulus of steel, respectively. Yield strength $f_{y,T}$ and Young modulus E_T could be determined by Equations (1) and (2), respectively:

$$\eta_T = \frac{f_{y,T}}{f_y} = \begin{cases} 1.0 & 20\,^{\circ}\mathrm{C} \leq T \leq 300\,^{\circ}\mathrm{C} \\ 1.24 \times 10^{-8}T^3 - 2.096 \times 10^{-5}T^2 + 9.288 \times 10^{-3}T - 0.2168 & 300\,^{\circ}\mathrm{C} < T < 800\,^{\circ}\mathrm{C} \\ 0.5 - \frac{T}{2000} & 800\,^{\circ}\mathrm{C} \leq T \leq 1000\,^{\circ}\mathrm{C} \end{cases} \tag{1}$$

$$\chi_T = \frac{E_T}{E} = \begin{cases} \frac{7T-4780}{6T-4760} & 20\,^{\circ}\text{C} \leq T < 600\,^{\circ}\text{C} \\ \frac{1000-T}{6T-2800} & 600\,^{\circ}\text{C} \leq T \leq 1000\,^{\circ}\text{C} \end{cases} \qquad (2)$$

where f_y is the yield strength and E is the Young's modulus of steel at ambient temperature, and T is the temperature in $^{\circ}$C.

Current standards do not cover the behavior of high-strength bolts in fire. Based on the study in [29], the reduction in the yield strength ($f_{ybolt,T}$) and Young's modulus ($E_{bolt,T}$) of high-strength bolts (10.9 grade) at elevated temperatures could be estimated by Equations (3) and (4), respectively:

$$\lambda_T = \frac{f_{ybolt,T}}{f_{ybolt}} = 4 \times 10^{-9}T^3 - 6 \times 10^{-6}T^2 + 0.0011T + 0.9603 \qquad (3)$$

$$\omega_T = \frac{E_{bolt,T}}{E_{bolt}} = 6 \times 10^{-9}T^3 - 8 \times 10^{-6}T^2 + 0.0016T + 0.9433 \qquad (4)$$

where f_{ybolt} and E_{bolt} are the yield strength and Young's modulus of high-strength bolts at ambient temperature. Figure 6 and Table 1 show the reduction factors for the yield strength and Young's modulus of steel and high-strength bolts at elevated temperatures.

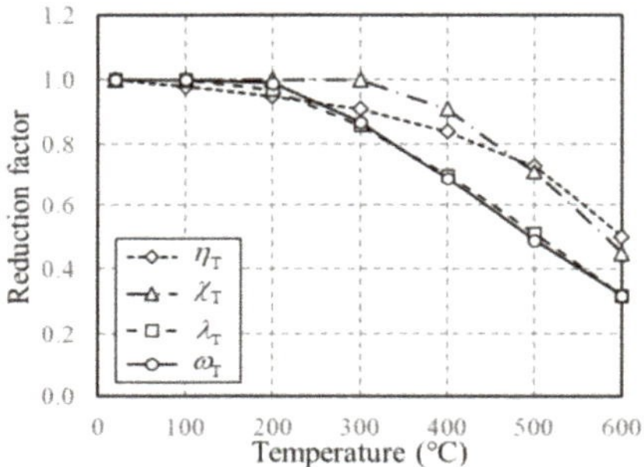

Figure 6. Reduction factors for steel and high-strength bolts at elevated temperatures.

Table 1. Reduction factors for the yield strength and Young's modulus of steel and high-strength bolts.

T (°C)	Steel		High-Strength Bolt	
	η_T	χ_T	λ_T	ω_T
20	1.00	1.00	1.00	1.00
100	0.98	1.00	1.00	1.00
200	0.95	1.00	0.97	0.99
300	0.91	1.00	0.86	0.87
400	0.84	0.91	0.70	0.69
500	0.73	0.71	0.51	0.49
600	0.50	0.45	0.32	0.32

3. Parametric Study

3.1. Specimens

As shown in Figure 7, four types of loads on BBC joints were investigated in the current study, which are compression (denoted as "CN" joint), tension ("TN"), positive bending ("BN") and negative bending ("RBN"). The key parameters affecting the structural behavior of BBC joints include the outer diameter (D), thickness (t) and height (H) of the circular tube, and the width (b_1) and height (H_0) of the tubular beam (Figure 1). Twenty-one FE joints were developed for parametric study to cover these key parameters, as shown in Table 2. For each joint, seven temperatures (i.e., 20 °C, 100 °C, 200 °C, 300 °C, 400 °C, 500 °C and 600 °C) and three steel strengths (i.e., Q235, Q345 and Q690 steel with f_y = 235 MPa, 345 MPa and 690 MPa, respectively) were selected, yielding a total of 441 FE models, to investigate the structural behavior of BBC joints at elevated temperatures. Due to the limitation of article length, the results of Q690 joints are reported in the following subsections and Q235 and Q345 joints exhibited a similar trend as that of Q690 joints.

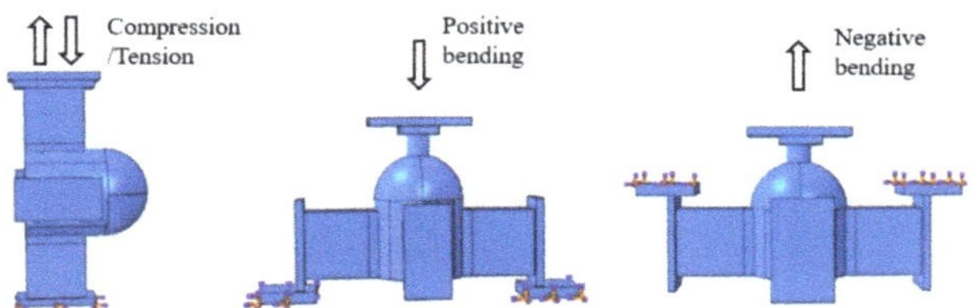

Figure 7. Loading scenarios for BBC joints.

Table 2. Joints for parametric study.

Joint No.	Variable	D (mm)	t (mm)	H (mm)	b_1 (mm)	b_2 (mm)	H_0 (mm)
N1	N/A	150	10	160	60	26	120
N2		110	10	160	60	26	120
N3	D	130	10	160	60	26	120
N4		170	10	160	60	26	120
N5		190	10	160	60	26	120
N6		150	8	160	60	26	120
N7	t	150	9	160	60	26	120
N8		150	11	160	60	26	120
N9		150	12	160	60	26	120
N10		150	10	120	60	26	120
N11	H	150	10	130	60	26	120
N12		150	10	140	60	26	120
N13		150	10	150	60	26	120
N14		150	10	160	50	26	120
N15	b_1	150	10	160	55	26	120
N16		150	10	160	65	26	120
N17		150	10	160	70	26	120
N18		150	10	160	60	26	130
N19	H_0	150	10	160	60	26	140
N20		150	10	160	60	26	150
N21		150	10	160	60	26	160

3.2. Joints under Compression

Load–displacement curves of a typical Q690 specimen N1 under compression with various temperatures are plotted in Figure 8. For joints subjected to temperatures less or equal than 300 °C, the curves are almost identical, indicating that the temperature effect is insignificant. With further increase in the temperature, the initial stiffness and ultimate compressive capacity of a joint decrease gradually. In general, the shapes of the load–displacement curves of the joints subjected to various temperatures are similar. Similar trends could be found for all the other joints (i.e., N2–N21).

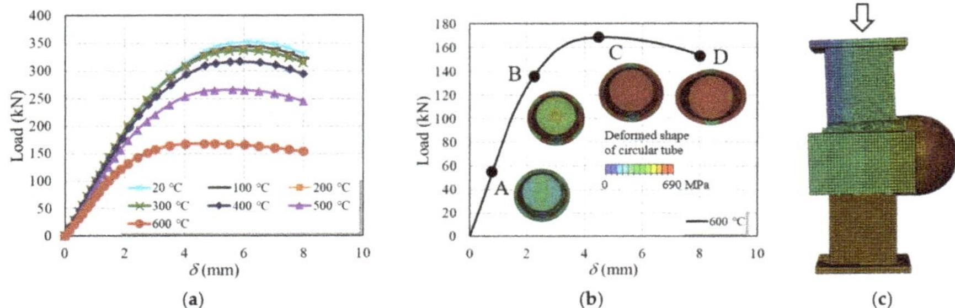

Figure 8. Effects of elevated temperatures on the load–displacement curves of specimen N1 under compression: (**a**) load–displacement curves; (**b**) failure process; (**c**) deformed shape corresponding to point C.

By examining the deformed shapes and stress contours of the joints under various temperatures, the failure mechanisms are similar to those under ambient temperature. Due to compression, the free end of the circular tube in a joint first yields, and the yield zone extends to the hemisphere. At the ultimate load, complete yielding of the free end is observed, and plastic hinges are formed at the tube faces connected to the tubular beams. After the peak load, the applied load decreases slowly. Figure 8b shows the failure process of specimen N1 at 600 °C. The deformed shape of the joint at ultimate load is shown in Figure 8c.

Based on FE analysis, key parameters affecting the structural behavior of a BBC joint include the outer diameter, thickness and height of the circular tube and the width and height of the tubular beam. Parametric studies were conducted to investigate the effects of these key parameters on the initial stiffness and ultimate capacity of Q690 BBC joints, as shown in Figures 9 and 10, respectively. The initial stiffness is defined as the slope of the load–displacement curve in the elastic region. In this paper, it is equal to the slope of the line between the origin and point A as shown in Figure 8b, where the load at point A is generally a quarter to a third of the peak load. The ultimate capacity is defined as the peak load during the loading process (i.e., point C in Figure 8b).

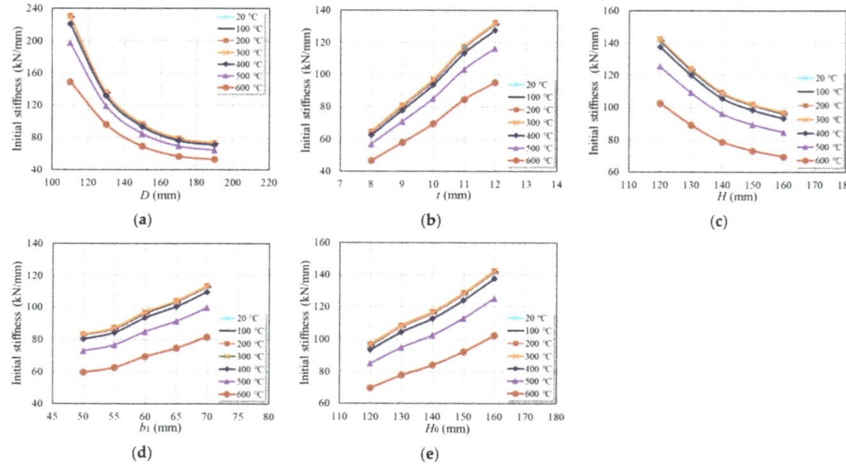

Figure 9. Effects of parameters on the initial stiffness of BBC joints under compression: (**a**) outer diameter of the tube; (**b**) thickness of the tube; (**c**) height of the tube; (**d**) width of the tubular beam; (**e**) height of the tubular beam.

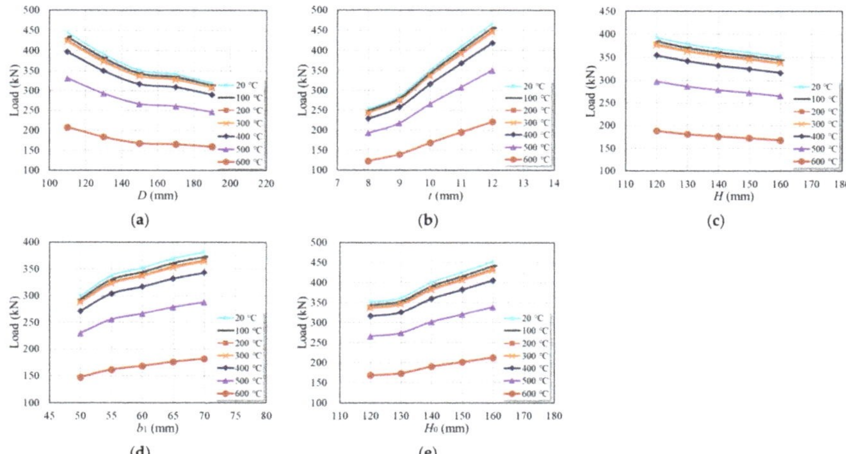

Figure 10. Effects of parameters on the load-carrying capacity of BBC joints under compression: (**a**) outer diameter of the tube; (**b**) thickness of the tube; (**c**) height of the tube; (**d**) width of the tubular beam; (**e**) height of the tubular beam.

With the increase in the outer diameter of the tube, both the initial stiffness and ultimate capacity decrease dramatically first, and then the decreasing rate slows down, as a tube with a larger diameter is more prone to local buckling. Comparison between Figures 9a and 10a indicates that the effect of the outer diameter is more prominent on the initial stiffness than that on the ultimate capacity. A thicker tube also leads to an enhancement in the initial stiffness and ultimate capacity. However, increasing the height of the tube has an opposite effect. Furthermore, an increase in the beam size (i.e., width and height) leads to a gradual increase in the initial stiffness and the ultimate capacity since the stress concentration on the tube wall is less severe. As expected, an elevated temperature reduces the initial stiffness and ultimate capacity of the joint due to the deterioration in material properties. Nevertheless, if the temperature is lower than 300 °C, the negative effect of temperature is insignificant.

3.3. Joints under Tension

Figure 11 plots the load–displacement curves of Q690 specimen N1 under tension with various temperatures. In general, obvious strain hardening is observed for joints under tension as the joint capacity is mainly controlled by material strength. When the temperature is higher than 400 °C, significant reductions in the initial stiffness and the capacity are observed. Nevertheless, the failure modes of the joint under ambient and elevated temperatures are similar, which is a successive yielding of the circular tube. As the applied load did not drop during the loading process, the load at point C in Figure 11b is defined as the ultimate capacity. The ultimate load corresponds to a complete yielding of the circular tube (Figure 11b). From this point, the slope of the load–displacement curve remains constant. The deformed shape of the joint at ultimate load is shown in Figure 11c.

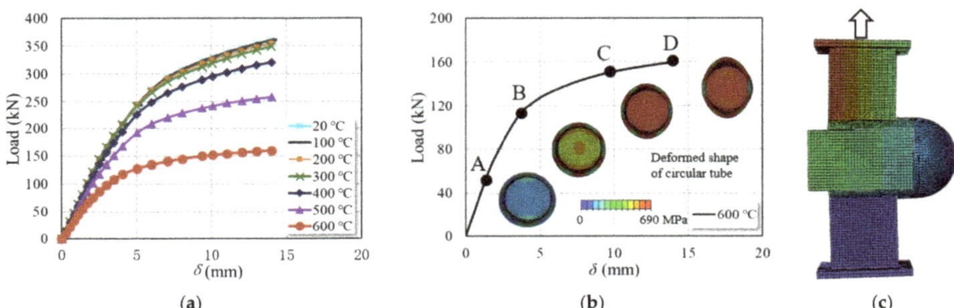

Figure 11. Effects of elevated temperatures on the load–displacement curves of specimen N1 under tension: (**a**) load–displacement curves; (**b**) failure process; (**c**) deformed shape corresponding to point C.

Effects of the key parameters on the initial stiffness and ultimate capacity of Q690 BBC joints are shown in Figures 12 and 13, respectively. With the increase in the outer diameter of the tube, the initial stiffness and ultimate load decrease first and keep almost consistent if the diameter is larger than 170 mm. The initial stiffness and ultimate capacity increase gradually with the increase in tube thickness or with the decrease in tube height. The width of tubular beams does not notably affect the structural behavior of a joint, as show in Figures 12d and 13d. With the increase in the tubular beam height, the initial stiffness increases slightly, but its effect on the ultimate capacity is insignificant.

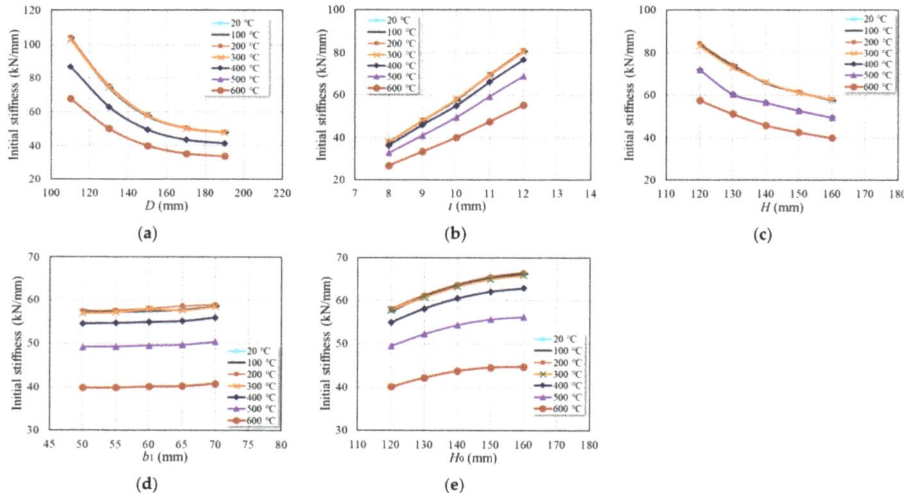

Figure 12. Effects of parameters on the initial stiffness of BBC joints under tension: (**a**) outer diameter of the tube; (**b**) thickness of the tube; (**c**) height of the tube; (**d**) width of the tubular beam; (**e**) height of the tubular beam.

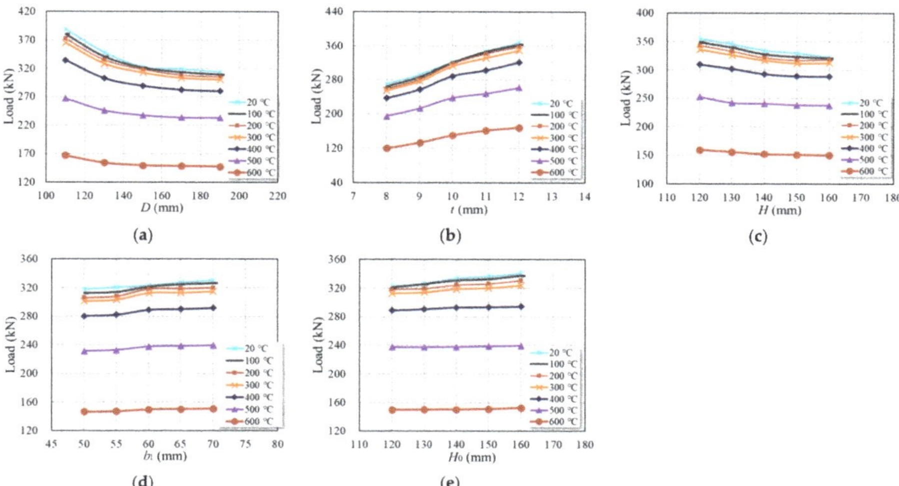

Figure 13. Effects of parameters on the load-carrying capacity of BBC joints under tension: (**a**) outer diameter of the tube; (**b**) thickness of the tube; (**c**) height of the tube; (**d**) width of the tubular beam; (**e**) height of the tubular beam.

3.4. Joints under Positive Bending

Figure 14 shows the effects of elevated temperatures on the load–displacement curves of Q690 specimen N1 under positive bending. Similar to the joints under tension, the applied load did not drop during the loading process, and the ultimate load is defined as the load at point C in Figure 14b. The deformed shape of the joint at ultimate load is shown in Figure 14c. If the temperature is less than 300 °C, the effect of temperature on joint behavior is insignificant. However, with a further increase in temperature, both initial stiffness and ultimate capacity deteriorate dramatically. Failure mode of the joint at an elevated temperature is similar as that at ambient temperature, in which plastic hinges are formed at the tube-to-beam junctions.

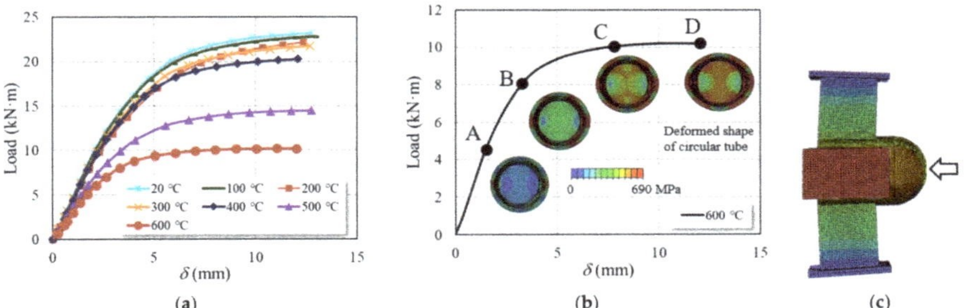

Figure 14. Effects of elevated temperatures on the load–displacement curves of specimen N1 under positive bending: (**a**) load–displacement curves; (**b**) failure process; (**c**) deformed shape corresponding to point C.

As shown in Figure 15, the initial stiffness of Q690 joints increases with the increase in tube thickness and the tubular beam height. Increasing the outer diameter of the tube leads to a reduction in the initial stiffness as the tube is more prone to buckling. Nevertheless, the effects of tube height and tubular beam width on the initial stiffness are insignificant. For the ultimate capacity shown in Figure 16, effects of tube diameter, tube height and beam width are insignificant. The ultimate capacity is enhanced by increasing the tube thickness and the height of the tubular beams.

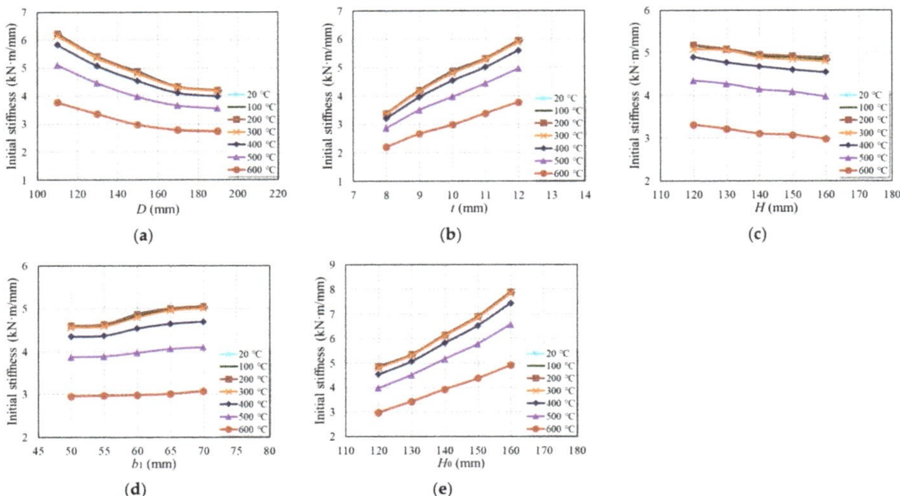

Figure 15. Effects of parameters on the initial stiffness of BBC joints under positive bending: (**a**) outer diameter of the tube; (**b**) thickness of the tube; (**c**) height of the tube; (**d**) width of the tubular beam; (**e**) height of the tubular beam.

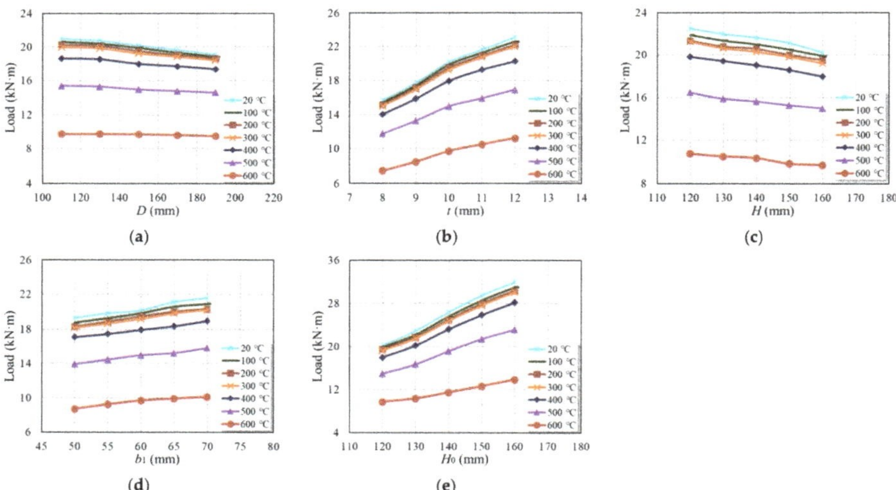

Figure 16. Effects of parameters on the load-carrying capacity of BBC joints under positive bending: (**a**) outer diameter of the tube; (**b**) thickness of the tube; (**c**) height of the tube; (**d**) width of the tubular beam; (**e**) height of the tubular beam.

3.5. Joints under Negative Bending

For Q690 joints under negative bending, the effects of elevated temperatures on the load–displacement curves are illustrated in Figure 17. The applied load almost keeps constant after reaching the peak load and the ultimate load is defined as the peak load (i.e., point C in Figure 17b). The deformed shape of the joint at ultimate load is shown in Figure 17c. Similar to joints under positive bending, the adverse effect of elevated temperatures on the joint behavior becomes obvious when the temperature is higher than 400 °C. For a given joint, the ultimate capacity under negative bending is lower than that under positive bending, as the free end of the circular tube is prone to local buckling.

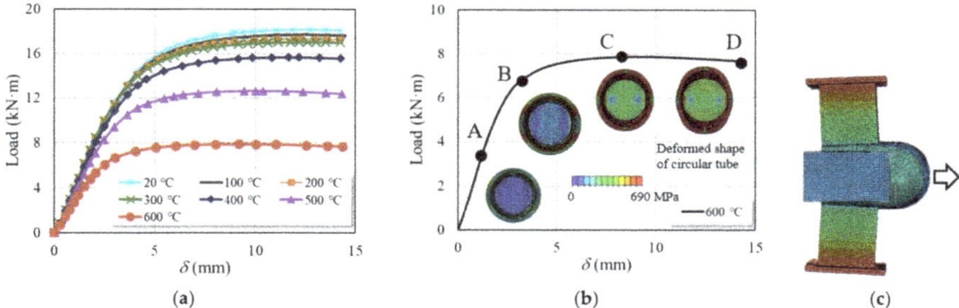

Figure 17. Effects of elevated temperatures on the load–displacement curves of specimen N1 under negative bending: (**a**) load–displacement curves; (**b**) failure process; (**c**) deformed shape corresponding to point C.

As shown in Figure 18, the initial stiffness decreases with the increase in the outer diameter of the tube or with the decrease in tube thickness. The effect of tube height on the initial stiffness is insignificant. A slight increase in the initial stiffness is observed when increasing the beam size (i.e., width and height). Figure 19 shows the effect of the key parameters on the ultimate capacity. In general, an increase in tube thickness of beam width leads to an increase in the ultimate capacity. However, the influences of other parameters such as tube diameter, tube height and beam height on the ultimate capacity are insignificant.

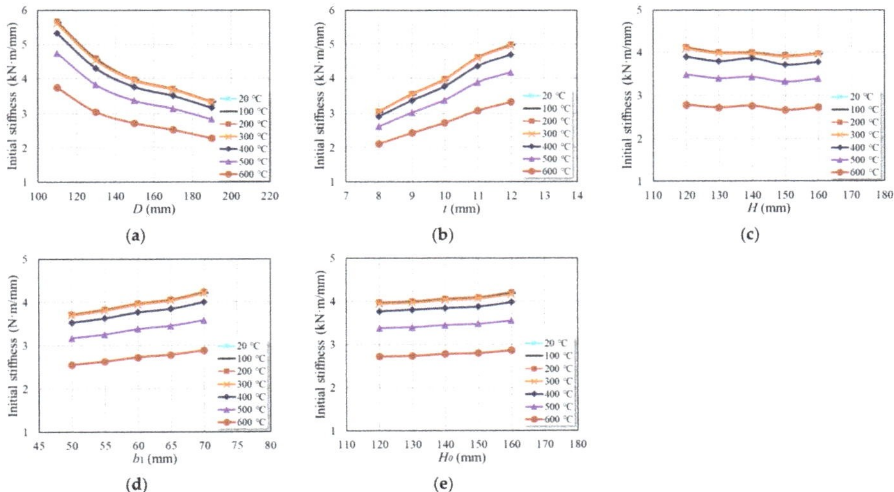

Figure 18. Effects of parameters on the initial stiffness of BBC joints under negative bending: (**a**) outer diameter of the tube; (**b**) thickness of the tube; (**c**) height of the tube; (**d**) width of the tubular beam; (**e**) height of the tubular beam.

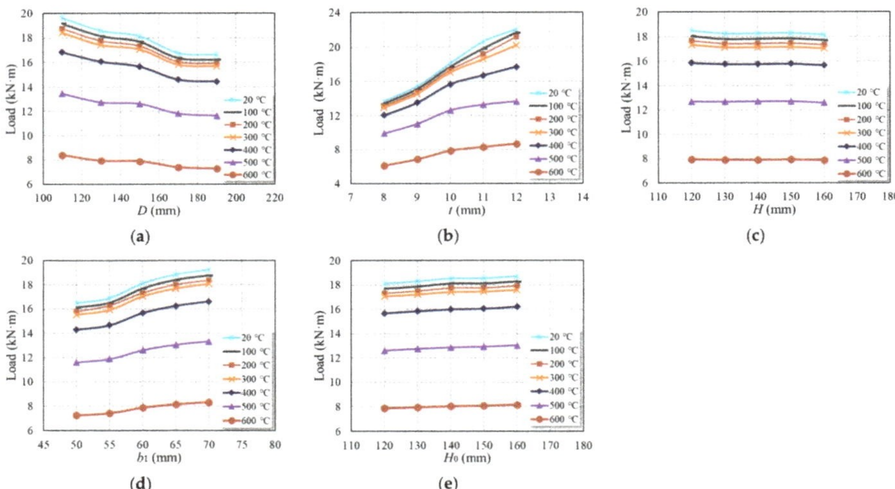

Figure 19. Effects of parameters on the load-carrying capacity of BBC joints under negative bending: (**a**) outer diameter of the tube; (**b**) thickness of the tube; (**c**) height of the tube; (**d**) width of the tubular beam; (**e**) height of the tubular beam.

4. Theoretical Analysis

4.1. Effects of Elevated Temperatures on the Capacities of BBC Joints

Based on parametric studies, the reduction in the ultimate capacity of a joint at elevated temperatures has a similar trend to the yield strength reduction of steel, regardless of the loading scenario. Therefore, the ultimate capacity of a joint at elevated temperatures could be predicted using the formulas for ambient temperature, but replacing the material properties at ambient temperature with those at elevated temperatures.

4.2. Design Method

Based on FE analysis, plastic hinges were formed in the circular tube. For a circular tube under compression or tension, plastic hinges first appeared at the loaded beam-to-tube junctions (location 1 in Figure 20) and then formed at location 2. The ultimate load was achieved when plastic hinges were formed at location 2.

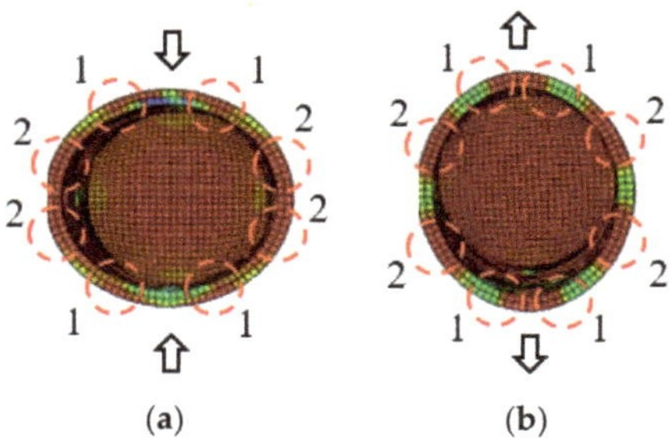

Figure 20. Plastic hinges formed in the circular tube (location 1 and 2): (**a**) under compression; (**b**) under tension.

For a joint, owing to the constraints provided by washers and end plates, rigid parts were formed in the circular tube, shown as the dark areas in Figure 21a. The length of the rigid part is equal to the width of the tubular beams (i.e., b_1 for the loaded beams and b_2 for

25. Ozyurt, E. Finite element study on axially loaded reinforced Square Hollow Section T-joints at elevated temperatures. *Thin-Walled Struct.* **2020**, *148*, 106582. [CrossRef]
26. Azari-Dodaran, N.; Ahmadi, H. Structural behavior of right-angle two-planar tubular TT-joints subjected to axial loadings at fire-induced elevated temperatures. *Fire Saf. J.* **2019**, *108*, 102849. [CrossRef]
27. Dassault Systemes Simulia Corp. *ABAQUS*; Dassault Systemes Simulia Corp.: Johnston, RI, USA, 2016.
28. *GB/T 228.1-2010*; Metallic Materials—Tensile Testing—Part 1: Method of Test at Room Temperature. China Quality and Standards Publishing: Beijing, China, 2010. (In Chinese)
29. Li, G.; Li, M.; Yin, Y.; Jiang, S. Experimental study on material properties of high strength bolt 20 Mn TiB steel at elevated temperature. *China Civ. Eng. J.* **2001**, *34*, 100–104. (In Chinese) [CrossRef]

metals

MDPI

Article

Numerical Study on Elastic Buckling Behavior of Diagonally Stiffened Steel Plate Walls under Combined Shear and Non-Uniform Compression

Yuqing Yang *, Zaigen Mu and Boli Zhu

School of Civil and Resource Engineering, University of Science and Technology Beijing, Beijing 100083, China; zgmu@ces.ustb.edu.cn (Z.M.); zboly@ustb.edu.cn (B.Z.)
* Correspondence: yqyang@ustb.edu.cn

Abstract: Unstiffened steel plate walls (SPWs) are prone to buckling in practical engineering and will invariably be subjected to vertical loads. The use of stiffeners can improve the buckling behavior of thin plates. Considering the effect of the torsional stiffness of C-shaped stiffeners, the elastic buckling of the diagonally stiffened steel plate wall (DS-SPW) under combined shear and non-uniform compression is investigated. The interaction curves for the DS-SPW under combined action are presented, as well as a proposed equation for the elastic buckling coefficient. In addition, the effects of the stiffener's flexural and torsional stiffness on the elastic buckling stress were investigated, and the threshold stiffness formulae were proposed. The results show that the interaction curve of the DS-SPW under combined shear and non-uniform compression is approximately parabolic. The critical buckling stress of the DS-SPW can be increased by increasing the stiffener's torsional-to-flexure stiffness ratio and the non-uniform compression distribution factor, while the buckling stress can be decreased by increasing the non-uniform compression-to-shear ratio. Simultaneous action of shear and axial compression will increase the threshold stiffness by approximately 40% when compared to the plate under pure shear action. Therefore, the safety threshold stiffness formula is suggested, considering the combined action of shear and non-uniform compression.

Keywords: steel plate wall; diagonal stiffener; torsional stiffness; elastic buckling; threshold stiffness

Citation: Yang, Y.; Mu, Z.; Zhu, B. Numerical Study on Elastic Buckling Behavior of Diagonally Stiffened Steel Plate Walls under Combined Shear and Non-Uniform Compression. *Metals* **2022**, *12*, 600. https://doi.org/10.3390/met12040600

Academic Editors: Carlos Garcia-Mateo and Angelo Fernando Padilha

Received: 22 February 2022
Accepted: 24 March 2022
Published: 31 March 2022

Publisher's Note: MDPI stays neutral with regard to jurisdictional claims in published maps and institutional affiliations.

Copyright: © 2022 by the authors. Licensee MDPI, Basel, Switzerland. This article is an open access article distributed under the terms and conditions of the Creative Commons Attribution (CC BY) license (https://creativecommons.org/licenses/by/4.0/).

1. Introduction

Steel plate walls (SPWs), as illustrated in Figure 1, are lateral force resistant elements with great ductility and energy dissipation capability. SPWs are widely used in high-rise and super high-rise buildings [1], such as the Los Angeles Hotel [2] and the 74-storey Tianjin Tower [3]. Thorburn et al. [4] found that thin steel plates formed tension fields after buckling and had a high shear capacity. However, unstiffened SPWs are prone to buckling and exhibit noticeable "pinching" hysteresis curves [5]. Thus, stiffeners are usually applied to the thin SPWs to improve their buckling behaviors [6]. According to certain experiments [7], it was found that the SPWs with flat-bar stiffeners had a considerable influence on the stiffeners after buckling, resulting in twisting and compressive buckling of the stiffeners themselves, which seriously harmed and even failed their stiffening effect. In the Technical Specification for Steel Plate Shear Walls (JGJ/T380-2015) [8], the SPW is designed to bear only horizontal loads, while the SPW will be subject to vertical load such as gravity or live load in actual works. For those reasons, some researchers advocated for the usage of closed form stiffeners to improve the strength and performance of the stiffeners, and allow the SPW to carry a portion of the vertical load [9–11].

In recent years, several researchers have investigated the behavior of SPWs while considering the effect of the torsional stiffness of C-shaped stiffeners. Xu et al. [10,12–14] analyzed the elastic and elastoplastic stability of vertically stiffened SPWs under uniaxial compression, shear, and their combined action. According to these studies, C-shaped stiffeners with higher flexural and torsional stiffness can significantly increase the elastic buckling stress of stiffened plates. Vu et al. [15] carried out the buckling analysis of transversely stiffened plates under combined bending and shear. The buckling of transversely stiffened corrugated plates was discussed by Feng et al. [16]. Tong and Guo [17] investigated the buckling behavior of vertically stiffened corrugated plates. Furthermore, some experts are concerned about plates that have been diagonally stiffened. Mu and Yang [9] tested the behavior of diagonally stiffened steel plate walls (DS-SPWs) with side openings. Additionally, the shear buckling behavior and threshold stiffness of DS-SPWs with C-shaped stiffeners are studied by the finite element method [18]. Yuan et al. [11], and Martins and Cardoso [19] investigated the elastic shear buckling coefficients for diagonally stiffened plates.

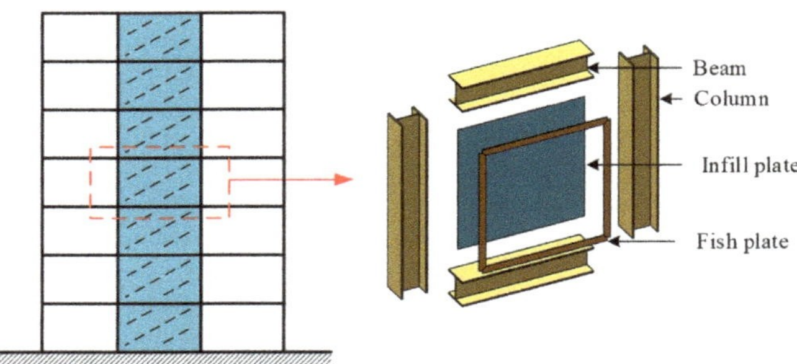

Figure 1. Unstiffened steel plate wall.

The current research focuses primarily on the buckling behavior of vertically or diagonally stiffened plates under pure compression or shear. However, there have been few investigations on the diagonally stiffened plate under combined action. In this paper, the influence of the aspect ratio of the plate, the torsional and flexural stiffness of the stiffeners, and a load of combined shear and non-uniform compression on the elastic buckling behavior of the DS-SPWs is investigated. In addition, the formula for the threshold stiffness of the DS-SPWs is established for engineering design, considering the torsional stiffness of stiffeners.

2. Elastic Buckling Coefficient and Interaction Formula

2.1. Elastic Buckling Stresses for Unstiffened Plate

Timoshenko and Gee [20] initially researched the elastic stability of unstiffened plates and found that the aspect ratio γ significantly has a substantial impact on the critical buckling stress τ_{cr} of the structure, as stated in Equations (1) and (2). The buckling coefficients for a four-edge simply supported plate under shear are calculated by Equations (3) and (4) [21]. The buckling coefficients of a four-edge simply supported plate under bending are approximated by Equation (5) [22].

$$\gamma = \frac{L}{H} \tag{1}$$

$$\tau_{cr} = \frac{k_{cr}\pi^2 D}{H^2 t}, \text{where } D = \frac{Et^3}{12(1-\nu^2)}, \nu = 0.3 \tag{2}$$

$$k_{cr} = 5.34 + 4/\gamma^2 \ (L \geq H) \tag{3}$$

$$k_{cr} = 4 + 5.34/\gamma^2 \ (L \leq H) \tag{4}$$

$$\begin{cases} k_{cr} = 15.87 + 1.87/\gamma^2 + 8.6\gamma^2 & (\gamma < 2/3) \\ k_{cr} = 23.9 & (\gamma \geq 2/3) \end{cases} \tag{5}$$

where L, H, and t is the length, height, and thickness of the plate, respectively; D is the flexural rigidity of the plate; E is elastic modulus of the plate; k_{cr} is the elastic buckling coefficient; and ν is Poisson's ratio.

2.2. Interaction Formula of Unstiffened Plate

Tong [23] analyzed the buckling stability of a four-edge simply supported plate under combined shear and non-uniform compression, giving the interaction formula for plate combined action of axial compression, bending, and shear as Equation (6). The non-uniform compression can be split into a combined action of bending and axial compression. The non-uniform compression distribution factor ζ and the compression-to-shear ratio δ are defined in Equations (8) and (9), respectively. Then, the elastic buckling coefficient k of the plate under combined action can be solved by Equation (10), which is obtained from Equations (2) and (6)–(9).

$$\frac{\sigma}{\sigma_{cr}} + \left(\frac{\sigma_b}{\sigma_{bcr}}\right)^2 + \left(\frac{\tau}{\tau_{cr}}\right)^2 = 1 \tag{6}$$

$$\begin{cases} \sigma = (1 - 0.5\zeta)\sigma_{max} \\ \sigma_b = 0.5\zeta\sigma_{max} \end{cases} \tag{7}$$

$$\zeta = \frac{\sigma_{max} - \sigma_{min}}{\sigma_{max}} \tag{8}$$

$$\delta = \frac{\sigma_{max}}{\tau} \tag{9}$$

$$\frac{(1 - 0.5\zeta)\delta k}{k_{\sigma,cr}} + \left(\frac{0.5\zeta\delta k}{k_{\sigma b,cr}}\right)^2 + \left(\frac{k}{k_{\tau,cr}}\right)^2 = 1 \tag{10}$$

2.3. Elastic Buckling Coefficient of Diagonally Stiffened Plate in Single Load Action

Mikami et al. [22] used the finite difference method to analyze the buckling behavior of the diagonally stiffened plates. The shear buckling coefficient for the diagonally stiffened plate, which is a four-edge simply supported plate stiffened over the compression and tension diagonal, is given as Equation (11). Moreover, the buckling coefficient of the diagonally stiffened plate in bending and axial compression can be calculated by Equations (12) and (13), respectively.

$$\begin{cases} k_{cr} = 11.9 + 10.1/\gamma + 10.9/\gamma^2 & \text{compression} \\ k_{cr} = 17.2 - 22.5/\gamma + 16.7/\gamma^2 & \text{tension} \end{cases} \tag{11}$$

$$k_{cr} = 22.5 + 4.23\gamma + 2.75\gamma^2 \tag{12}$$

$$k_{cr} = -1.2 + 5.5\gamma + 5.5\gamma^2 \tag{13}$$

3. Description of Numerical Model

A diagonally stiffened plate with C-shaped stiffeners, as illustrated in Figure 2, where b is the height of the flange of C-shaped stiffener, b_s is the width of the web of C-shaped stiffener, and t_s is the thickness of C-shaped stiffener.

3.1. Definition of Parameters

The stiffener-to-plate flexural stiffness ratio η is used to measure the relationship between the flexural stiffness of the stiffener and the plate, as shown in Equation (14). The stiffener's torsional-to-flexural stiffness ratio K is defined as Equation (15). When the C-shaped stiffeners are arranged on both sides of the plate, the moment of inertia I_s and torsional constant J_s of the stiffeners are calculated by Equations (16) and (17), respectively. Thus, K in Equation (15) is a function with factors of b_s and b. In this paper, b_s is taken as a constant of 100 mm, and K is changed by varying b, as shown in Table 1.

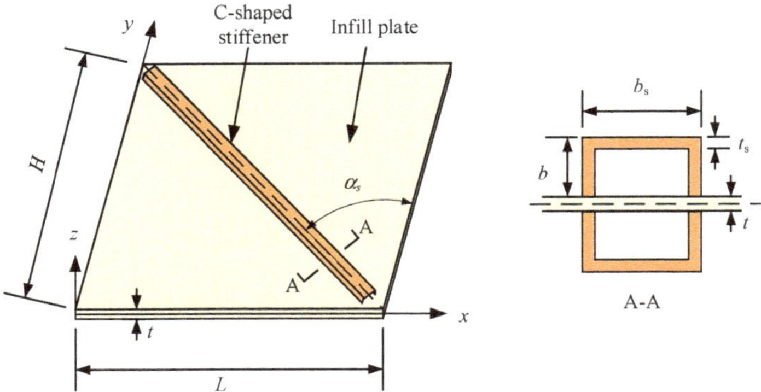

Figure 2. Diagonally stiffened steel plate walls.

$$\eta = \frac{E_s I_s}{D L_e}, \text{ where } L_e = H \sin \alpha_s + L \cos \alpha_s \tag{14}$$

$$K = \frac{G_s J_s}{E_s I_s} = \frac{4 b_s^2}{2(1+v)(b_s + 2b)\left(b_s + \frac{2b}{3}\right)} \tag{15}$$

$$I_s = \frac{2 b^3 t_s}{3} + b^2 b_s t_s \tag{16}$$

$$J_s = 2\frac{b^2 (2 b_s)^2 t_s^2}{2 b t_s + b_s t_s} \tag{17}$$

where α_s is the inclination angle of the plate's diagonal; G_s and E_s are the shear modulus and elastic modulus of the stiffeners, respectively.

Table 1. The Values of Stiffener's Torsional-to-Flexural Stiffness Ratio K.

Parameter	Value					
b/mm	25	50	75	100	125	150
K	0.879	0.577	0.410	0.307	0.240	0.192

This paper investigates the effects of four parameters on the buckling behavior of the diagonally stiffened plates, including the non-uniform compression distribution factor ζ, the compression-to-shear ratio δ, the stiffener's torsional-to-flexural stiffness ratio K, and the stiffener-to-plate flexural stiffness ratio η. SPWs will inevitably be subjected to vertical pressure in actual projects [24], but still dominated by the shear load. Therefore, ζ ranges from 0 to 1, δ ranges from 0 to 2, η ranges from 1 to 70, and K is taken according to Table 1.

3.2. Finite Element Model

Three finite element models are established, as displayed in Figure 3. An unstiffened SPW is used as a validation model in Figure 3a. The stiffened edge of the DS-SPW is assumed to be simply supported Figure 3b. When the stiffeners are rigid, the out-of-plate displacement of the stiffened edge is constrained, and the critical buckling stress value of local buckling can be solved by this model. Figure 3c shows a DS-SPW with C-shaped stiffeners, which is used for the parametric analysis.

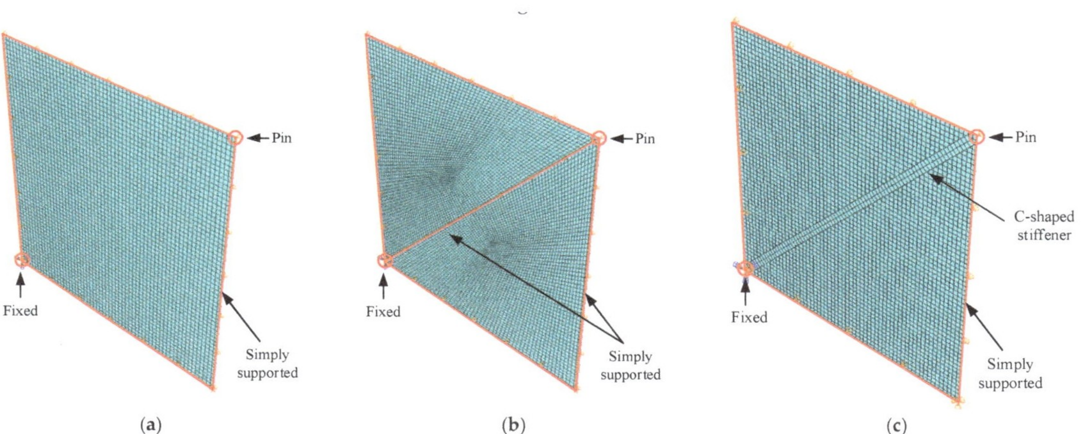

Figure 3. Finite models of SPWs: (**a**) SPW, (**b**) SPW with diagonal simply supported, and (**c**) SPW with C-shaped stiffeners.

The plate and the stiffeners are both made of shell element S4R, which is suitable for analyzing the buckling problems of thin panels with out-of-plane deformations. The stiffeners are connected to the steel plate by TIE constraints, which are simulated as a welded connection. To avoid rigid body displacement of the plate, one corner of the plate is fixed and another corner is simply supported. Out-of-plane deformation of four edges is constrained to simulate the case of simply supported. The size of the plate is 3000 mm × 3000 mm × 10 mm ($L \times H \times t$), the size of the C-shaped stiffener is 100 mm × 50 mm × 6 mm ($b_s \times b \times t_s$), and the mesh grid is 50 mm × 50 mm to ensure accuracy and less time-consuming calculation. The modulus of elasticity of the steel is 206,000 MPa, and Poisson's ratio is 0.3. Tangential line loads are applied on the four edges of the plate to simulate the shear stress. Normal line loads are applied on the top and bottom edges of the plate to simulate the compressive stress, and the action form of the non-uniform compression stress can be changed by adjusting the distribution function. The buckling eigenvalue analysis is carried out to solve the buckling stress of the structure.

3.3. Verification of Boundary Conditions

The plate in an SPW is connected to the boundary elements by the fishplate, and it is usually assumed that the plate is simply supported on four edges. The elastic shear buckling coefficient of the plates can be calculated by the equations in Sections 2.1 and 2.3.

Table 2 presents the results of the FE analysis for the unstiffened plate when it is subjected to shear, axial compression, bending, or their combined action. The theoretical buckling coefficient k can be calculated from Section 2, then k is substituted into Equation (2) to obtain the theoretical value of critical buckling stress. Moreover, the critical buckling stresses of DS-SPWs under single load action are similarly verified by comparing the results with Equations (11)–(13). The results indicate that the modelling approach is valid and that it can be utilized to analyze the elastic buckling behavior of the stiffened plate.

Table 2. Comparison Buckling Stress of Finite Element and Theoretical Formula.

Force Conditions	Type	ζ	δ	σ_{FEA} /MPa	σ_{TH} /MPa	σ_{TH}/σ_{FEA}
Shear	Unstiffened	-	0	19.30	19.32	1.001
Axial compression	Unstiffened	0	1.0	8.12	8.27	1.018
Non-uniform compression	Unstiffened	1.0	1.0	16.05	16.12	1.004
Bending	Unstiffened	2.0	1.0	52.83	52.90	1.001
Combined shear and non-uniform compression	Unstiffened	1.0	1.0	15.99	16.16	1.011
Combined shear and non-uniform compression	Unstiffened	1.0	0.5	14.37	14.43	1.004
Shear (compression)	Diagonally stiffened	-	0	63.6	68.0	1.069
Shear (tension)	Diagonally stiffened	-	0	23.5	23.8	1.013
Bending	Diagonally stiffened	2.0	1.0	64.5	63.1	0.978
Axial compressoion	Diagonally stiffened	0	1.0	19.6	20.9	1.066

Note: σ_{FEA} is buckling stress from FE analysis and σ_{TH} is critical buckling stress from theoretical equation.

4. Interaction Curves for Diagonally Stiffened Plate

Using the model in Figure 3b as the base model for the analysis in this section, shear stress τ is applied on four edges of the diagonally stiffened plate, and non-uniform compression stress $\sigma = \zeta\tau$ is applied on the top and bottom edges. The critical shear stress τ_{cr} of the diagonally stiffened plate under combined action is determined by analysis results, and the critical non-uniform compression stress is calculated as $\sigma_{cr} = \zeta\tau_{cr}$. Then, one point on the interaction curve is obtained as $(\tau_{cr}/\tau_{cr0}, \sigma_{cr}/\sigma_{cr0})$, where τ_{cr0} is the critical stress of the plate under pure shear and σ_{cr0} is the critical stress of the plate under pure non-uniform compression. The interaction curves for the diagonally stiffened plate under combined shear and non-uniform compression can be obtained by varying the compression-to-shear ratio δ and the non-uniform compression distribution factor ζ.

4.1. Combined Shear and Axial Compression $\zeta = 0$

The non-uniform compression is equal to axial compression when the factor ζ is 0. By varying the compression-to-shear ratio δ, the interaction curves of the diagonally stiffened plate under combined shear and axial compression are shown in Figure 4. The aspect ratio γ of the plate affects the shape of the curves. The interaction curves with different aspect ratios are close to parabolas. When $\gamma < 1.2$, the curve rises and moves away from the origin point with increasing γ. When $1.2 < \gamma < 2.0$, the curve reaches the outermost circle and changes insignificantly. When $\gamma > 2.0$, the curve moves closer to the origin point with increasing γ.

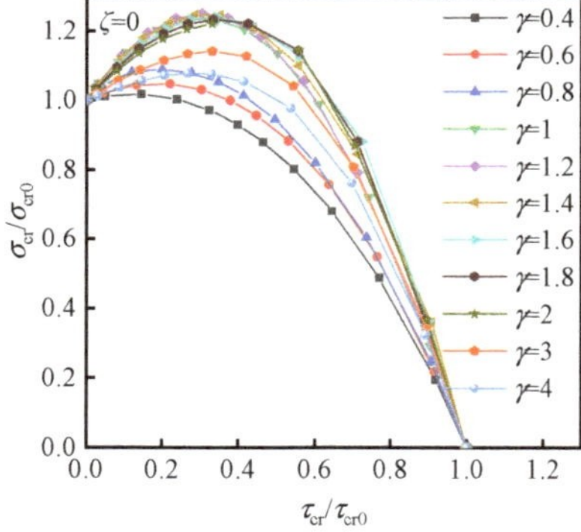

Figure 4. Interaction curves of diagonally stiffened plate under combined shear and axial compression $\zeta = 0$.

4.2. Combined Shear and Non-Uniform Compression $0 < \zeta < 2$

When the non-uniform compression distribution factor $0 < \zeta < 2$, the interaction curves of the diagonally stiffened plate under combined shear and non-uniform compression are shown in Figure 5. The aspect ratio γ and non-uniform compression distribution factor ζ affect the shape of the curves. The interaction curves with different aspect ratios are also close to parabolas. The variation of the interaction curves with different γ is similar to that of factor $\zeta = 0$. As the factor ζ increases, the peak of the interaction curve gradually decreases, and the axis of symmetry of the parabola is shifted towards the Y-axis.

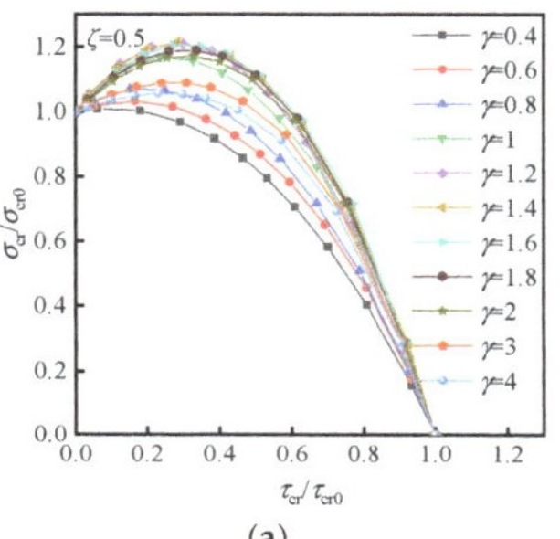
(a)

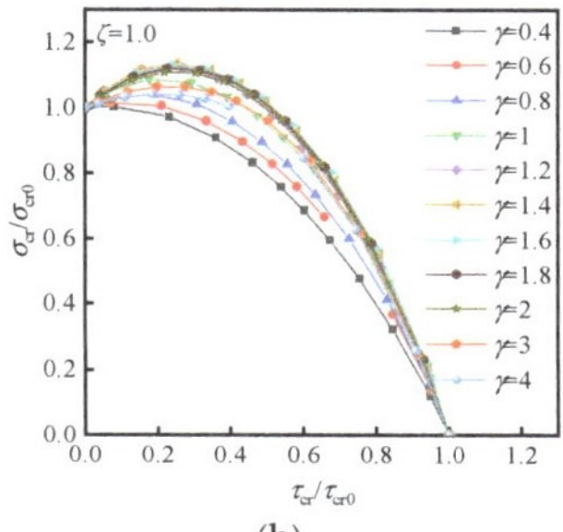
(b)

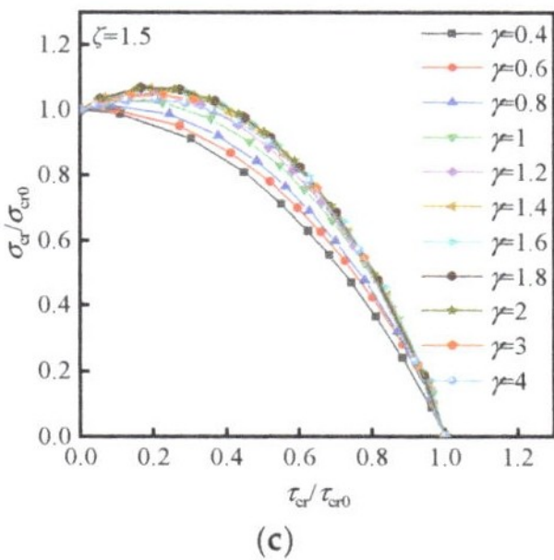
(c)

Figure 5. Interaction curves of diagonally stiffened plate under combined shear and non-uniform compression: (**a**) $\zeta = 0.5$, (**b**) $\zeta = 1.0$, and (**c**) $\zeta = 1.5$.

4.3. Combined Shear and Bending $\zeta = 2.0$

Mikami et al. [22] used Equation (18) to approximate the interaction curves of the diagonally stiffened plate under combined shear and bending when the aspect ratio γ of plate was 1.0. Figure 6 illustrated the interaction curves for various aspect ratios. The spacing between each curve is small, and the shape of curves is closer to a parabola curve of which the axis of symmetry is close to the Y-axis. There is still a large difference in using Equation (18) to approximate the interaction curve. Therefore, a more accurate approximation curve equation will be given in the following section.

$$\left(\frac{\tau}{\tau_{cr}}\right)^2 + \frac{\sigma_b}{\sigma_{bcr}} = 1, \gamma = 1.0 \tag{18}$$

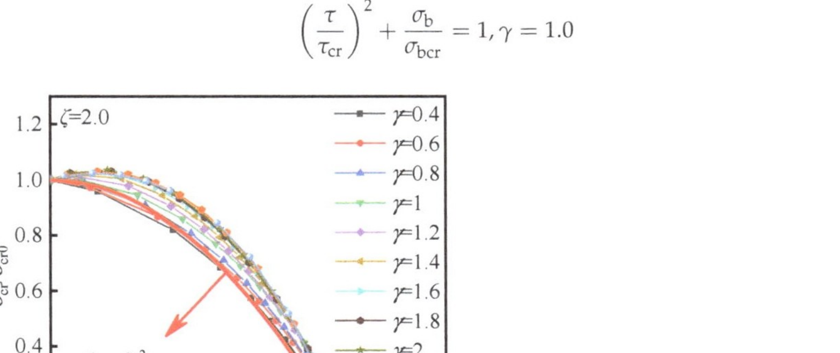

Figure 6. Interaction curves of diagonally stiffened plate under combined shear and bending.

4.4. Effect of ζ on Interaction Curves

Taking the plate with an aspect ratio γ of 1.0 as an example, the interaction curves of the diagonally stiffened plate under combined shear and non-uniform compression are shown in Figure 7. It can be seen that as ζ increases, the curve gradually converges to a parabola with the axis of symmetry as the Y-axis, and the curve is closer to the origin point. The approximate calculation formula Equation (19) is given according to the variation of the interaction curves. Equation (19) agrees well with the finite element analysis results, indicating that it can reflect the relationship of the diagonally stiffened plate under combined shear and non-uniform compression, i.e., the combined action of shear-bending-axial compression.

$$\left[\frac{5}{2} - \frac{2}{3}\zeta\right]\left(\frac{\tau}{\tau_{cr}}\right)^2 - \left[\frac{3}{2} - \frac{2}{3}\zeta\right]\left(\frac{\tau}{\tau_{cr}}\right) + \frac{\sigma}{\sigma_{cr}} + \left(\frac{\sigma_b}{\sigma_{bcr}}\right)^{\frac{5}{4}} = 1, \gamma = 1.0 \tag{19}$$

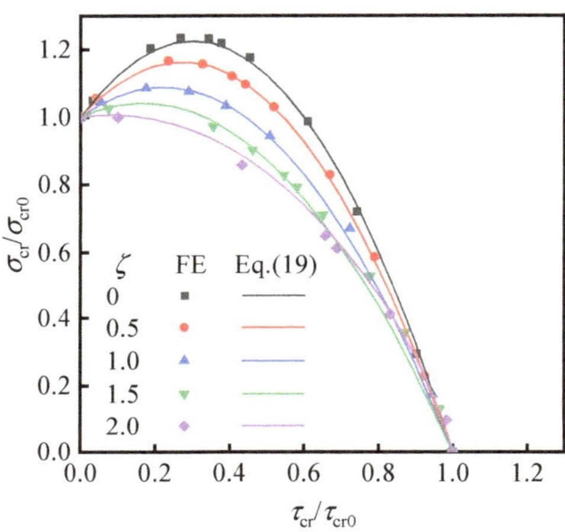

Figure 7. Interaction curves of diagonally stiffened plate under combined shear, axial compression, and bending ($\gamma = 1.0$).

Equation (19) can be written as Equation (20), where $k_{\tau,cr}$, $k_{\sigma,cr}$, and $k_{\sigma b,cr}$ are calculated by Equations (11)–(13). The elastic buckling coefficient k of the diagonally stiffened plate under combined shear and non-uniform compression can be calculated by solving the one-variable function of Equation (20).

$$\left[\frac{5}{2} - \frac{2}{3}\zeta\right]\left(\frac{k}{k_{\tau,cr}}\right)^2 - \left[\frac{3}{2} - \frac{2}{3}\zeta\right]\left(\frac{k}{k_{\tau,cr}}\right) + \frac{(1 - 0.5\zeta)\delta k}{k_{\sigma,cr}} + \left(\frac{0.5\zeta\delta k}{k_{\sigma b,cr}}\right)^{\frac{5}{4}} = 1, \gamma = 1.0 \tag{20}$$

When $\delta = 0$, the diagonally stiffened plate is under the pure shear, at which point the theoretical elastic buckling coefficient k is 32.9, as obtained by Equation (11). The result calculated from Equation (20) is also 32.9. When $0 < \delta < 1.0$, varying the value of ζ, the corresponding elastic buckling coefficient k can be found in Table 3.

Table 3. Buckling Coefficients k of the Diagonally Stiffened Plates Under Combined Action ($\gamma = 1.0$).

δ	$\zeta = 0$	$\zeta = 0.5$	$\zeta = 1.0$	$\zeta = 1.5$	$\zeta = 2.0$
0.25	25.4	26.3	27.3	28.5	30.0
0.5	19.3	20.8	22.4	24.3	26.6
0.75	14.8	16.5	18.4	20.7	23.5
1.0	11.6	13.3	15.3	17.7	20.7

5. Threshold Stiffness of DS-SPW

Section 4 gives the interaction curves and elastic buckling coefficients of the diagonally stiffened plate in which the out-of-plane displacement of the stiffened edge is restrained. In actual engineering, the stiffener is not an ideally rigid body. Thus, the flexural stiffness of stiffeners is a significant factor in the design of buildings with DS-SPW.

SPWs are inevitably subjected to shear, axial compression, and bending in the building structures. Their buckling behaviors can be significantly different from those under pure shear action. The performance of the diagonally stiffened plate under pure shear has been studied [18]. Furthermore, the effects of the torsional and flexural stiffnesses of stiffeners on the buckling behavior of the structure under different combined actions are analyzed in this section.

5.1. Buckling Mode of DS-SPW

By changing the modulus of elasticity of the stiffeners, the value of the stiffener-to-plate flexural stiffness ratio η is changed without affecting any other parameters. Figure 8 illustrates the curves of τ-η, and the critical local buckling stress is indicated by the blue horizontal line. The shear buckling stress curves of the plate stiffened over the compression and tension diagonal are not the same. Combined with the first-order buckling modes corresponding to different η in Figures 9 and 10, the buckling modes of the tensile and compressive types can be divided into global buckling, local buckling, and global–local interaction buckling.

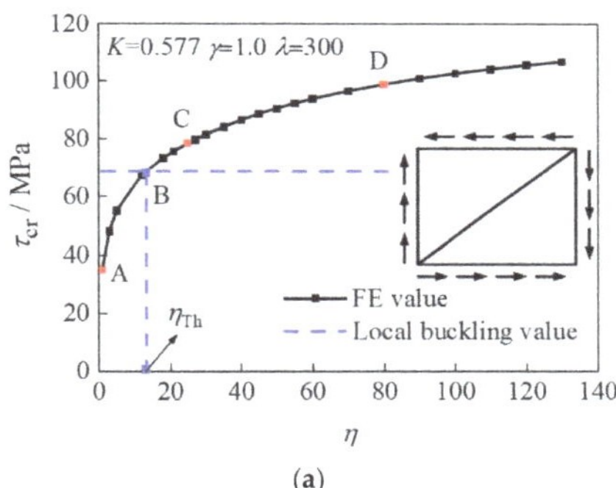

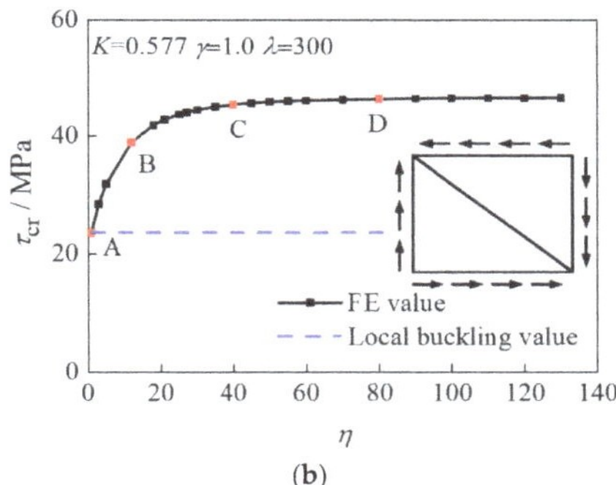

Figure 8. τ-η curve for DS-SPW with C-shaped stiffeners under shear: (**a**) compressive type and (**b**) tensile type.

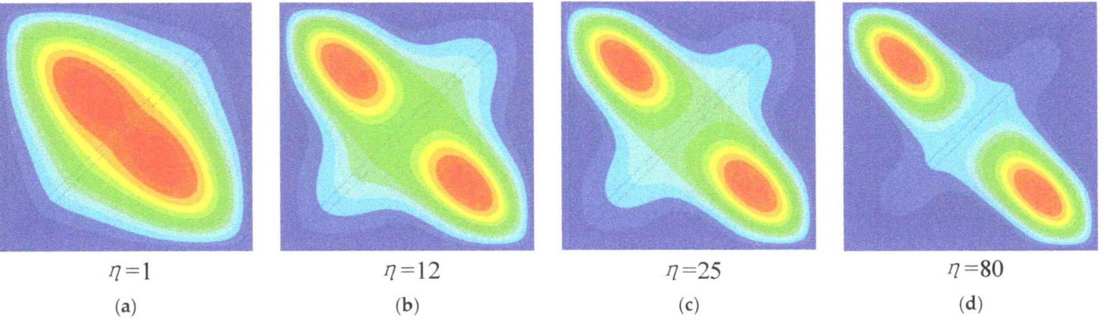

Figure 9. Shear buckling mode of SPW with diagonal compression: (**a**) global buckling at Point A, (**b**) local buckling at Point B, (**c**) local buckling at Point C, and (**d**) local buckling at Point D.

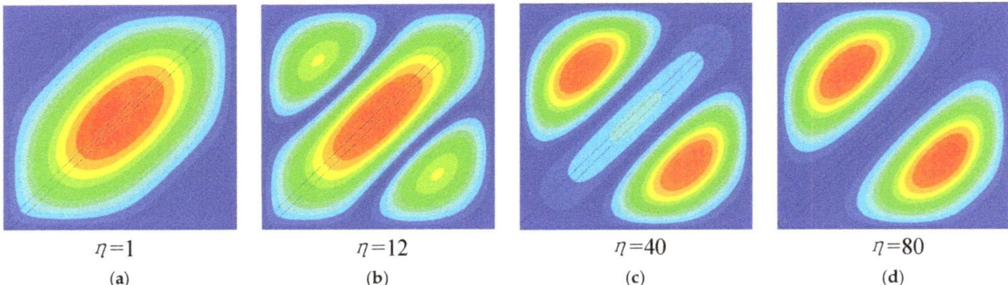

Figure 10. Shear buckling mode of SPW with diagonal tension: (**a**) global buckling at Point A, (**b**) global–local interaction buckling at Point B, (**c**) local buckling at Point C, and (**d**) local buckling at Point D.

The buckling stress of the plate stiffened over the compression and tension diagonal increases rapidly as η increases, such as Figure 9a where the global buckling occurs. When η exceeds the critical local buckling load, local buckling of the steel plate occurs, at which time the η is called the threshold stiffness η_{TH}, such as in Figure 9b. It is indicated that the stiffened plate reaches the minimum η corresponding to local buckling. As η continues to increase, the buckling stress of the structure has increased, and this increased part of value is provided by the stiffeners, such as in Figure 9c,d. Therefore, in a DS-SPW, it is very important to find out the appropriate threshold stiffness η_{TH} to ensure that the global buckling of the structure does not occur, and also to make an economical choice of stiffeners.

When the η is very small, the global buckling of the plate stiffened over the compression and tension diagonal occurs, such as in Figure 10a. With the increase of the η, the local buckling occurs between the two nodal lines such as Figure 10b. Then, when η exceeds a certain value, the nodal lines of the plate with $\gamma = 1.0$ overlap, decreasing from two to one, and the local buckling occurs in sub-plate such as Figure 10c. After that, as η continues to increase, the buckling stress of the structure does not increase.

The τ-η curves of the diagonally stiffened plate under combined action are presented in Figure 11. Figure 11a shows the plate under combined shear and axial compression with δ of 0.25, and the key points on the curve are shown in Figure 12. Figure 11b shows the plate under combined shear and bending with δ of 1.0, and the key points on the curve are shown in Figure 13. The development of τ-η curves is similar to the plate stiffened over the compression and tension diagonal under pure shear, and the buckling mode still changes from global buckling to local buckling with increasing η. Under combined shear and axial compression action, the local buckling half-wave shifts and is not symmetrical along the diagonal line. Under combined shear and bending action, the maximum deformation point shifts to the upper-left corner. The local buckling has only one half-wave and no antisymmetric half-wave. This is due to the fact that the axial tension force on the right of plate affects the formation of the local buckling half-wave.

5.2. Buckling Stress of Stiffened Plate under Combined Action

Changing the modulus of elasticity of the stiffeners and the height of flange of the C-shaped stiffener b varied the value of K without changing other parameters. The C-shaped stiffener has a large torsional stiffness, which obviously affects the elastic buckling stress of the stiffened plate. Figure 14 shows the normalized buckling stress $\tau_{\mathrm{cr}}/\tau_{\mathrm{cr0}}$ corresponding to different K, ζ, and δ. The following regularities can be obtained from Figure 14: (1) When ζ and δ remain constants, the buckling stress of the structure keeps increasing as K increases. (2) When ζ remains constant, the buckling stress of the structure keeps decreasing as δ increases. (3) When δ remains constant, the buckling stress of the structure keeps increasing as ζ increases. In brief, the K and ζ are positive for improving the elastic buckling stress of stiffened plate, while δ is negative.

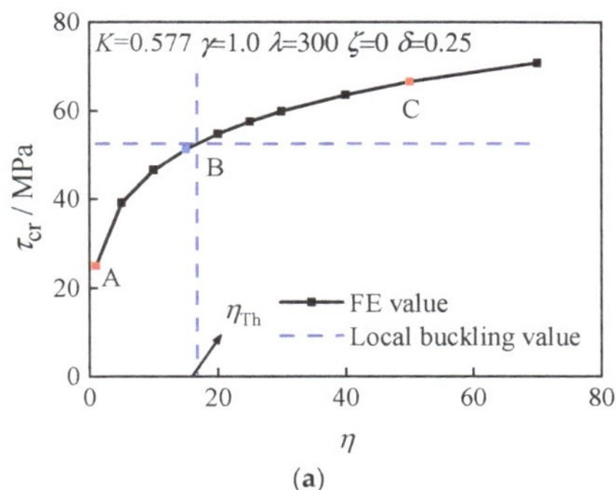

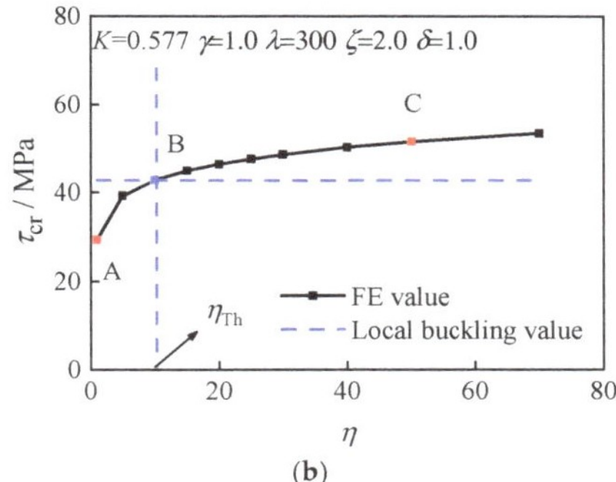

Figure 11. τ-η curve for DS-SPW with C-shaped stiffeners under combined action: (**a**) $\zeta = 0$ $\delta = 0.25$ and (**b**) $\zeta = 2.0$ $\delta = 1.0$.

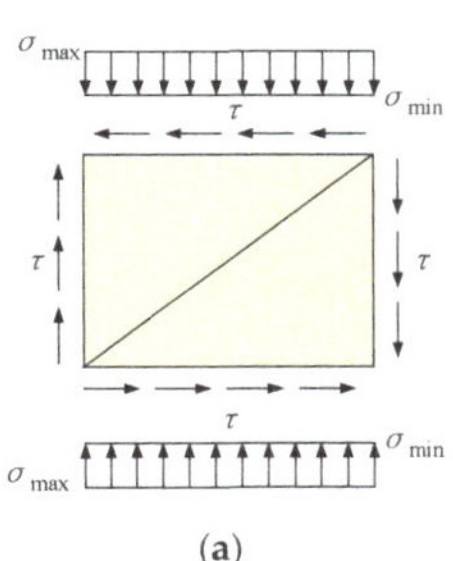

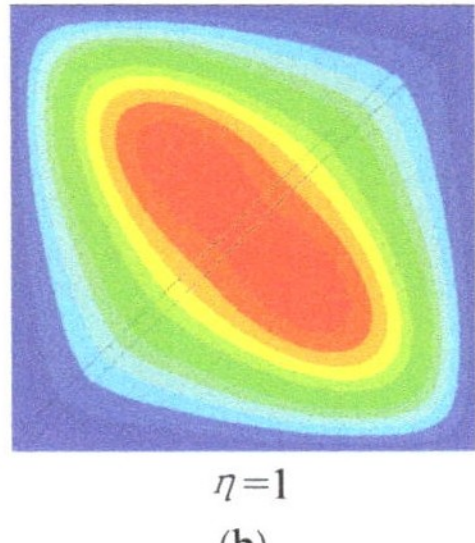

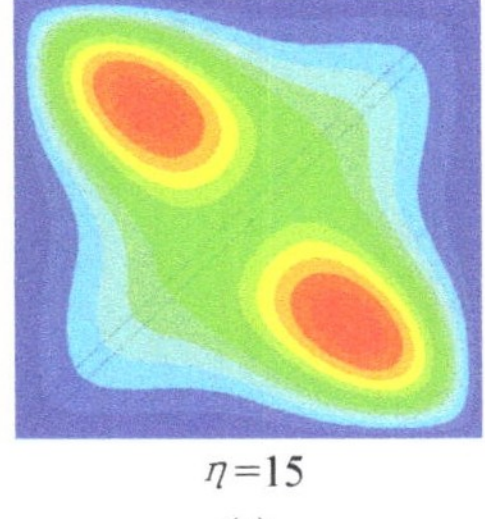

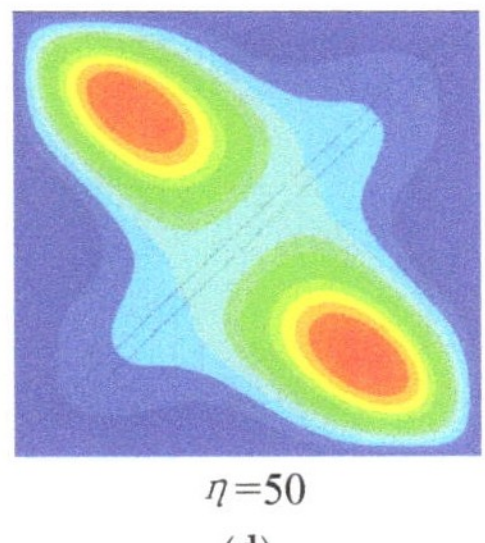

Figure 12. Combined shear and axial compression buckling mode of SPW with diagonal compression: (**a**) $\zeta = 0$ $\delta = 0.25$, (**b**) global buckling at Point A, (**c**) local buckling at Point B, and (**d**) local buckling at Point C.

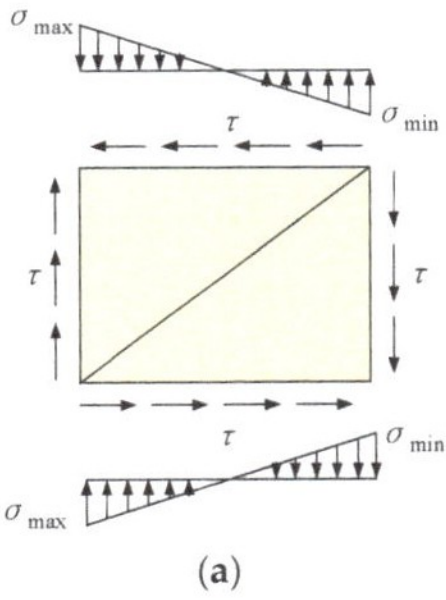

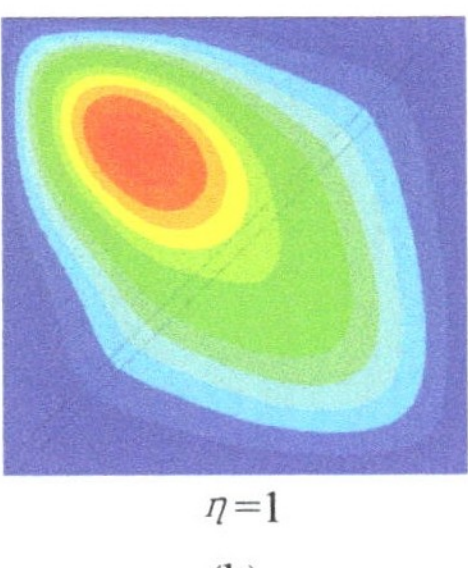

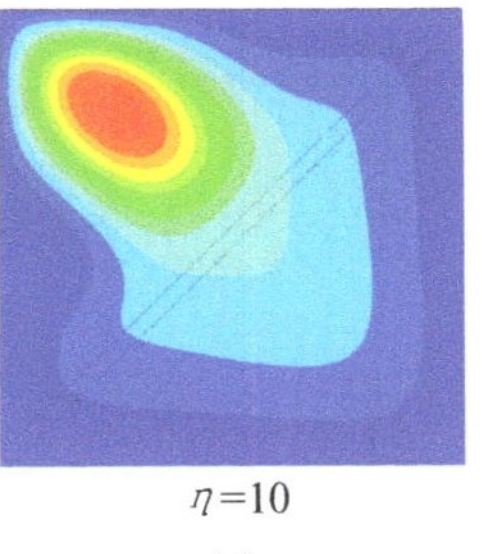

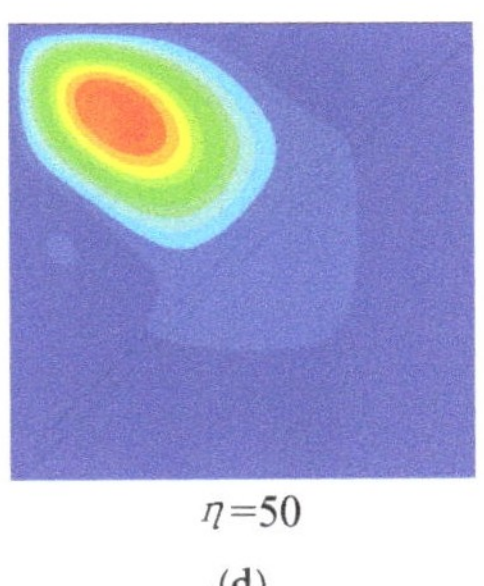

Figure 13. Combined shear and bending buckling mode of SPW with diagonal compression: (**a**) $\zeta = 2.0$ $\delta = 1.0$, (**b**) global buckling at Point A, (**c**) local buckling at Point B, and (**d**) local buckling at Point C.

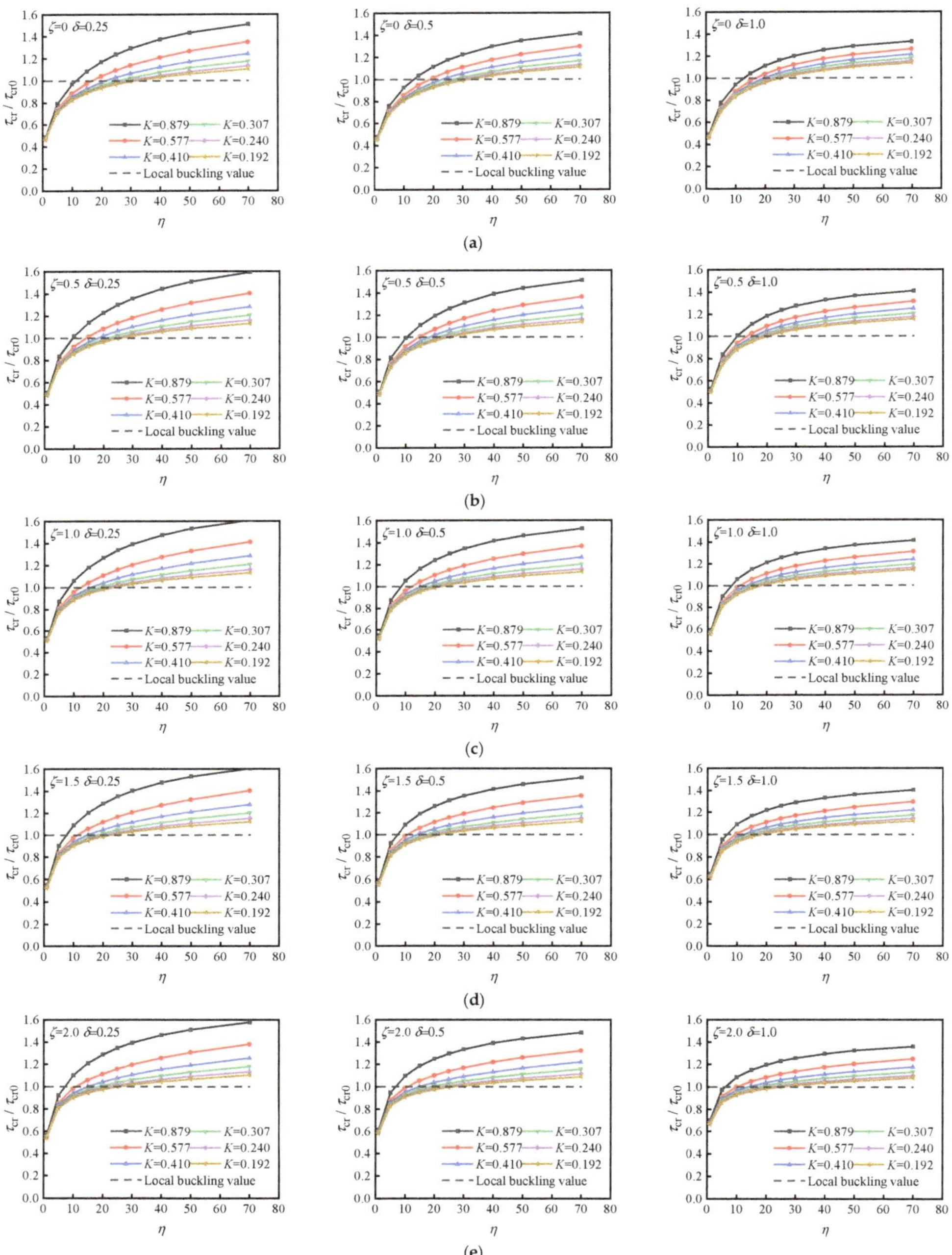

Figure 14. η-τ_{cr} curves of DS-SPW ($\gamma = 1.0$): (**a**) $\zeta = 0$, (**b**) $\zeta = 0.5$, (**c**) $\zeta = 1.0$, (**d**) $\zeta = 1.5$, and (**e**) $\zeta = 2.0$.

5.3. Threshold Stiffness

The intersection of the critical buckling stress and the η-τ_{cr} curve is the threshold stiffness η_{TH}. Then, the threshold stiffness corresponding to different K, ζ, and δ is shown in Figure 15.

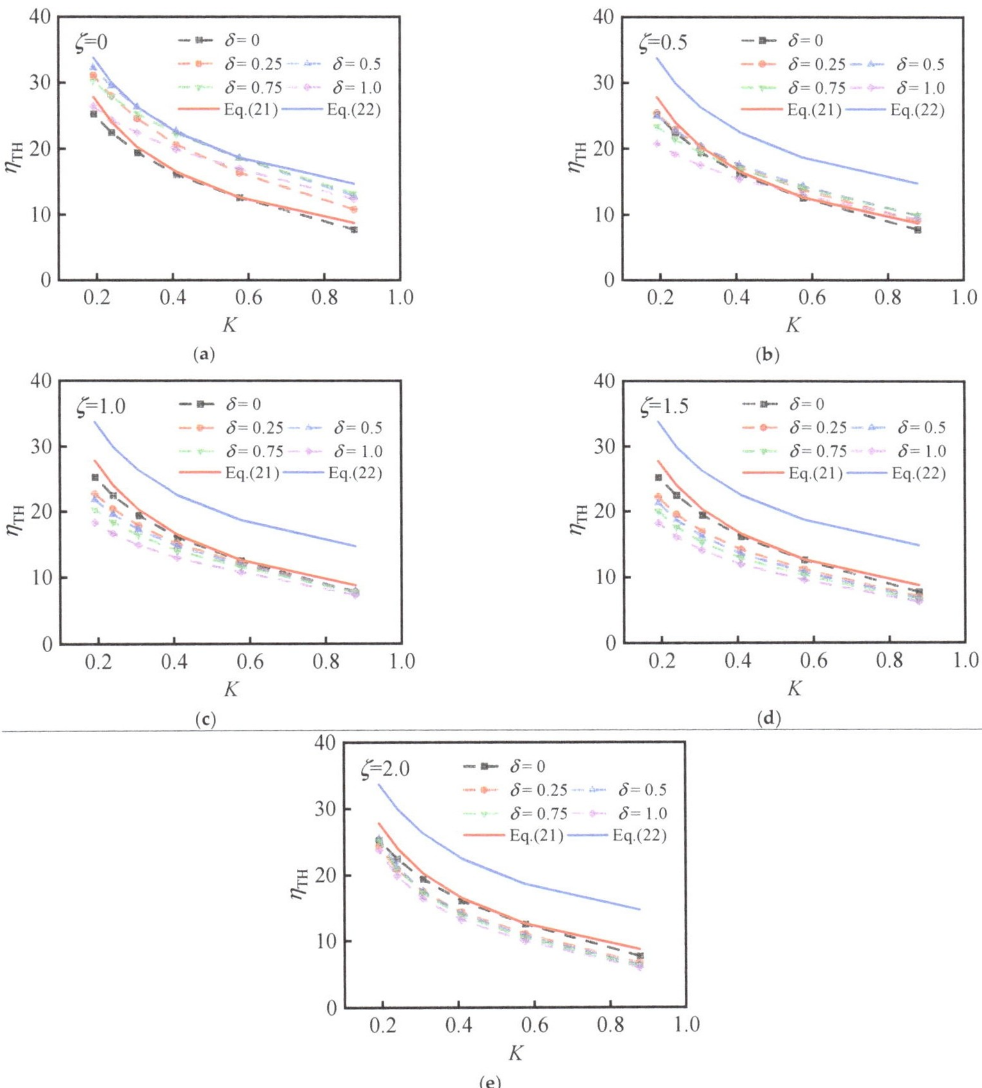

Figure 15. K-η_{TH} curves of DS-SPW ($\gamma = 1.0$): (**a**) $\zeta = 0$, (**b**) $\zeta = 0.5$, (**c**) $\zeta = 1.0$, (**d**) $\zeta = 1.5$, and (**e**) $\zeta = 2.0$.

With the increase of K, the threshold stiffness tends to decrease, indicating that increasing the K of the stiffeners can reduce the requirement of flexural stiffness of the stiffeners. The height of the stiffeners is higher with a lower value of K, so the stiffeners themselves are more likely to twist. Therefore, a higher flexural stiffness of the stiffener is required. According to the analysis under pure shear, the K of the stiffener is recommended to be not less than 0.3 [18].

Using $\delta = 0$ as a reference line (i.e., the threshold stiffness of the stiffened plate under pure shear), as δ increases, the threshold stiffness of the structure does not change in the same trend. When $\zeta = 0$, the threshold stiffness is maximum at δ of 0.5. That is, the threshold stiffness of the structure is maximum when the plate is under combined shear and axial compression. At this time, the threshold stiffness is increased by about 40% compared to the pure shear state. Therefore, if an SPW is subjected to a combination of shear and axial compression in practice, the threshold stiffness obtained from plate under pure shear is on the low side and the structure will occur global buckling instead of local buckling. The threshold stiffness curve of the combined action tends to decrease as ζ increases. When $\zeta = 0.5$, it basically overlaps with the threshold stiffness curve of pure shear. When $1.0 < \zeta < 2.0$, the threshold stiffness curve of joint action is below the threshold stiffness curve of pure shear. In this case, taking the plate under pure shear as the threshold stiffness is on the safe side.

The most unfavorable action is the combination of axial compression and shear. The threshold stiffness requirement can be satisfied by considering this case only during the design. For diagonally stiffened plates in pure shear or $1.0 < \zeta < 2.0$, the threshold stiffness can be calculated according to Equation (21). For $0 < \zeta < 1.0$, the threshold stiffness can be calculated according to Equation (22), and the results obtained are safe.

$$\eta_{TH} = -8 + 15.7/K^{0.5}, \zeta = 0 \tag{21}$$

$$\eta_{TH} = -2 + 15.7/K^{0.5}, \zeta \leq 1.0 \tag{22}$$

6. Conclusions

In this study, a diagonally stiffened steel plate wall (DS-SPW) with C-shaped stiffeners is presented and elastic eigenvalue buckling calculations are performed using ABAQUS software to analyze the buckling stresses and buckling modes of DS-SPW under combined shear and non-uniform compression. The accuracy of the performance of the numerical model is first verified with the four-sided simply supported rectangular plate and the diagonally stiffened plate. The parameters, such as the compression-to-shear ratio, the non-uniform compression distribution factor and the stiffener's torsional-to-flexural stiffness ratio, and the elastic buckling performance of the stiffened plates, are changed to carry out a large number of numerical models for DS-SPWs, and the elastic buckling performance of diagonally stiffened plates under combined shear and non-uniform compression is investigated. Considering the torsional stiffness of C-shaped stiffeners, the formula of threshold stiffness, which is the minimum stiffness of stiffeners required for a stiffened plate from overall buckling to local buckling, is proposed. The main results of the study are as follows.

(1) The elastic buckling behavior of the diagonally stiffened plate under combined shear and non-uniform compression is analyzed. The interaction curves of a plate under combined action are parabolic, and the aspect ratio γ and the non-uniform compression distribution factor ζ obviously affect the shape of the parabolas. The larger the factor ζ is, the more the parabola moves toward the origin point and the axis of symmetry of the parabola is closer to the Y-axis. As the aspect ratio increases $0.4 < \gamma < 1.2$, the interaction curve moves away from the origin point; when the aspect ratio continues to increase $1.2 < \gamma < 2.0$, the interaction curve is closer to the origin point, and the change is not obvious; as the aspect ratio continues to increase $2.0 < \gamma < 4.0$, the curve moves closer to the origin point. When $\gamma = 1.0$, the interaction equations of diagonally stiffened plate combined shear and non-uniform compression, considering the non-uniform compression distribution factor ζ and the compression-to-shear ratio δ, are given; moreover, the equation of critical elastic buckling coefficient of that are also given;

(2) The effect of stiffener's torsional-to-flexural stiffness ratio K on the elastic buckling stresses of stiffened plates is investigated. With the increase of the stiffener-to-plate

flexural stiffness ratio η, the buckling stress of the stiffened plate with diagonal tension increases rapidly and then tends towards a certain value. The buckling stress of the stiffened plate with diagonal compression increases with η, which is due to the diagonal stiffeners assuming the role of compression bars. As K and ζ increase, the buckling stress of the structure increases, whereas, an increase in δ decreases the buckling stress of the structure;

(3)   The effects of ζ, K, and δ on the threshold stiffness η_{TH} are analyzed. The threshold stiffness of the structure decreases when K and ζ increase. When $1.0 < \zeta < 2.0$, increasing δ can reduce the threshold stiffness of the structure, at which time the threshold stiffness under combined action is lower than that under pure shear action. When $\zeta = 0$, the threshold stiffness of the structure under the combined axial compression and shear is 40% higher than that under pure shear. At this time, if the threshold stiffness in pure shear will not ensure that the structure meets the requirements of the critical local buckling stress, the global buckling will occur and it is unsafe. Therefore, the calculation of threshold stiffness is given in favor of safety.

Author Contributions: Conceptualization, Y.Y. and Z.M.; methodology, Y.Y.; software, Y.Y.; validation, Y.Y.; formal analysis, Y.Y.; investigation, Y.Y.; resources, Y.Y., Z.M. and B.Z.; data curation, Y.Y.; writing—original draft preparation, Y.Y.; writing—review and editing, Z.M. and B.Z.; visualization, Y.Y.; supervision, Z.M. and B.Z.; project administration, Z.M.; funding acquisition, Z.M. All authors have read and agreed to the published version of the manuscript.

Funding: This research was funded by [National Natural Science Foundation of China] grant number [51578064]; [Natural Science Foundation of Beijing Municipality] grant number [8172031].

Institutional Review Board Statement: Not applicable.

Informed Consent Statement: Not applicable.

Data Availability Statement: All the data supporting the results were provided within the article.

Conflicts of Interest: The authors declare no conflict of interest.

References

1.   Aghayari, R.; Dardaei, S. Evaluating the effect of the thickness and yield point of steel on the response modification factor of RC frames braced with steel plate. *KSCE J. Civ. Eng.* **2018**, *22*, 1865–1871. [CrossRef]
2.   Youssef, N.; Wilkerson, R.; Fischer, K.; Tunick, D. Seismic performance of a 55-storey steel plate shear wall. *Struct. Des. Tall Spec. Build.* **2010**, *19*, 139–165. [CrossRef]
3.   Lee, S.; Wang, D.; Liao, Y.; Mathias, N. Performance based seismic design of a 75 story buckling restrained slender steel plate shear wall tower. In *Structures Congress, Proceedings of Structures Congress 2010, Orlando, FL, USA, 12–15 May 2010*; American Society of Civil Engineers: Reston, VA, USA, 2010; pp. 2871–2884.
4.   Thorburn, L.J.; Kulak, G.L.; Montgomery, C.J. Analysis of steel plate shear walls. In *Structural Engineering Report. No.107*; Department of Civil Engineering, University of Alberta: Edmonton, AB, Canada, 1983.
5.   Lubell, A.S.; Prion, H.G.L.; Ventura, C.E.; Rezai, M. Unstiffened steel plate shear wall performance under cyclic loading. *J. Struct. Eng.-ASCE* **2000**, *126*, 453–460. [CrossRef]
6.   Wang, M.; Zhang, X.; Yang, L.; Yang, W. Cyclic performance for low-yield point steel plate shear walls with diagonal T-shaped-stiffener. *J. Constr. Steel Res.* **2020**, *171*, 106163. [CrossRef]
7.   Alavi, E.; Nateghi, F. Experimental study of diagonally stiffened steel plate shear walls. *J. Struct. Eng.* **2013**, *139*, 1795–1811. [CrossRef]
8.   *JGJ/T380-2015 Technical Specification for Steel Plate Shear Walls*; China Architecture & Building Press: Beijing, China, 2015.
9.   Mu, Z.; Yang, Y. Experimental and numerical study on seismic behavior of obliquely stiffened steel plate shear walls with openings. *Thin-Walled Struct.* **2020**, *146*, 106457. [CrossRef]
10.   Xu, Z.; Tong, G.; Zhang, L. Design of horizontal stiffeners for stiffened steel plate walls in compression. *Thin-Walled Struct.* **2018**, *132*, 385–397. [CrossRef]
11.   Yuan, H.X.; Chen, X.W.; Theofanous, M.; Wu, Y.W.; Cao, T.Y.; Du, X.X. Shear behaviour and design of diagonally stiffened stainless steel plate girders. *J. Constr. Steel Res.* **2019**, *153*, 588–602. [CrossRef]
12.   Xu, Z.; Tong, G.; Zhang, L. Elastic and elastic-plastic threshold stiffness of stiffened steel plate walls in compression. *J. Constr. Steel Res.* **2018**, *148*, 138–153. [CrossRef]
13.   Xu, Z.; Tong, G.; Zhang, L. Design of vertically stiffened steel plate walls under combined uniaxial compression and shear loads. *Structures* **2020**, *26*, 348–361. [CrossRef]

14. Xu, Z.; Tong, G.; Zhang, L. Stiffness demand on stiffeners of vertically stiffened steel plate shear wall. *J. Struct. Eng.-ASCE* **2020**, *146*, 06020005. [CrossRef]
15. Vu, Q.; Papazafeiropoulos, G.; Graciano, C.; Kim, S. Optimum linear buckling analysis of longitudinally multi-stiffened steel plates subjected to combined bending and shear. *Thin-Walled Struct.* **2019**, *136*, 235–245. [CrossRef]
16. Feng, L.; Sun, T.; Ou, J. Elastic buckling analysis of steel-strip-stiffened trapezoidal corrugated steel plate shear walls. *J. Constr. Steel Res.* **2021**, *184*, 106833. [CrossRef]
17. Tong, J.; Guo, Y. Elastic buckling behavior of steel trapezoidal corrugated shear walls with vertical stiffeners. *Thin-Walled Struct.* **2015**, *95*, 31–39. [CrossRef]
18. Yang, Y.; Mu, Z. Elastic shear buckling of steel plate shear walls diagonally stiffened by channel stiffeners. *J. Build. Struct.* **2019**, *41*, 182–190. [CrossRef]
19. Martins, J.P.; Cardoso, H.S. Elastic shear buckling coefficients for diagonally stiffened webs. *Thin-Walled Struct.* **2022**, *171*, 108657. [CrossRef]
20. Timoshenko, S.P.; Gere, J.M. *Theory of Elastic Stability*, 2nd ed.; McGraw-Hill: New York, NY, USA, 1961.
21. Bleich, F. Buckling strength of metal structures. In *Engineering Societies Monographs*; McGraw-Hill: New York, NY, USA, 1952.
22. Mikami, I.; Matsushita, S.; Nakahara, H.; Yonezawa, H. Buckling of plate girder webs with diagonal stiffener. *Proc. Jpn. Soc. Civ. Eng.* **1971**, *192*, 45–54. (In Japanese) [CrossRef]
23. Tong, G. *Out-of-Plane Stability of Steel Structures*; China Architecture & Building Press: Beijing, China, 2007.
24. Li, J.; Lv, J.; Sivakumar, V.; Chen, Y.; Huang, X.; Lv, Y.; Chouw, N. Shear strength of stiffened steel shear walls with considering the gravity load effect through a three-segment distribution. *Structures* **2021**, *29*, 265–272. [CrossRef]

Article

Development of Closed-Form Equations for Estimating Mechanical Properties of Weld Metals according to Chemical Composition

Jeong-Hwan Kim, Chang-Ju Jung, Young IL Park and Yong-Taek Shin *

Department of Naval Architecture and Offshore Engineering, Dong-A University, Busan 49315, Korea; jhkim81@dau.ac.kr (J.-H.K.); qwe100301@gmail.com (C.-J.J.); parkyi1973@dau.ac.kr (Y.I.P.)
* Correspondence: ytshin@dau.ac.kr

Abstract: In this study, data analysis was performed using an artificial neural network (ANN) approach to investigate the effect of the chemical composition of welds on their mechanical properties (yield strength, tensile strength, and impact toughness). Based on the data collected from previously performed experiments, correlations between related variables and results were analyzed and predictive models were developed. Sufficient datasets were prepared using data augmentation techniques to solve problems caused by insufficient data and to make better predictions. Finally, closed-form equations were developed based on the predictive models to evaluate the mechanical properties according to the chemical composition.

Keywords: data analysis; artificial neural network (ANN); chemical composition of welds; data augmentation technique

Citation: Kim, J.-H.; Jung, C.-J.; Park, Y.I.; Shin, Y.-T. Development of Closed-Form Equations for Estimating Mechanical Properties of Weld Metals according to Chemical Composition. *Metals* **2022**, *12*, 528. https://doi.org/10.3390/met12030528

Academic Editors: Zhihua Chen and Alberto Campagnolo

Received: 10 February 2022
Accepted: 16 March 2022
Published: 21 March 2022

Publisher's Note: MDPI stays neutral with regard to jurisdictional claims in published maps and institutional affiliations.

1. Introduction

As welding is applied in almost all industrial fields, it is crucial in modern industries. Numerous studies have been conducted to improve the quality of weldments. The quality of a weld depends on its mechanical properties such as yield and tensile strengths, impact toughness, and hardness. These properties are determined by parameters such as chemical composition, microstructure, heat input, interpass temperature, and preheating temperature [1].

In terms of the microstructure, acicular ferrite (AF) is formed at a low heat input with a fast cooling rate, improving the low-temperature toughness. It grows in the form of laths and plates, and is formed in an interlocking structure, which prevents crack propagation. As the AF fraction increases, the strength also increases. Grain boundary ferrite (GBF) with large grain sizes is frequently generated at a slow cooling rate, which is a condition of high heat input. Ferrite side plates (FSPs) consist only of the boundary between laths grown in the same direction at the austenite grain boundary. Both GBF and ESPs have a significant adverse effect on toughness owing to their low crack resistance [1,2]. In addition, martensite–austenite (M–A) constituents, which are frequently generated in high heat input welding, are microstructures that adversely affect the transition temperature as their fraction is increased [3].

In terms of welding conditions, the higher the interpass temperature, the slower the cooling rate and the lower the AF fraction. Therefore, the tensile and yield strengths decrease, and the transition temperature also tends to decrease. In the case of a large heat input, the AF fraction decreases, and the tensile and yield strengths decrease. However, in the case of an excessively low heat input, the impact toughness is adversely affected; therefore, an appropriate amount of heat should be inputted [4–6].

Thus far, numerous studies have been conducted to determine the effect of chemical composition on mechanical properties.

Shao et al. [7] studied the effect of chemical composition on the fracture toughness of bulk metallic glasses. Balaguru et al. [8] studied the effect of weld metal composition on the impact toughness properties of SMAW-welded ultrahigh hard armor steel joints. Glover et al. [4] studied the effect of cooling rate and chemical composition on the microstructure of weld joints of C–Mn and HSLA steels. Takashima and Minami [9] predicted the Charpy absorbed energy of steel for welded structures in the ductile–brittle transition temperature (DBTT) range. Jorge et al. [10] reviewed the relationship between the microstructure and the impact toughness of C–Mn and high-strength low-alloy steel weld metals based on the work of Evans and Bailey [1]. Khalaj and Poraliakbar [11,12] predicted the effects of chemical composition and heat treatment on the phase transformation of microalloy steel using ANN models to estimate the bainite fraction using the austenitization temperature as a parameter. Pak et al. [13] predicted the impact toughness change owing to the interlayer temperature and Ni and Mn concentrations using ANN models. Jung et al. [14] predicted the yield strength, ultimate tensile strength, and high-strength yield for various microstructures by means of soft magnetic wave linear regression and ANN-based algorithms. He et al. [15] devised a physical model to predict the yield stress of bainitic steel by dislocation strengthening and lath boundary strengthening based on the correlation between dislocation density, lath thickness, and yield stress.

Whereas numerous studies have been carried out on the effect of chemical composition on mechanical properties of weld metals, most of the studies were limited to specific conditions or chemical components. There are few studies that formulate experimental results for various chemical components so that their effects can be simply presented. Therefore, it is worthwhile to propose closed-form equations developed by a statistical method.

In this study, artificial neural networks (ANNs), a field of machine learning, were used to determine the effect of chemical composition on mechanical properties (yield strength, tensile strength, and impact toughness). Based on the collected data, correlations between the related variables and the results were analyzed, and a predictive model was developed. The experimental results were extracted from Evans and Bailey [1], and a sufficient dataset was prepared using a data augmentation technique to make better predictions. Finally, closed-form equations were developed based on the predictive models to evaluate the mechanical properties according to the chemical composition.

2. Data Collection and Augmentation

2.1. Data Collection

In this study, experimental data were extracted from Evans and Bailey [1] to understand changes in the mechanical properties with respect to the chemical composition. Mn, which is known to improve the strength significantly, was used as the base. Next, results of experiments on the yield strength, tensile strength, and impact toughness according to the increase in the content of each alloying element were analyzed. To evaluate the impact toughness, the Charpy V-notch (CVN) transition temperature at 100 J was applied, hereinafter referred to as the CVN temperature.

Figure 1a,b shows the shape of the weld used in this study and the sampling method of the specimen. Shielded metal arc welding (SMAW) was applied as the welding technique. The base material and the electrode were welded by using the arc heat generated between a coated electrode and a metal. Herein, ISO 2560:2020, the welding standard for mild steel and low-alloy steel, was applied [16].

The welding electrode was manufactured using the standard technique for a 25% iron powder-coated electrode, and all components were kept constant, except for the components used in the investigation. The automatic spectrographic (ICP-AES) technique was applied for chemical composition analysis. All the core wires of the welding electrode had similar typical compositions of 0.07 C, 0.50 Mn, 0.008 Si, 0.006 S, 0.008 P, 0.02 Cr, 0.003 Mo, 0.03 Ni, 0.02 Cu, 0.0004 Ti, 0.0015 Al, 0.0005 Nb, 0.0005 V, 0.0002 B, 0.02 O, and 0.0025 N (wt%).

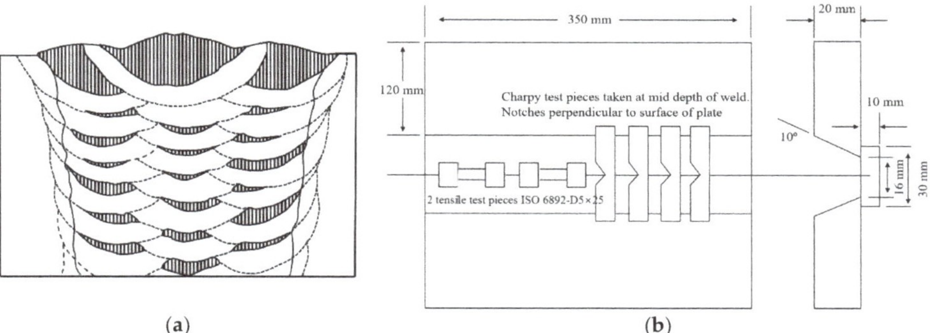

Figure 1. Test assembly for extraction of test pieces: (**a**) disposition of weld runs; (**b**) sample extraction overview adapted from [1].

The weldment was performed in three beads per layer, as shown in Figure 1a. Two 20 mm-thick mild steel plates were attached to a backing strip, and a jig in a flat (downhand) position with a stringer bead was welded. The root gap size was designed to be 16 mm as shown in Figure 1b. Welding conditions were applied such that the current was 170 A, the voltage was 21 V, the interpass temperature was maintained at 200 °C, and the welding speed was adjusted so that the heat input was 1 kJ/mm. For the interpass temperature, the maximum interpass temperature suitable for welding short blocks in ISO 2560:2020 was applied.

The cross section of the panel after welding is shown in Figure 1b. The mechanical properties were tested under 'as-welded' conditions, and the specimens for tensile testing were heated at 250 °C for 14 h to remove diffusive hydrogen. Duplicate tests were conducted with ISO 6892 standard specimens using a gauge with a length of 500 mm and a diameter of 5 mm in a direction parallel to the welding direction.

In the case of metals, the fracture behavior varies from ductile to brittle as the temperature changes from high to low. This change in the fracture mode depends on the material ability to absorb fracture energy. Evans and Bailey [1] conducted an impact toughness test using a CVN impact test, and the specimen was manufactured by machining an axis perpendicular to the welding direction and a notch perpendicular to the plate surface, as shown in Figure 1b. The CVN impact test followed the E23 test procedure of the American Society for Testing Materials (ASTM). Figure 2 shows a typical CVN impact energy curve with respect to the temperature. As shown in Figure 2, the curve is divided into three regions: the upper shelf, lower shelf, and transition. The upper shelf region is a ductile region, the lower shelf is a brittle region, and the transition region is a reference temperature for defining the failure mode as the DBTT. The Charpy impact toughness is generally evaluated based on the transition temperature when absorbing 28 J and 100 J impact energies. In this study, we applied only 100 J of impact energy from a conservative point of view. The 100 J transition temperature data used in this study were obtained from the impact energy and temperature curves obtained through 36 impact tests for each composition condition.

Table 1 shows the composition of the weld metal with varying alloy and Mn content used in this study. Nominal Mn contents of 0.6%, 1.0%, 1.4%, and 1.8% were used, and for each Mn content, three to five varying contents of each element were tested. Table 2 shows typical compositions, except for the elements under investigation.

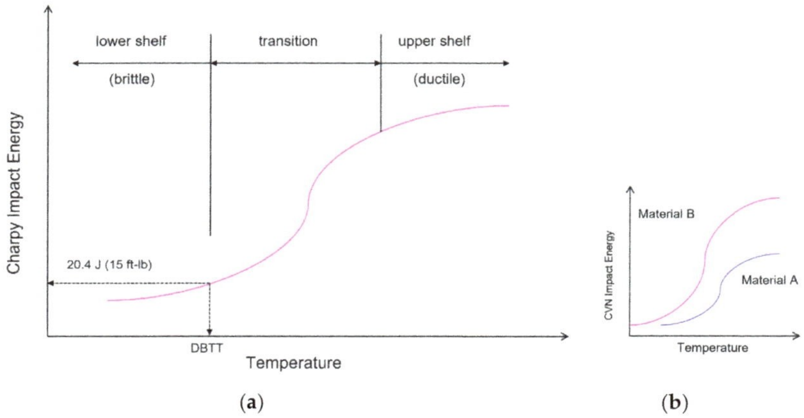

Figure 2. (**a**) CVN impact energy versus temperature and ductile–brittle transition temperature (DBTT); (**b**) comparison of materials A and B adapted from [17].

Table 1. Composition of the weld metal with varying alloy and Mn contents data adapted from [1].

Mn	C	Si	Cr	Ni	Mo	O	V	Nb
	0.04	0.20	0.25	0.50	0.0	0.03	0.0004	0.0004
	0.06	0.40	0.50	1.0	0.25	0.037	0.02	0.01
0.6	0.10	0.60	1.0	2.25	0.50	0.045	0.04	0.02
	0.15	0.90	2.3	3.5	1.1	-	0.06	0.045
	-	-	-	-	-	-	0.08	0.09
	0.04	0.20	0.25	0.50	0.0	0.03	0.0004	0.0004
	0.06	0.40	0.50	1.0	0.25	0.037	0.02	0.01
1.0	0.10	0.60	1.0	2.25	0.50	0.045	0.04	0.02
	0.15	0.90	2.3	3.5	1.1	-	0.06	0.045
	-	-	-	-	-	-	0.08	0.09
	0.04	0.20	0.25	0.50	0.0	0.03	0.0004	0.0004
	0.06	0.40	0.50	1.0	0.25	0.037	0.02	0.01
1.4	0.10	0.60	1.0	2.25	0.50	0.045	0.04	0.02
	0.15	0.90	2.3	3.5	1.1	-	0.06	0.045
	-	-	-	-	-	-	0.08	0.09
	0.04	0.20	0.25	0.50	0.0	0.03	0.0004	0.0004
	0.06	0.40	0.50	1.0	0.25	0.037	0.02	0.01
1.8	0.10	0.60	1.0	2.25	0.50	0.045	0.04	0.02
	0.15	0.90	2.3	3.5	1.1	-	0.06	0.045
	-	-	-	-	-	-	0.08	0.09

Table 2. Typical composition except for the elements under investigation data adapted from [1].

	C	Si	Cr	Ni	Mo	O	V	Nb	S	P	N	Cu	Ti	Al	B
Mn–C	-	0.30	0.03	0.03	0.005	0.049	0.012	0.002	0.006	0.013	0.007	0.03	0.0055	0.0005	0.0002
Mn–Si	0.066	-	0.03	0.03	0.005	0.04	0.012	0.002	0.006	0.013	0.006	0.03	0.0055	0.0005	0.0002
Mn–Cr	0.046	0.32	-	0.03	0.005	0.04	0.012	0.002	0.006	0.013	0.006	0.03	0.0055	0.0005	0.0002
Mn–Ni	0.045	0.32	0.03	-	0.005	0.04	0.012	0.002	0.006	0.013	0.006	0.03	0.0055	0.0005	0.0002
Mn–Mo	0.043	0.33	0.03	0.03	-	0.04	0.012	0.002	0.006	0.013	0.006	0.03	0.0055	0.0005	0.0002
Mn–O	0.078	0.32	0.03	0.03	0.005	-	0.012	0.002	0.006	0.008	0.007	0.03	0.0005	0.0005	0.0002
Mn–V	0.076	0.31	0.03	0.03	0.005	0.042	-	0.0006	0.007	0.006	0.008	0.03	0.0032	0.0005	0.0002
Mn–Nb	0.076	0.32	0.03	0.03	0.005	0.042	0.0007	-	0.007	0.006	0.009	0.03	0.0036	0.0005	0.0002

2.2. Data Augmentation

A sufficiently large dataset is required for accurately determining the relationship between the input and the output because the ANN is a data-driven approach. However, the amount of publicly available data on changes in mechanical properties with respect to weld chemical composition is small. Although experiments have been conducted in a large number of studies, it is difficult to create a single dataset owing to different welding methods and conditions. Although the data by Evans and Bailey [1] applied in this study are relatively diverse and provide many experimental results, the amount of data for individual experiments is still insufficient, which may lead to inaccurate predictions. Therefore, in this study, the amount of initial input data was increased by applying a data augmentation technique.

Figure 3 shows an example of the data augmentation applied in this study. There are only four Mn test data points for the 0.04 C condition. Because the amount of data is small and the trends are different, training the ANN directly may cause an inaccurate fitting or overfitting. Therefore, the test data (asterisks in Figure 3) were first plotted for each condition, and then, nonlinear regression was performed with a line that expressed the test data as good as possible. Finally, ten additional data points (circles in Figure 3) were added to the regression line. Although scatter exists owing to the nature of the experimental data, it is expected that a predictive model with better performance can be generated because the finely tuned data after direct fitting are used for model training.

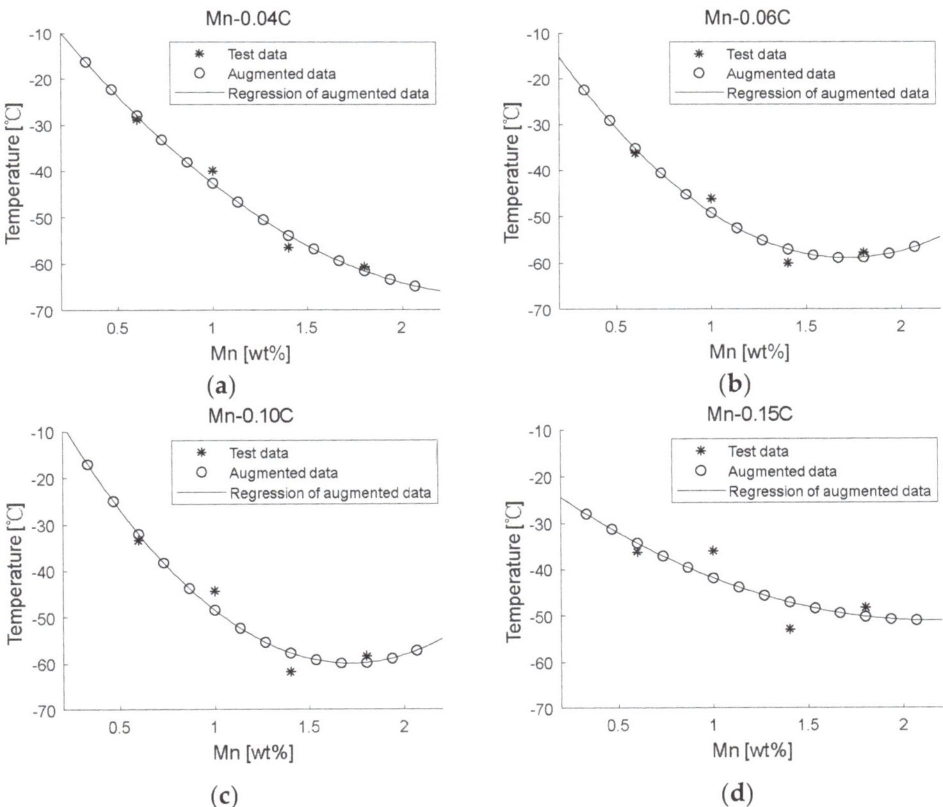

Figure 3. Example of data augmentation: (**a**) Mn–0.04C; (**b**) Mn–0.06C; (**c**) Mn–0.10C; and (**d**) Mn–0.15C.

3. Development of Closed-Form Equations Using the ANN Model

3.1. ANN Model

As shown in Table 1, in this study, we intended to create ANN models that can predict mechanical properties of welds using two variables (chemical compositions). If only one output is obtained from two inputs, multivariate nonlinear regression can be applied because it can be expressed in three dimensions. However, it is difficult to use the existing statistical method to create a model that calculates three outputs (yield strength, tensile strength, and 100 J Charpy temperature) simultaneously. Therefore, it is required to apply the ANN, which is a machine-learning approach that can effectively express multivariate nonlinear systems. The ANN has no limit to the number of input or output variables. If an appropriate structure is used, multiple outputs can be simultaneously derived from multiple inputs.

Figure 4 shows the structure of the ANN model used in this study. This neural network had one input layer, one hidden layer, and one output layer. In the figure, the circle represents the node constituting each layer, and the connecting line represents the weight and the bias representing the relationship between each node. Two input vectors (for example, the content of Mn and the content of C) are fed into the input layer and the output vectors (yield strength, tensile strength, and 100 J Charpy transition temperature) are fed into the output layer. Weights and biases between each node are calculated according to the predetermined ANN model structure. Because the number of hidden layer nodes determines the nonlinearity of the entire system, it should be optimized for the corresponding system. That is, too few nodes result in high computation speed but poor accuracy, and many nodes can lead to overfitting and slow computation.

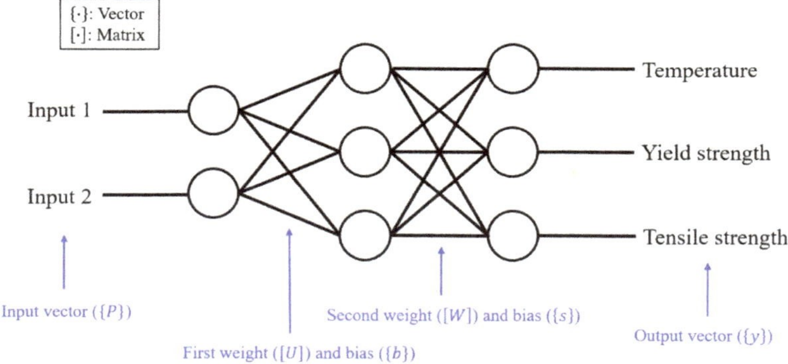

Figure 4. Basic structure of the ANN model applied in this study.

In this study, the number of nodes in the hidden layer was determined using a case study, as shown in Figure 5. For each model, from two to five hidden layer nodes were tested based on the prediction performance. It was observed that the results almost converged if the number of hidden layer nodes exceeded two. Therefore, in consideration of efficiency, three nodes were selected for the hidden layer. The prediction performance was evaluated using the Pearson correlation coefficient between the predicted and target values. That is, the closer it is to one, the better the prediction result.

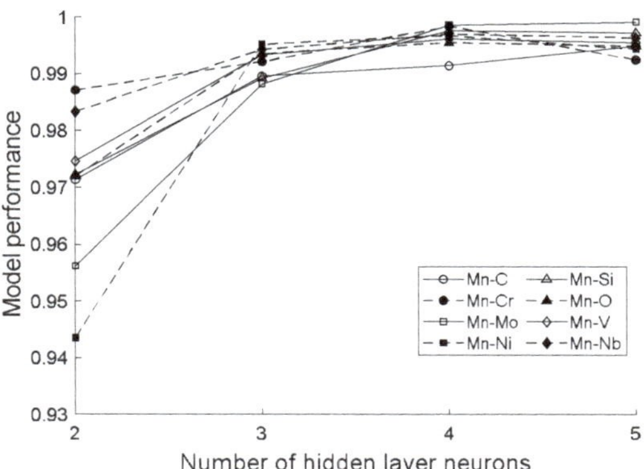

Figure 5. Sensitivity study for determining the number of hidden layer neurons.

Equation (1) presents the ANN model applied in this study in the matrix form: 'tanh' was used as the activation function for the nonlinearity of the model and the backpropagation algorithm was applied for weight and bias optimization. For the calculation efficiency, all values applied to the calculation were normalized between −1 and 1, as in Equation (2). Therefore, the final derived outputs should be denormalized using Equation (1).

$$\{y\} = \{s\} + [W]^T tanh([U]\{P\} + \{b\}),\tag{1}$$

$$X_N = 2 \times \left[\frac{X_R - X_{min}}{X_{max} - X_{min}}\right] - 1,\tag{2}$$

where X_N is the normalized value and X_R is the original value. X_{max} and X_{min} are the maximum and minimum values of X_R, respectively.

To evaluate the performance of the trained model, 15% of the original dataset was used as a test dataset. That is, 85% of the total dataset was used for model training and 15% was used for model evaluation. Moreover, out of 85%, 15% was again separated and used for model validation to prevent data overfitting [18]. In summary, the original dataset was divided into training, validation, and testing sets, as illustrated in Figure 6.

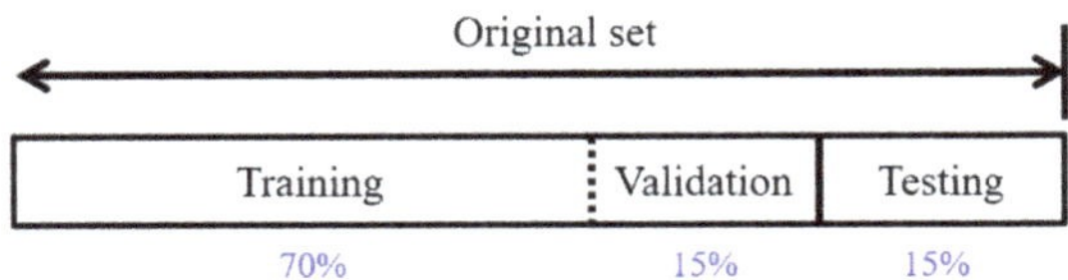

Figure 6. Data splitting ratio.

Figure 7a,b shows the model performances at the training and testing stages in the Mn–C model, respectively, and Figure 7c presents the performance after combing these two stages. Here, the x-axis represents the target value, and the y-axis represents the predicted value. R represents the Pearson correlation coefficient. Table 3 shows R values for all models. As a result, it can be seen that all models show high prediction accuracy.

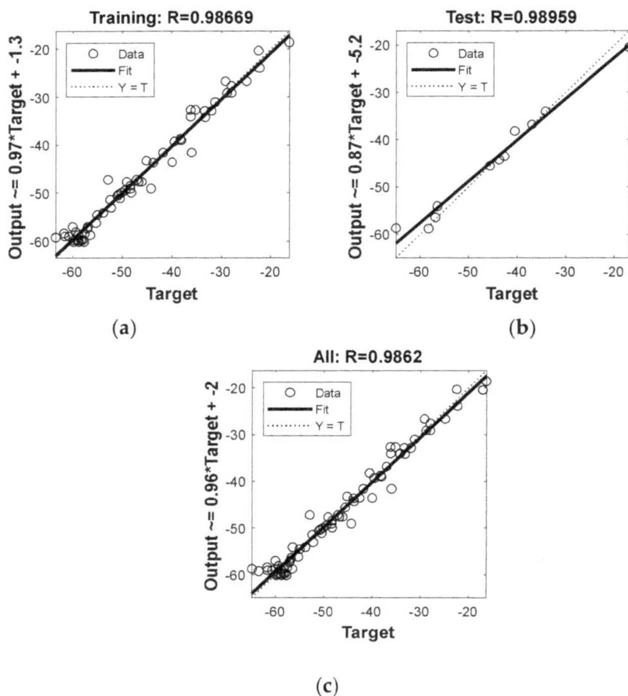

Figure 7. Prediction performances: (**a**) training stage; (**b**) validation stage; (**c**) combination of the training and validation stages.

Table 3. Tested R value for each model.

Mn–C	Mn–Si	Mn–Cr	Mn–Ni	Mn–Mo	Mn–O	Mn–V	Mn–Nb
0.98959	0.98922	0.99215	0.99525	0.98821	0.99358	0.99334	0.99430

3.2. Closed-Form Equations

The previous section discussed the ANN model development for predicting the mechanical properties according to changes in the content of Mn and other compositions. Developing an ANN model is the process of deriving the connection between each node, that is, weight and bias, as shown in Equation (1), through the 'learning' process based on the given data. Therefore, it is possible to use Equation (1) with the calculated weights and biases as a closed-form equation. We prepared the ANN models for each case derived in this study using Equation (3). The models for all eight cases, as summarized in Table 1, are presented in the matrix form. For example, $n = 1$ represents a model for Mn–C, and $n = 2$ represents Mn–Cr. To calculate the CVN temperature, yield strength, and tensile strength according to the Mn–C content, the Mn and C contents are entered into the P_1 vector. Each value of the y_1 vector is then calculated through a matrix operation. However, because these values are normalized, as mentioned in Section 3.1, the input values should be normalized first before applying Equation (3). The values required for normalization are listed in Table 4. The final calculated y_1 must be denormalized again to convert it to the original scale.

$$\{y_n\} = \{s_n\} + [W_n]^T tanh([U_n]\{P_n\} + \{b_n\}), n = 1 \sim 8 \tag{3}$$

$$\{y_1\} = \left\{ \begin{array}{c} \text{Temp}_{\text{MnC}} \\ \text{YS}_{\text{MnC}} \\ \text{TS}_{\text{MnC}} \end{array} \right\}, \{s_1\} = \left\{ \begin{array}{c} 1.049636 \\ 0.233576 \\ 0.241714 \end{array} \right\}, [W_1] = \left[\begin{array}{ccc} 1.010525 & 0.635715 & 0.773552 \\ -0.284486 & 1.199921 & -0.18018 \\ -0.406651 & 1.102196 & -0.02591 \end{array} \right],$$

$$[U_1] = \left[\begin{array}{cc} -0.86956 & -0.64436 \\ 0.51526 & 0.45486 \\ -1.05571 & 0.46409 \end{array} \right], \{P_1\} = \left\{ \begin{array}{c} \text{Mn} \\ \text{C} \end{array} \right\}, \{b_1\} = \left\{ \begin{array}{c} -0.93884 \\ -0.69465 \\ -0.66777 \end{array} \right\}$$

$$\{y_2\} = \left\{ \begin{array}{c} \text{Temp}_{\text{MnCr}} \\ \text{YS}_{\text{MnCr}} \\ \text{TS}_{\text{MnCr}} \end{array} \right\}, \{s_2\} = \left\{ \begin{array}{c} 1.311486 \\ -0.06367 \\ -0.05746 \end{array} \right\}, [W_2] = \left[\begin{array}{ccc} 1.872796 & 0.87106 & -2.98703 \\ 0.000750 & -0.014840 & 3.639551 \\ 0.004089 & -0.000017 & 3.643273 \end{array} \right],$$

$$[U_2] = \left[\begin{array}{cc} 1.186769 & 0.603857 \\ -1.36505 & 0.914433 \\ 0.163576 & 0.113679 \end{array} \right], \{P_2\} = \left\{ \begin{array}{c} \text{Mn} \\ \text{Cr} \end{array} \right\}, \{b_2\} = \left\{ \begin{array}{c} -0.92949 \\ -0.71739 \\ 0.016649 \end{array} \right\}$$

$$\{y_3\} = \left\{ \begin{array}{c} \text{Temp}_{\text{MnNi}} \\ \text{YS}_{\text{MnNi}} \\ \text{TS}_{\text{MnNi}} \end{array} \right\}, \{s_3\} = \left\{ \begin{array}{c} 1.951395 \\ -0.02477 \\ -0.99595 \end{array} \right\}, [W_3] = \left[\begin{array}{ccc} 1.72015 & -0.25232 & -2.50412 \\ -0.55666 & -1.45942 & -0.39684 \\ -0.75407 & -0.81762 & -0.54086 \end{array} \right],$$

$$[U_3] = \left[\begin{array}{cc} -0.63435 & -0.28357 \\ -0.34123 & 0.025948 \\ -0.77397 & -0.35558 \end{array} \right], \{P_3\} = \left\{ \begin{array}{c} \text{Mn} \\ \text{Ni} \end{array} \right\}, \{b_3\} = \left\{ \begin{array}{c} -0.69519 \\ 0.014862 \\ 0.921754 \end{array} \right\}$$

$$\{y_4\} = \left\{ \begin{array}{c} \text{Temp}_{\text{MnSi}} \\ \text{YS}_{\text{MnSi}} \\ \text{TS}_{\text{MnSi}} \end{array} \right\}, \{s_4\} = \left\{ \begin{array}{c} 0.024837 \\ -0.01735 \\ 0.168692 \end{array} \right\}, [W_4] = \left[\begin{array}{ccc} 1.039145 & -0.09436 & 0.047961 \\ -0.22634 & -1.59469 & -0.35168 \\ -0.01153 & -2.00920 & -0.17194 \end{array} \right],$$

$$[U_4] = \left[\begin{array}{cc} -1.08566 & 0.581902 \\ -0.32260 & -0.14604 \\ -0.32434 & -1.23222 \end{array} \right], \{P_4\} = \left\{ \begin{array}{c} \text{Mn} \\ \text{Si} \end{array} \right\}, \{b_4\} = \left\{ \begin{array}{c} -0.38397 \\ 0.140286 \\ -1.27282 \end{array} \right\}$$

$$\{y_5\} = \left\{ \begin{array}{c} \text{Temp}_{\text{MnMo}} \\ \text{YS}_{\text{MnMo}} \\ \text{TS}_{\text{MnMo}} \end{array} \right\}, \{s_5\} = \left\{ \begin{array}{c} 1.057515 \\ -0.03976 \\ -0.03826 \end{array} \right\}, [W_5] = \left[\begin{array}{ccc} -0.99300 & -1.49291 & 1.051739 \\ -0.78631 & 0.499370 & 0.429281 \\ -0.85297 & 0.472669 & 0.406191 \end{array} \right],$$

$$[U_5] = \left[\begin{array}{cc} -0.03755 & -0.49507 \\ 0.992052 & 0.111495 \\ 1.084931 & 0.121627 \end{array} \right], \{P_5\} = \left\{ \begin{array}{c} \text{Mn} \\ \text{Mo} \end{array} \right\}, \{b_5\} = \left\{ \begin{array}{c} -0.00999 \\ 0.779431 \\ -0.82102 \end{array} \right\}$$

$$\{y_6\} = \left\{ \begin{array}{c} \text{Temp}_{\text{MnO}} \\ \text{YS}_{\text{MnO}} \\ \text{TS}_{\text{MnO}} \end{array} \right\}, \{s_6\} = \left\{ \begin{array}{c} 1.077799 \\ 0.067542 \\ 0.067552 \end{array} \right\}, [W_6] = \left[\begin{array}{ccc} 2.043723 & -0.59919 & -0.19960 \\ 0.768526 & 1.142815 & 0.213232 \\ 0.768559 & 1.142852 & 0.213249 \end{array} \right],$$

$$[U_6] = \left[\begin{array}{cc} 1.049968 & 0.121991 \\ 0.833696 & 0.110389 \\ -0.13802 & -0.71512 \end{array} \right], \{P_6\} = \left\{ \begin{array}{c} \text{Mn} \\ \text{O} \end{array} \right\}, \{b_6\} = \left\{ \begin{array}{c} -1.08092 \\ 0.556272 \\ -0.18993 \end{array} \right\}$$

$$\{y_7\} = \left\{ \begin{array}{c} \text{Temp}_{\text{MnV}} \\ \text{YS}_{\text{MnV}} \\ \text{TS}_{\text{MnV}} \end{array} \right\}, \{s_7\} = \left\{ \begin{array}{c} 3.559261 \\ -0.315284 \\ -0.319956 \end{array} \right\}, [W_7] = \left[\begin{array}{ccc} 1.958502 & -2.09365 & 2.507940 \\ -0.18359 & 1.548604 & 0.480616 \\ -0.22796 & 1.513895 & 0.500507 \end{array} \right],$$

$$[U_7] = \left[\begin{array}{cc} -0.81832 & 0.358623 \\ 0.335151 & 0.280673 \\ 0.865442 & 0.267847 \end{array} \right], \{P_7\} = \left\{ \begin{array}{c} \text{Mn} \\ \text{V} \end{array} \right\}, \{b_7\} = \left\{ \begin{array}{c} -0.94493 \\ 0.428865 \\ -1.24693 \end{array} \right\}$$

$$\{y_8\} = \left\{ \begin{array}{c} \text{Temp}_{\text{MnNb}} \\ \text{YS}_{\text{MnNb}} \\ \text{TS}_{\text{MnNb}} \end{array} \right\}, \{s_8\} = \left\{ \begin{array}{c} -0.22090 \\ -0.31089 \\ -0.28607 \end{array} \right\}, [W_8] = \left[\begin{array}{ccc} 0.769446 & -1.06027 & 0.724919 \\ 0.724100 & -1.66486 & -0.02200 \\ 0.667899 & -1.79086 & 0.005047 \end{array} \right],$$

$$[U_8] = \left[\begin{array}{cc} -0.00234 & 1.101230 \\ -0.29114 & -0.08185 \\ -1.43752 & 0.558518 \end{array} \right], \{P_8\} = \left\{ \begin{array}{c} \text{Mn} \\ \text{Nb} \end{array} \right\}, \{b_8\} = \left\{ \begin{array}{c} 1.016240 \\ 0.006815 \\ -0.26343 \end{array} \right\}$$

Table 4. Reference values for normalization and denormalization.

		Min. Value	Max. Value	Unit
	Mn	0.6	1.8	wt%
	C	0.04	0.15	wt%
	Si	0.2	0.9	wt%
	Cr	0.25	2.3	wt%
	Ni	0.5	3.5	wt%
	Mo	0	1.1	wt%
	O	0.03	0.045	wt%
	V	0.0004	0.08	wt%
	Nb	0.0004	0.09	wt%
Mn–C	Temp.	−64.91	−16.29	°C
	YS	389	626	MPa
	TS	435	711.67	MPa
Mn–Si	Temp.	−66.25	−2.56	°C
	YS	364	587	MPa
	TS	427	665	MPa
Mn–Cr	Temp.	−53.31	33.19	°C
	YS	416	743	MPa
	TS	482	797	MPa
Mn–Ni	Temp.	−62.62	69.39	°C
	YS	391	589	MPa
	TS	458	683	MPa
Mn–Mo	Temp.	−54.93	16.16	°C
	YS	372	736	MPa
	TS	430	796	MPa
Mn–O	Temp.	−65.47	25.86	°C
	YS	378	468	MPa
	TS	467	560	MPa
Mn–V	Temp.	−69.74	−21.27	°C
	YS	366	648	MPa
	TS	459	685	MPa
Mn–Nb	Temp.	−66.45	15.50	°C
	YS	384	682	MPa
	TS	475	729	MPa

4. Model Performance Estimation and Discussion

The results calculated using the ANN model (Equation (3)) were compared with the original test data to estimate the model performance. Figures 8–10 show the comparison between the calculated and experimental results using one of the eight models (Mn–Ni) in terms of 100 J Charpy temperature, yield strength, and tensile strength, respectively. The comparison results for the rest of the models for temperature are shown in Figures A1–A7 in Appendix A. In the case of yield and tensile strengths, we only included one case (Figures 9 and 10) because all other cases showed a nearly linear relationship and were almost consistent with the experimental results. In the figure, the lines represent the estimation (E), and the symbols represent the tested data (T).

It can be seen that the trend of the fitting results for the experimental results varies greatly depending on the type or content of the ingredients contained. For example, in Figure 8a, the temperature difference is not large depending on the Mn content in 0.5 Ni, and the temperature does not increase even when the Mn content is increased. However, as the Ni content is increased, the temperature increases rapidly with the Mn content, particularly for 3.5 Ni (Figure 8d). As shown in the other figures in the Appendix A, the trends are all different for the other elements. This indicates that the nonlinearity is large, depending on the type or content of the chemical component. In contrast, in terms of yield and tensile strengths, although the slopes are slightly different, they show an almost perfect linear relationship with the Mn content, regardless of the type or content of the component.

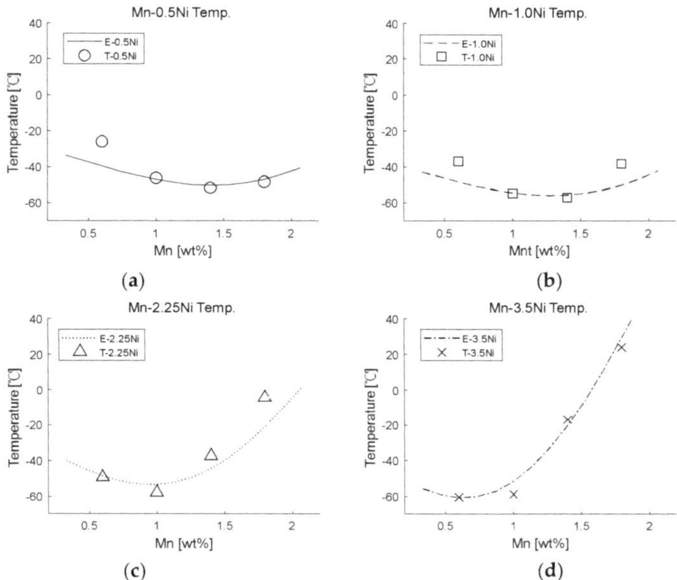

Figure 8. Comparison of the estimation and test data in terms of CVN temperature: (**a**) Mn–0.5Ni; (**b**) Mn–1.0Ni; (**c**) Mn–2.25Ni; (**d**) Mn–3.5Ni.

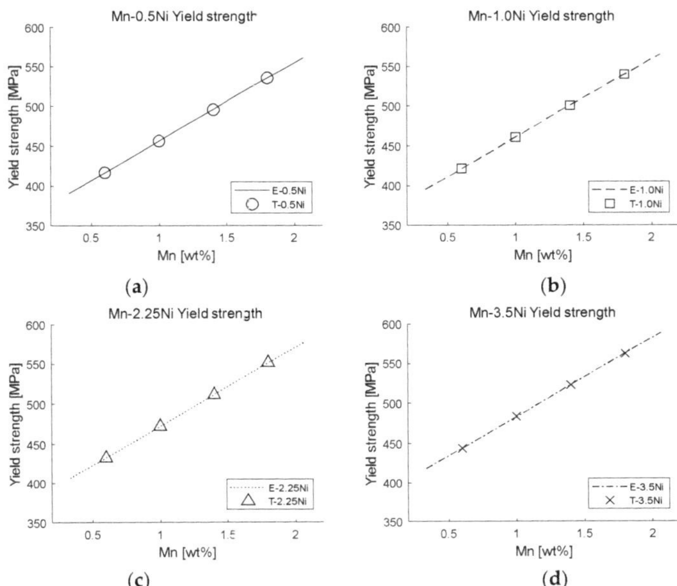

Figure 9. Comparison of the estimation and test data in terms of yield strength: (**a**) Mn–0.5Ni; (**b**) Mn–1.0Ni; (**c**) Mn–2.25Ni; (**d**) Mn–3.5Ni.

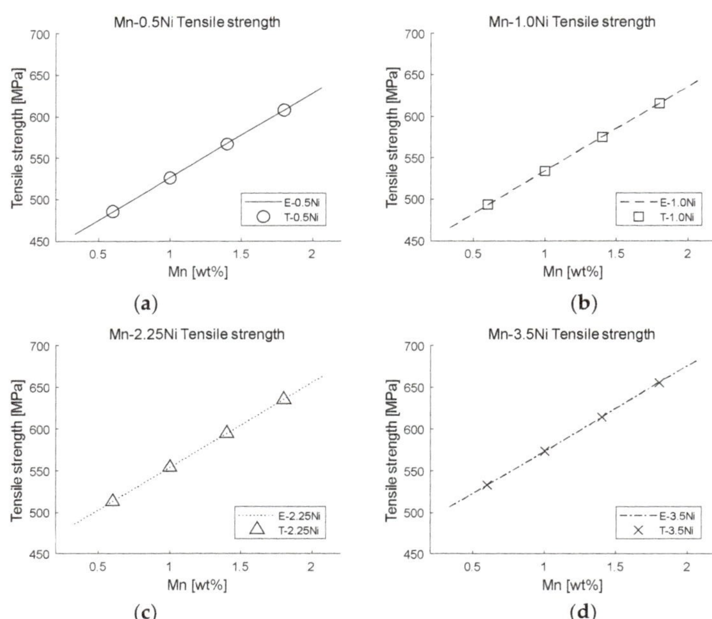

Figure 10. Comparison of the estimation and test data in terms of tensile strength: (**a**) Mn–0.5Ni; (**b**) Mn–1.0Ni; (**c**) Mn–2.25Ni; (**d**) Mn–3.5Ni.

As a result, in most cases, the estimated and experimental values show good agreement. However, in some cases, for example, in the case of the CVN temperature of Mn–1.1Mo in Figure A4, an estimation error occurs. This seems to be because, in this case, the experimental results have greater nonlinearity than in the other cases. When the nonlinearity of the data used in the ANN model is high, the fitting accuracy can be improved by increasing the number of hidden layers. However, if the number of hidden layers is increased, there is a risk of overfitting. Therefore, the decision should be made by considering the overall data trend. In the case of the CVN temperature of Mn–1.1Mo, this error seems to be unavoidable because only this case has a large nonlinearity. It is expected that as the amount of experimental data is increased, the accuracy can be increased further. In contrast, for all cases of yield and tensile strengths, the prediction results show high accuracy, as compared with the test results, because the data used for the model training show a strong linearity.

5. Conclusions

In this study, existing experimental data were collected to investigate changes in the following mechanical properties: 100 J Charpy temperature, yield strength, and tensile strength, depending on the chemical composition. Trends were analyzed by applying an ANN to the data. Data augmentation was performed to solve the problem of insufficient data caused by the dependence of experimental results on specific conditions. Finally, closed-form equations were developed based on the coefficients derived from the ANN models to facilitate a prediction. Based on these results, the following conclusions were drawn:

- By increasing the amount of data through data augmentation, the performance of the ANN model improved. Inaccurate regression that may occur due to the insufficient number of experimental results was prevented in advance, and efficient ANN model training was made. However, some cases of CVN temperature showed an estimation error owing to the large nonlinearity in the data used for the ANN training. Because each condition has a different tendency, accurate regression could not be made in this case with relatively large nonlinearity. For a better predictive model, securing more experimental results is essential. In contrast, the yield and tensile strengths showed high accuracy, as the data showed a linear relationship.

- The developed ANN models are presented in the form of vectors and matrices. Therefore, the three mechanical properties considered as targets in this study were calculated by inputting the content of each component through a simple matrix operation.
- However, because each ANN model developed in this study only considered changes in the content of two elements, there is a limitation in that an accurate prediction cannot be performed if any element with a content different from that of the specimen used for the ANN model is included. That is, the results of this study can be mainly used to predict the relative increase or decrease according to the change in the content of two elements, including Mn.
- Further studies are recommended to develop an ANN model based on data collected in a variety of ranges, including other factors. It will then be possible to efficiently estimate the exact mechanical properties with respect to the contents of chemical compositions.

Author Contributions: Conceptualization, Y.-T.S.; methodology, J.-H.K.; validation, Y.I.P.; formal analysis, C.-J.J.; investigation, C.-J.J.; writing—original draft preparation, J.-H.K.; writing—review and editing, Y.I.P.; visualization, C.-J.J.; supervision, Y.-T.S.; project administration, Y.-T.S.; funding acquisition, Y.-T.S. All authors have read and agreed to the published version of the manuscript.

Funding: This work was supported by the Dong-A University research fund.

Institutional Review Board Statement: Not applicable.

Informed Consent Statement: Not applicable.

Data Availability Statement: The data presented in this study are available on request from the corresponding author.

Conflicts of Interest: The authors declare no conflict of interest.

Appendix A

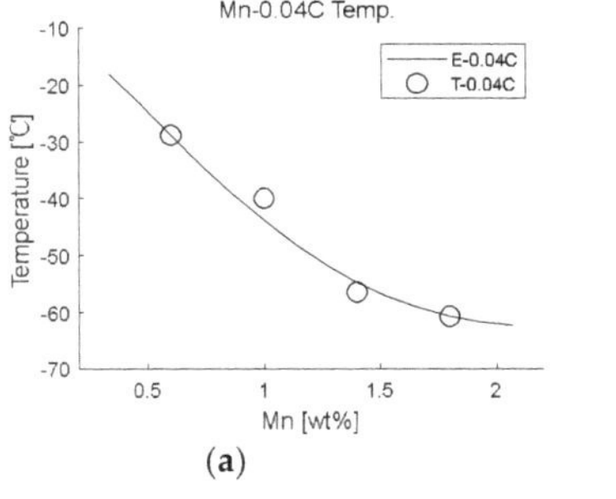

(a)

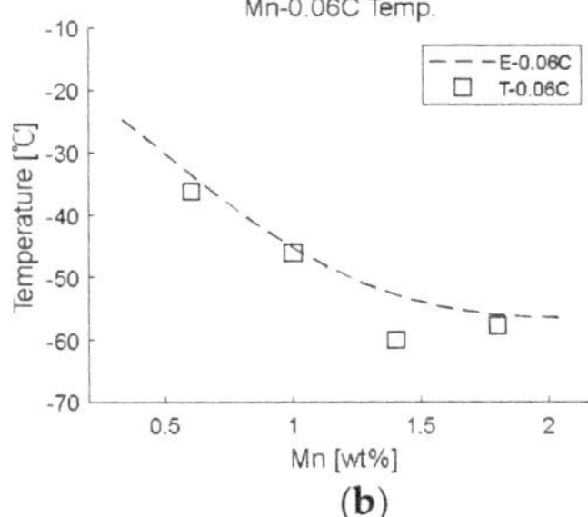

(b)

Figure A1. *Cont.*

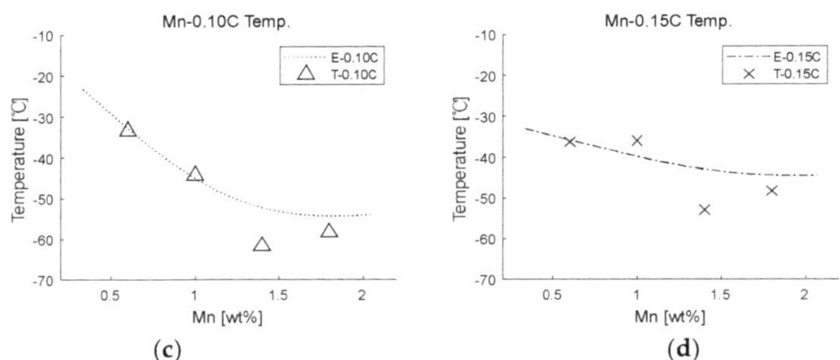

Figure A1. Comparison of the estimation and test data in terms of CVN temperature: (**a**) Mn–0.04C; (**b**) Mn–0.06C; (**c**) Mn–0.10C; (**d**) Mn–0.15C.

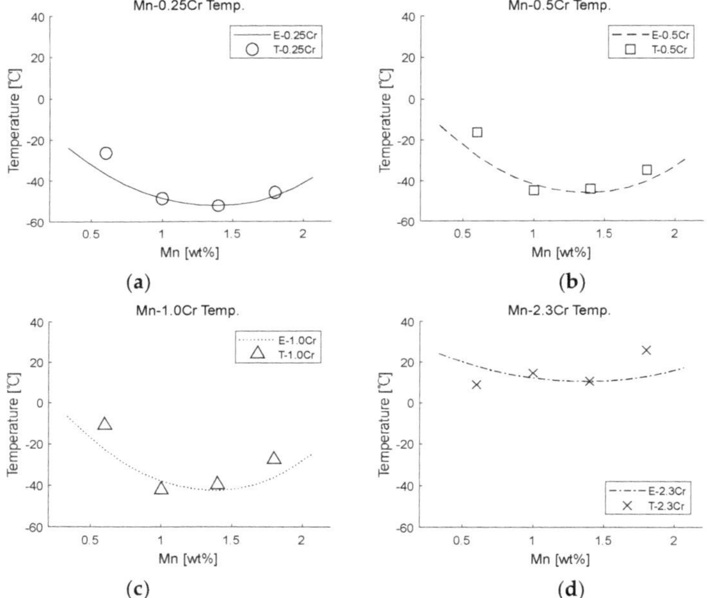

Figure A2. Comparison of the estimation and test data in terms of CVN temperature: (**a**) Mn–0.25Cr; (**b**) Mn–0.5Cr; (**c**) Mn–1.0Cr; (**d**) Mn–2.3Cr.

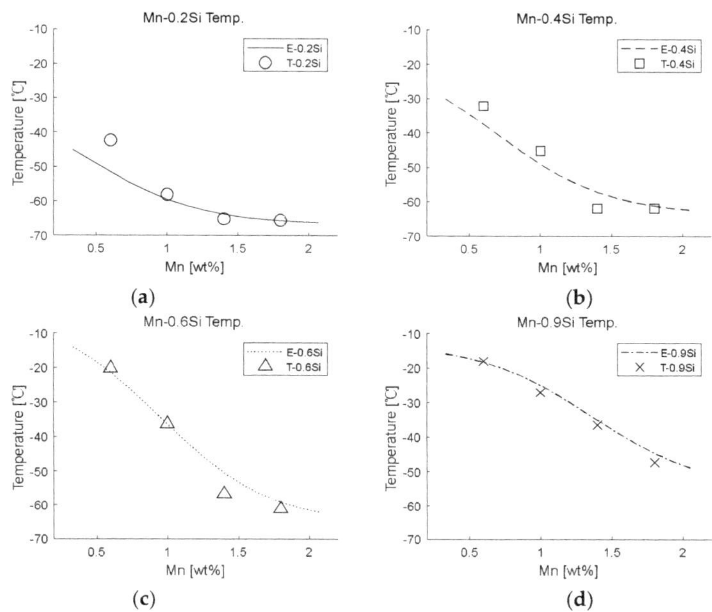

Figure A3. Comparison of the estimation and test data in terms of CVN temperature: (**a**) Mn–0.2Si; (**b**) Mn–0.4Si; (**c**) Mn–0.6Si; (**d**) Mn–0.9Si.

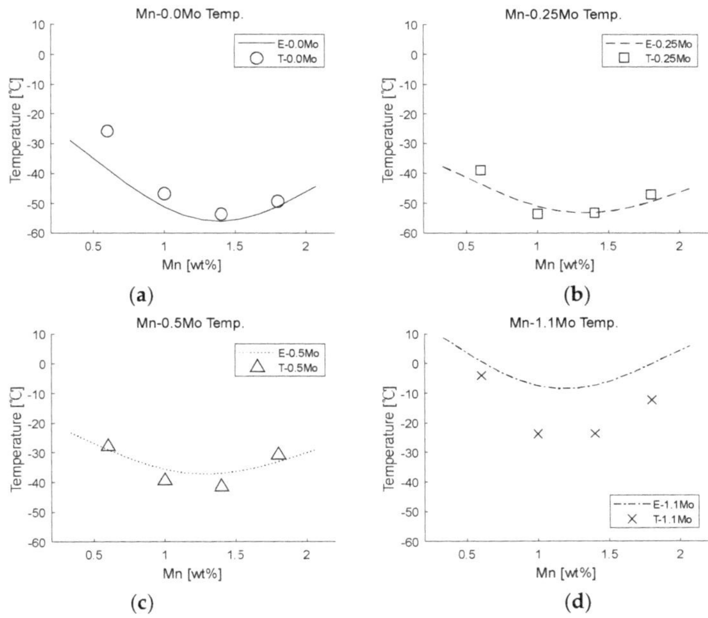

Figure A4. Comparison of the estimation and test data in terms of CVN temperature: (**a**) Mn–0.0Mo; (**b**) Mn–0.25Mo; (**c**) Mn–0.5Mo; (**d**) Mn–1.1Mo.

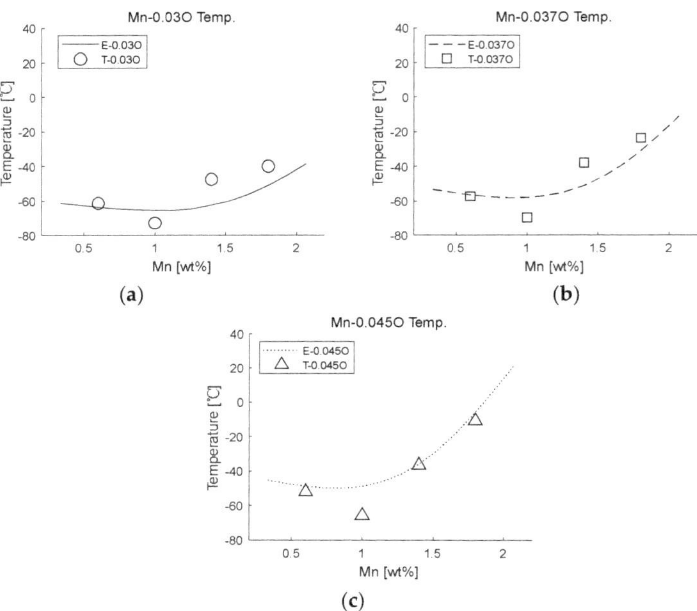

Figure A5. Comparison of the estimation and test data in terms of CVN temperature: (a) Mn–0.03O; (b) Mn–0.037O; (c) Mn–0.045O.

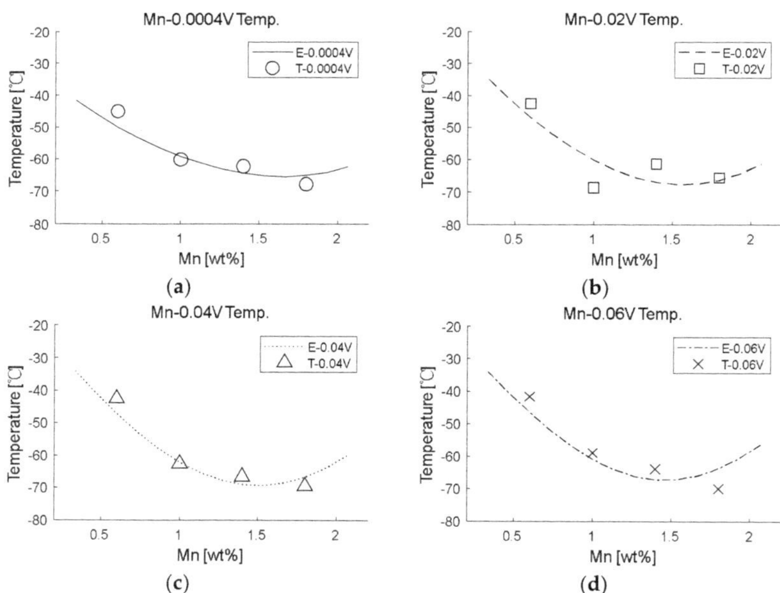

Figure A6. *Cont.*

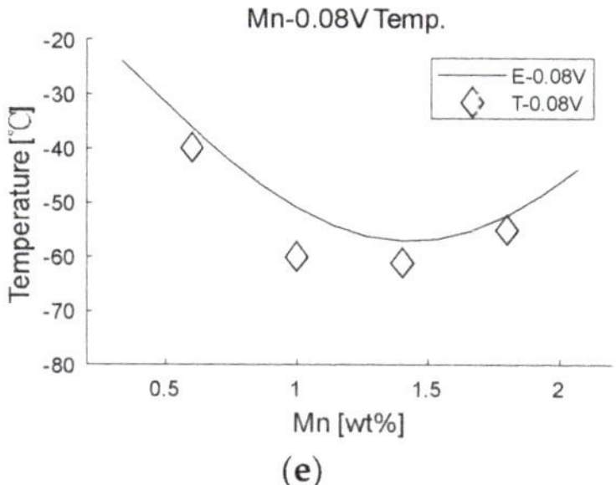

(**e**)

Figure A6. Comparison of the estimation and test data in terms of CVN temperature: (**a**) Mn–0.0004V; (**b**) Mn–0.02V; (**c**) Mn–0.04V; (**d**) Mn–0.06V; (**e**) Mn–0.08V.

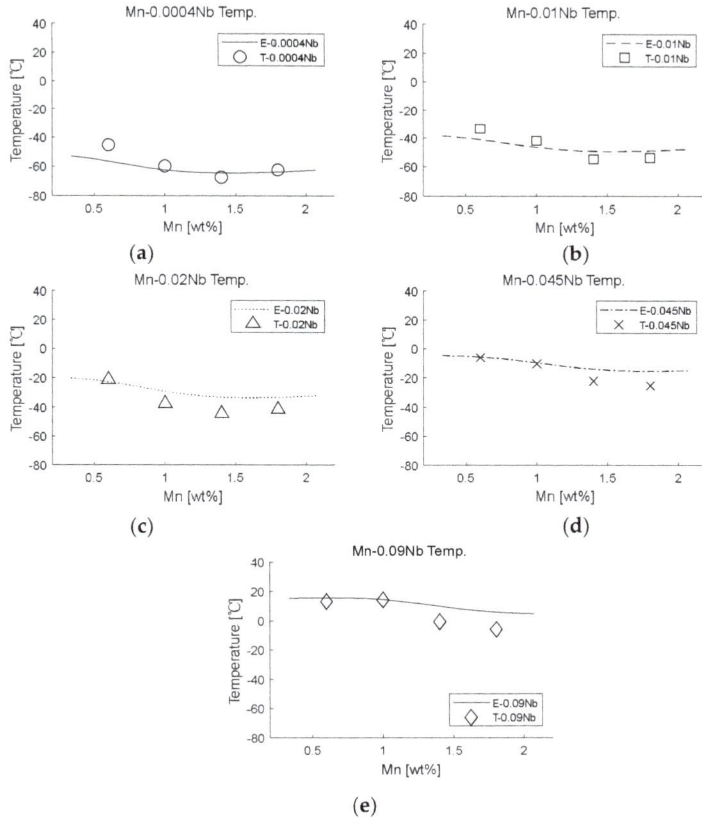

Figure A7. Comparison of the estimation and test data in terms of CVN temperature: (**a**) Mn–0.0004Nb; (**b**) Mn–0.01Nb; (**c**) Mn–0.02Nb; (**d**) Mn–0.045Nb; (**e**) Mn–0.09Nb.

References

1. Evans, G.M.; Bailey, N. *Metallurgy of Basic Weld Metal*; Woodhead publishing: Sawston, UK, 1997.
2. Fleck, N.A.; Grong, Ö.; Edwards, G.R.; Matlock, D.K. The role of filler metal wire and flux composition in submerged arc weld metal transformation kinetics. *Weld. J.* **1986**, *65*, 1135–1215.
3. Shin, Y.T.; Kang, S.W.; Lee, H.W. Fracture characteristics of TMCP and QT steel weldments with respect to crack length. *Mater. Sci. Eng. A* **2006**, *434*, 365–371. [CrossRef]

4. Glover, A.G.; McGrath, J.T.; Tinkler, M.J.; Weatherly, G.C. The influence of cooling rate and composition on weld meta microstructures in a C/Mn and a HSLA steel. *Simulation* **1977**, *60*, 80.

5. Smith, N.J.; McGrath, J.T.; Gianetto, J.A.; Orr, R.F. Microstructure/mechanical property relationships of submerged arc welds in HSLA 80 steel. *Weld. J.* **1989**, *68*, 11.

6. McGrath, J.T.; Gianetto, J.A.; Orr, R.F.; Letts, M.W. Factors Affecting the Notch Toughness Properties of High Strength HY80 Weldments. *Can. Metall. Q.* **1986**, *25*, 349–356. [CrossRef]

7. Shao, L.; Ketkaew, J.; Gong, P.; Zhao, S.; Sohn, S.; Bordeenithikasem, P.; Datye, A.; Mota, R.M.O.; Liu, N.; Kube, S.A.; et al. Effect of chemical composition on the fracture toughness of bulk metallic glasses. *Materialia* **2020**, *12*, 100828. [CrossRef]

8. Balaguru, V.; Balasubramanian, V.; Sivakumar, P. Effect of weld metal composition on impact toughness properties of shielded metal arc welded ultra-high hard armor steel joints. *J. Mech. Behav. Mater.* **2020**, *29*, 186–194. [CrossRef]

9. Takashima, Y.; Minami, F. Prediction of Charpy absorbed energy of steel for welded structure in ductile-to-brittle fracture transition temperature range. *Q. J. Jpn. Weld. Soc.* **2020**, *38*, 103s–107s. [CrossRef]

10. Jorge, J.C.F.; Souza, L.; Mendes, M.C.; Bott, I.S.; Araújo, L.S.; Santos, V.; Rebello, J.M.A.; Evans, G.M. Microstructure characterization and its relationship with impact toughness of C–Mn and high strength low alloy steel weld metals—A review. *J. Mater. Res. Technol.* **2021**, *10*, 471–501. [CrossRef]

11. Khalaj, G.; Pouraliakbar, H.; Mamaghani, K.R.; Khalaj, M.J. Modeling the correlation between heat treatment, chemical composition and bainite fraction of pipeline steels by means of artificial neural networks. *Neural Netw. World* **2013**, *23*, 351. [CrossRef]

12. Khalaj, G.; Nazari, A.; Pouraliakbar, H. Prediction of martensite fraction of microalloyed steel by artificial neural networks. *Neural Netw. World* **2013**, *23*, 117. [CrossRef]

13. Pak, J.; Jang, J.; Bhadeshia, H.K.D.H.; Karlsson, L. Optimization of neural network for Charpy toughness of steel welds. *Mater. Manuf. Process.* **2008**, *24*, 16–21. [CrossRef]

14. Jung, I.D.; Shin, D.S.; Kim, D.; Lee, J.; Lee, M.S.; Son, H.J.; Reddy, N.S.; Kim, M.; Moon, S.K.; Kim, K.T.; et al. Artificial intelligence for the prediction of tensile properties by using microstructural parameters in high strength steels. *Materialia* **2020**, *11*, 100699. [CrossRef]

15. He, S.H.; He, B.B.; Zhu, K.Y.; Huang, M.X. On the correlation among dislocation density, lath thickness and yield stress of bainite. *Acta Mater.* **2017**, *135*, 382–389. [CrossRef]

16. *ISO 2560-2020*; Covered Electrodes for Manual Metal Arc Welding of Non-Alloy and Fine Grain Steels—Classification. ISO: Geneva, Switzerland, 2020.

17. Chao, Y.J.; Ward, J.D., Jr.; Sands, R.G. Charpy impact energy, fracture toughness and ductile–brittle transition temperature of dual-phase 590 Steel. *Mater. Des.* **2007**, *28*, 551–557. [CrossRef]

18. Kim, J.H.; Kim, Y.; Lu, W. Prediction of ice resistance for ice-going ships in level ice using artificial neural network technique. *Ocean Eng.* **2020**, *217*, 108031. [CrossRef]

Article

Experimental Study on Mechanical Properties of Shear Square Section Steel Tube Dampers

Li Xiao [1], Yonggang Li [2], Cun Hui [2,*], Zhongyi Zhou [3] and Feng Deng [2]

[1] CNOOC Petroleum and Gas Power Group Co., Ltd., Beijing 100020, China; xiaoli6@cnooc.com.cn
[2] School of Architecture and Civil Engineering, Zhongyuan University of Technology, Zhengzhou 450007, China; 2020009306@zut.edu.cn (Y.L.); fengdeng@zut.edu.cn (F.D.)
[3] Key Laboratory of Earthquake Engineering and Engineering Vibration, Institute of Engineering Mechanics, China Earthquake Administration, Harbin 150080, China; zhouzy@iem.ac.cn
* Correspondence: hcun@zut.edu.cn

Abstract: Based on the excellent performance of shear metal dampers in building seismic capacity, the traditional shear metal damper was optimized. A double-sided shear steel tube damper with simple structure, easy replacement, and wide application is proposed. In order to study the influence of different design parameters on its seismic performance, taking the steel tube length, height, width, thickness, and connection mode as variables, five groups of 15 specimens were designed for experimental research, and the failure modes, characteristic loads and displacements, hysteretic curves, skeleton curves, stiffness degradation curves, and energy dissipation capacity of each specimen were analyzed in detail. The test results showed that the hysteretic curves of each specimen were full and that the energy dissipation capacity was good. The greater the thickness of the steel tube was, the greater the load-bearing capacity of the damper and the larger the hysteresis loop area were. The greater the width of the steel tube was, the greater the equivalent stiffness was. As displacement amplitude increased, the equivalent stiffness of the specimen showed a downward trend. The two connection modes had their own advantages and disadvantages, and a damper with reasonable connection form would need to be selected according to actual engineering needs.

Keywords: shear dampers; steel tube dampers; low cycle cyclic load; mechanical properties; energy dissipation capacity

Citation: Xiao, L.; Li, Y.; Hui, C.; Zhou, Z.; Deng, F. Experimental Study on Mechanical Properties of Shear Square Section Steel Tube Dampers. *Metals* **2022**, *12*, 418. https://doi.org/10.3390/met12030418

Academic Editors: Zhihua Chen, Hanbin Ge and Siu-lai Chan

Received: 21 January 2022
Accepted: 25 February 2022
Published: 26 February 2022

Publisher's Note: MDPI stays neutral with regard to jurisdictional claims in published maps and institutional affiliations.

1. Introduction

Shear metal dampers can consume seismic energy with plastic deformation by themselves. They can significantly enhance the comprehensive seismic capacity of high buildings. The replaceable property of the metal dampers can promote the rapid postearthquake recovery of building functions [1–4]. Traditional shear metal dampers are mainly used in high-rise or super-high-rise buildings and are not suitable for low multistory buildings because of their complex structure and high cost [5–7].

Since dampers can significantly improve the seismic capacity of buildings, many experts and scholars have studied them. Refs. [8–10] introduced the I-shaped steel damper with shear yield mechanism, which was economical and applicable, had a simple structure, and was easy to construct and replace after an earthquake, and evaluated the parameters of the proposed damper. The results showed that the I-shaped steel damper acted as a ductile fuse, which could prevent the buckling of diagonal elements of concentric braced frame systems and improve the hysteretic performance of concentric braced frames. The necessary relationship for designing this type of damper and a simple method for predicting the nonlinear behavior of the support installed with the damper and estimating the load displacement model of the damper without finite element modeling were also proposed. Lin et al. [11] developed a buckling restrained shear plate damper with removable steel–concrete composite restraint, which simplified the manufacturing process and made the limiter structure light and firm. Test results showed that the composite restraint plate of the damper effectively limited the local and overall out-of-plane deformation of the inner steel shear panel. Deng et al. [12] conducted a quasistatic test on a new buckling restrained shear plate damper (BRSPD) specimen. The results showed that a restrained plate with sufficient stiffness and strength could effectively restrain the out-of-plane buckling of the energy dissipation plate. The BRSPD was numerically analyzed by using the general finite element program. According to the test and analysis results, a design method for the restrained plate and bolt was proposed. Yao et al. [13] used a square steel tube as an out-of-plane stiffener of a steel core plate with low yield point, reduced the flange plate section to alleviate the end fracture, and used the combined hardening constitutive model with memory surface to characterize the plasticity of the square steel tube. They established a numerical model of the shear panel damper and proposed a simplified calibration method. Kamaludin et al. [14] conducted incremental dynamic analysis (IDA) of reinforced concrete frame structures with viscoelastic dampers, friction dampers, and BRB dampers, taking both far- and near-field seismic scenarios into account. The results showed that compared with other energy dissipation devices, viscoelastic dampers had better performance in reducing seismic damage. In contrast, Aydin et al. [15] considered the effect of soil–structure interaction in the optimal design of cohesive dampers and applied a damper optimization method based on the target damping ratio and interlayer displacement ratio from the literature to a building structure model considering different types of sandy soils. The first- and second-order modal responses were considered separately using the time–distance analysis method, and the results showed that the negative effects of sandy soils on the dynamic performance of the superstructure could be overcome by optimizing the dampers in the buildings. It was shown that the effect of soil should be considered when solving the damper optimization problem.

Cruze et al. [16,17] studied magnetorheological dampers with high energy dissipation capacity, low power consumption, perfect damping mechanism, good stability, and fast response time and proposed a method to generate magnetic fields using multiple coils. The magnetorheological dampers were simulated under different currents to simulate the seismic effects, and the results showed that such dampers could effectively reduce the structural response in regions of medium and high seismic activity. Numerical hybrid simulations were performed using OpenSees, and the results showed a significant reduction in displacement and an increase in energy dissipation capacity under major seismic events. Khalili et al. [18] proposed a hysteretic damper for energy dissipation of beam–column steel connections through the bending deformation of an internal hourglass-shaped steel pin. Through experiments and finite element simulations, it was shown that the damper had a high energy dissipation capacity and that the resistance did not decrease significantly during the loading cycle, which improved the ductility and seismic resilience of the whole building structure.

Several scholars have studied curved damped truss moment frame (CDTMF) systems designed using equivalent energy design methods [19,20]. The seismic performance of buildings with different story heights was evaluated using three different analysis methods: nonlinear static analysis, nonlinear time-dependent analysis, and incremental dynamic analysis, and the results showed that CDTMF had excellent ductility and energy dissipation capacity and could effectively control the top plate displacement, interstory displacement, and top plate acceleration. CDTMF could also be used as an effective seismic resisting system. Barbagallo et al. [21] proposed a design method for flexural restrained braced seismic reinforcement measures. Based on a numerical study, the parameters controlling the design method were calibrated to ensure that the near-collapse performance objectives specified in Eurocode 8 were achieved. Nuzzo et al. [22] proposed an effective and easy-to-use displacement-based procedure considering bracing flexibility for the seismic design or retrofit of frame structures with hysteresis dampers. The effectiveness of the proposed procedure was demonstrated in two case studies, for new and existing buildings, with high agreement between the analytical objectives and the numerical capacity curves. Finally, the reliability of the designed frames was evaluated by static and dynamic nonlinear analysis. Mazza [23] considered structural seismic degradation and proposed a displacement-based design (DBD) method using a hysteresis model based on plasticity and damage mechanisms to describe the inelastic response of reinforced concrete frame members. A nonlinear seismic analysis of a single-degree-of-freedom system, equivalent to a multi-degree-of-freedom model of the structure, was performed to generate capacity boundary curves using a hysteresis model defined from the initial backbone curve. Different strengthening schemes were compared, and the effects of different damage levels and different damage evolution patterns were investigated. Finally, pushover curves are calculated for unsupported and damped braced structures with and without consideration of the hysteretic response degradation of the RC frame members. Improvements will be made by inserting hysteretic damping supports to achieve the performance levels specified by the current Italian regulations in high-risk seismic zones.

Other scholars have also carried out experimental research and numerical simulations on the aspects of constitutive relationship and defect form. Golmoghany et al. [24] proposed a new hybrid control system composed of a friction damper and a vertical shear plate in series. In moderate earthquakes, the friction damper (FD) works as the first fuse, while in strong earthquakes, the FD and the vertical shear plate consume energy together. Based on a test, the model was established by software, and the results showed that the seismic effect of this hybrid damper was better than that of a standard reference. Kiani et al. [25] proposed a new type of two-stage yield vertical bar system hybrid damper. The applicability and cyclic response of the proposed system were tested by using the nonlinear finite element modeling protocol in the finite element software. The cyclic secant stiffness, energy dissipation capacity, and equivalent viscous damping ratio of the structure using the hybrid damper were significantly increased. A minimum free span height ratio was also proposed. Lin et al. [26] designed a new type of adaptive shear thickened fluid full-scale damper (STFD) and applied it to suppress the bad vibrations of cables. According to the experimental phenomenon, a phenomenon model in line with the observed performance of the STFD was proposed, the general differential equation of the cable system in the STFD was established, and the eigenvalue problem of the STFD cable system was solved by the finite difference method. This provided a tool for accurately measuring the attenuation process and damping force. Suzuki [27] proposed a method to search for the shape of defects to reproduce a target load displacement relationships and then used the modal iterative error correction method to change each modal coefficient to fit the target load displacement relationship. It was proved that the load displacement relationship of shear plate dampers could be controlled by the shape of defects. Xu et al. [28] conducted repeated load tests with load conditions and size parameters as test variables, proposed a kinematic isotropic composite hardening model, considered the loading history effect, and accurately simulated the complex hysteretic performance of significant strain hardening.

Although many experts and scholars have studied metal dampers, the form of these dampers is not simple enough and the cost is not low enough, so there is still much left to research. In this paper, traditional shear damping devices were optimized, and a kind of double-sided shear steel tube damper with low cost is proposed. The energy dissipation due to elastoplastic deformation caused by in-plane shearing of steel tube side plates was used to reduce vibration. The proposed dampers had the advantages of easy availability of materials, low cost, simple structure, and ability to be used in few-storied buildings in the areas of villages and small towns. Five groups with 15 square section seamless steel tube dampers were designed, and low-cycle repeated load tests were carried out. The influence of different design parameters on the mechanical properties of shear steel tube dampers was analyzed, and the seismic mechanism was revealed.

2. Experiments Details

The shear-type steel tube damper proposed was made of Q235 steel and consisted of steel tube, a T-shaped plate, and an L-shaped connecting plate, as shown in Figure 1.

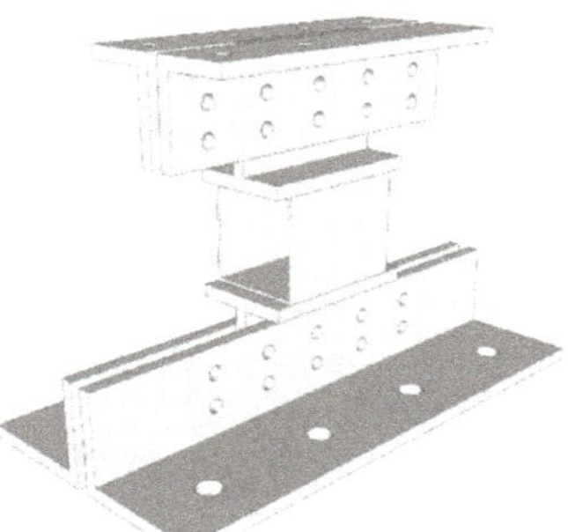

Figure 1. Schematic diagram of shear square section steel tube damper.

The upper and lower sides of the steel tube were welded to the T-shaped plate, and the L-shaped connectors were assembled to the T-shaped plate by bolts. In addition, a set of dampers connected by bolts between the steel tubes and T-shaped plate were set to study the influence of different connection modes on the mechanical properties of the dampers. The thickness of the T-shaped plate was 4 times that of the energy-dissipating part of the steel tube, which was much greater than the thickness of the steel tube. This ensured that only the energy-dissipating part of the steel tube would be consumed when the damper worked and that the steel plate connected to it remained elastic.

2.1. Design of the Specimens

The steel plates on the upper and lower sides of the shear square section steel tube damper were connected with other components by welding, and the L-shaped connector was assembled with the T-shaped plate by bolts. This constituted the nonenergy consuming part of the damper. The side steel plates bore the energy dissipation function and were the main working part of the dampers. In this paper, 5 groups of shear square section steel tube dampers were designed. Each group of dampers was designed with three widths, and the total number of specimens was 15. The influences of steel tube length L, steel tube height H, steel tube width B, steel tube thickness T, and connection mode on the shear seamless steel tube dampers with different design parameters were compared and studied. The structure is shown in Figure 2, and the design parameters are shown in Table 1. The connection forms were divided into welding and bolt connections. The effective thickness of the weld of the welded specimen was the same as that of the steel plate. The diameter of the bolt hole of the bolted specimen was 15 mm, and hexagon-head bolts were adopted.

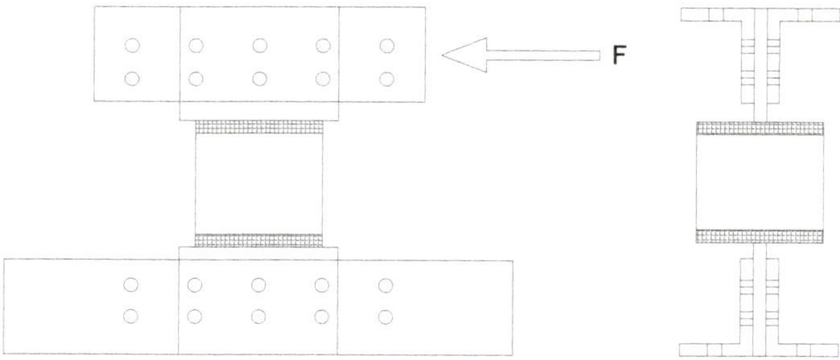

Figure 2. Structure of shear square section steel tube damper (L × H × B was 200 mm × 200 mm × 200 mm).

Table 1. Design parameters of the specimens.

Specimens	L/mm	H/mm	B/mm	t/mm	Connection Type
150-A1	150	150	150	3	Welded
150-A2	150	150	200	3	Welded
150-A3	150	150	250	3	Welded
150-B1	150	150	150	4	Welded
150-B2	150	150	200	4	Welded
150-B3	150	150	250	4	Welded
150-C1	150	150	150	5	Welded
150-C2	150	150	200	5	Welded
150-C3	150	150	250	5	Welded
150-D1	150	150	150	5	Bolted
150-D2	150	150	200	5	Bolted
150-D3	150	150	250	5	Bolted
200-A1	200	200	150	5	Welded
200-A2	200	200	200	5	Welded
200-A3	200	200	250	5	Welded

2.2. Material Characteristics

The steel used in the specimens was Q235. The standard tensile specimens were cut from the same batch of steel as the dampers. The size of the standard tensile specimens is shown in Figure 3. There were three standard tensile specimens in each group. The test results are shown in Table 2.

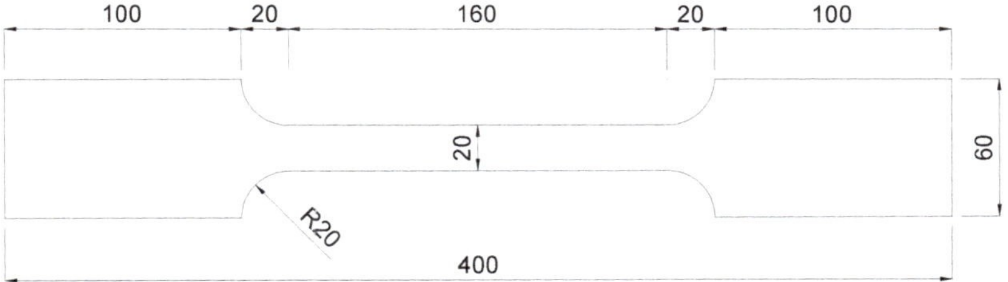

Figure 3. Drawing of standard tensile specimen (unit: mm).

Table 2. Measurement results of the steel plates.

Steel Plate	Yield Strength/MPa	Ultimate Tensile Strength/MPa	Yield Strength Ratio	Elongation/%	Elastic Modulus/MPa
3 mm	238.45	354.72	0.67	27.71	1.92×105
4 mm	242.08	359.76	0.67	30.43	2.12×105
5 mm	243.23	358.25	0.68	29.6	2.07×105

2.3. Loading and Measurement Scheme

The loading device is shown in Figure 4. The specimens were loaded with a displacement loading system. Suppose the yield displacement of the specimen was Δ. The target displacements of the specimens were 0.5Δ, 1Δ, 2Δ, 3Δ, 4Δ, 5Δ, 6Δ, 7Δ, 8Δ, 9Δ, and 10Δ. Each displacement was loaded 3 times until the load decreased to 85% of the ultimate load-bearing capacity. The loading scheme is shown in Figure 5.

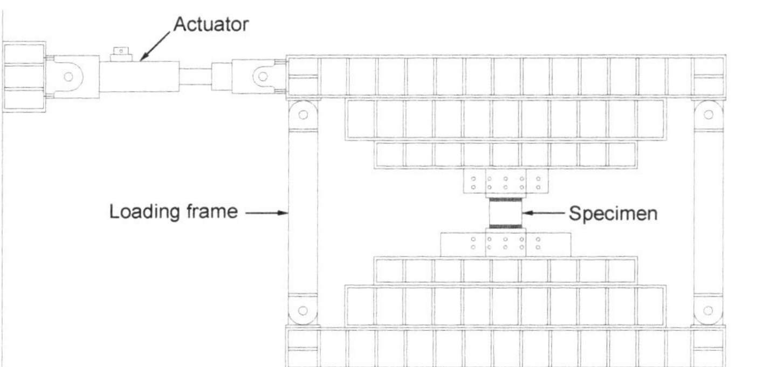

Figure 4. Loading device.

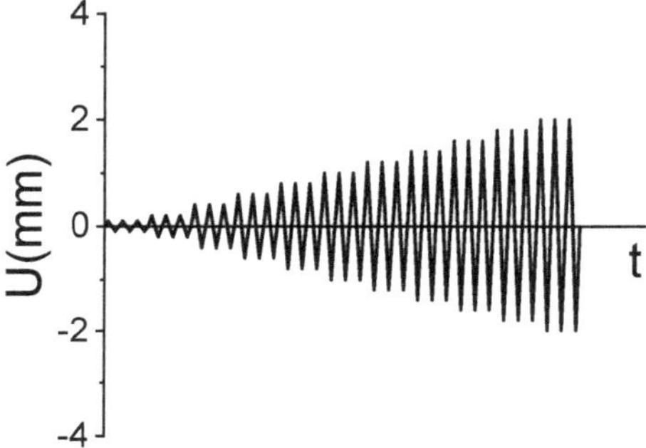

Figure 5. Loading scheme.

Displacement meters with Nos. H1, H2, H3 and H4 were set on the upper and lower sides of the T-shaped connectors and the upper and lower sides of the steel tube, respectively. A 3-directional strain was pasted on the central position of the side of the steel tube damper and an erect VIC-3D full-field strain measurement system (Beijing Reituo Technology Co., Ltd., Beijing, China) was placed on the other side of the damper. The equipment was manufactured by Beijing Reituo Technology Co., Ltd. of China. The measurement scheme is shown in Figure 6.

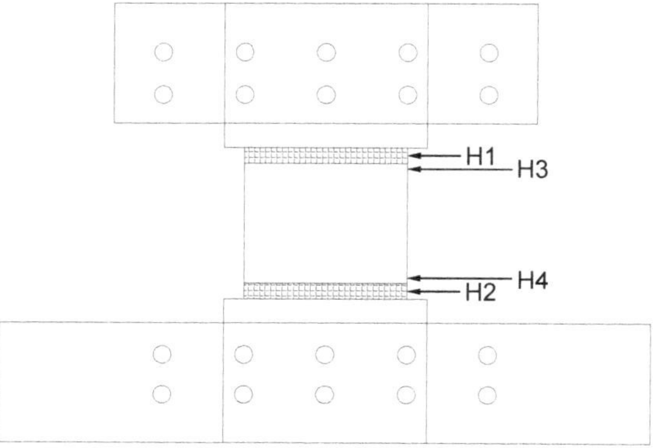

Figure 6. Measurement scheme.

3. Results and Analysis

3.1. Failure Modes

The typical failure modes of the welded steel tube dampers are shown in Figure 7, and the typical failure modes of the bolted steel tube dampers are shown in Figure 8.

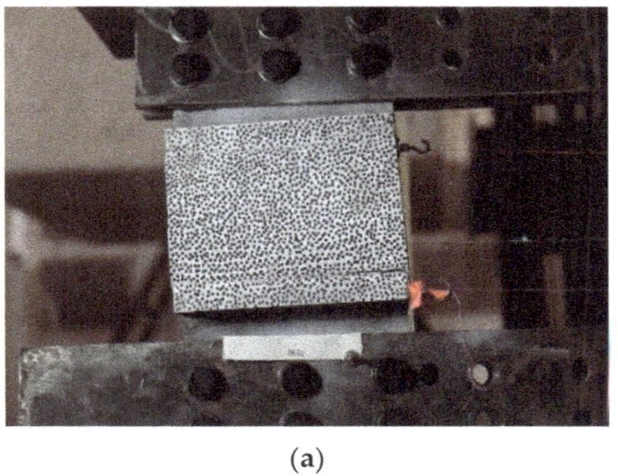
(**a**)

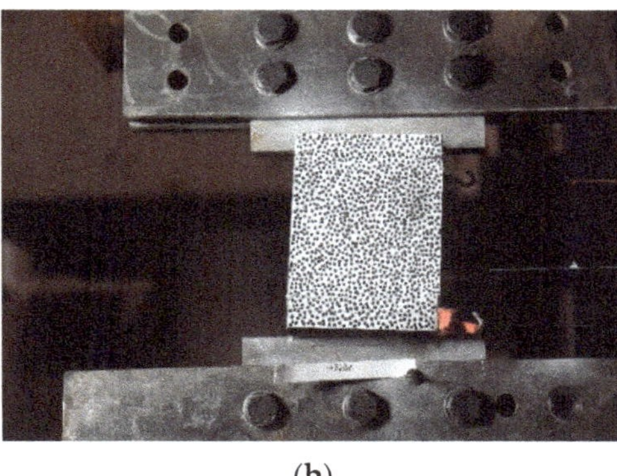
(**b**)

Figure 7. Failure modes of the welded steel tube dampers: (**a**) steel tube end of 150-C3 tearing; (**b**) weld of 200-A1 cracking.

(**a**)

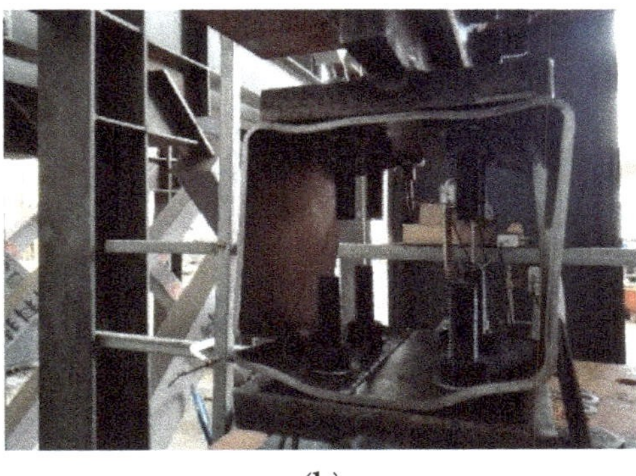
(**b**)

Figure 8. Failure modes of the bolted steel tube dampers: (**a**) corner of 150-D1 warping; (**b**) steel tube end of 150-D1 tearing.

It can be observed from Figure 7 that:

1. The failure modes of welded seamless steel tube dampers were divided mainly into two types: steel tube end tearing and weld cracking.
2. Taking the specimen 150-C3 as an example, the damper had no obvious changes at the initial loading stage. With increases in displacement and load, the side of the steel tube began to tilt slightly, and the steel tube tilt became more and more obvious. When the horizontal displacement was 4.42 mm, the corresponding load was 265.92 kN, and the damper yielded with no obvious cracks. When the horizontal displacement reached 7.16 mm, the peak load-bearing capacity of the damper reached 335.80 kN. Cracks began to appear at the corner of the steel tube at last, and the damper was damaged.

It can be observed from Figure 8 that:

1. At the beginning of loading, the connection between the steel tube and the T-shaped plate became warped. With increasing displacement of the damper, the warping of the corners of the steel tube became much more obvious.

2. When the deformation of the damper was 8.08 mm, the damper began to yield, and the yield strength was 54 kN. When the relative displacement was 17.49 mm, the specimen reached to its peak load-bearing capacity, which was 64.1 kN. With increasing displacement, the bottom of the steel tube was torn, the load declined, and the damper was damaged.

3.2. Characteristic Load and Displacement

The yield displacement, peak displacement, yield load-bearing capacity, peak load-bearing capacity, and initial stiffness of the seamless steel tube dampers are listed in Table 3. The ductility coefficient is the ratio of the peak displacement to the yield displacement of the damper.

Table 3. Characteristic load and displacement of seamless steel tube dampers.

Specimens	Initial Stiffness (kN/mm)	Yield Displacement (mm)	Yield Load (kN)	Peak Displacement (mm)	Peak Load (kN)
150-A1	80.34	2.34	109.56	3.31	122.01
150-A2	114.65	1.87	162.46	2.64	174.47
150-A3	165.81	1.66	166.01	3.72	207.75
150-B1	87.60	2.72	159.99	4.86	184.18
150-B2	124.14	3.15	218.35	5.31	268.07
150-B3	180.42	4.17	269.25	7.13	339.28
150-C1	93.31	3.76	161.04	7.07	191.43
150-C2	123.75	4.84	242.03	8.35	292.92
150-C3	199.28	4.42	265.92	7.16	335.80
150-D1	13.57	8.08	53.37	17.49	64.10
150-D2	24.58	5.87	88.02	9.93	104.97
150-D3	32.15	5.15	141.90	8.45	156.17
200-A1	33.27	5.05	111.63	10.53	131.00
200-A2	56.34	5.08	175.02	9.93	203.76
200-A3	89.19	5.14	233.95	9.87	280.53

It can be observed from Table 3 that:

1. As the thickness of the damper increased, the initial stiffness gradually increased. As the width of the shear surface increased, the initial stiffness of the damper gradually increased, and the proportion of the increase was much larger.
2. The yield loads of specimens 200-A1, 200-A2, and 200-A3 were 111.63 kN, 175.02 kN, and 233.95 kN, respectively. The peak loads were 131.00 kN, 203.76 kN, and 280.53 kN, respectively. The initial stiffnesses were 33.27 kN/mm, 56.34 kN/mm, and 89.19 kN/mm, respectively. As steel tube width increased, the yield load, peak load, and initial stiffness gradually increased.
3. For the three specimens 150-A1, 150-B1, and 150-C1, the yield displacements were 2.34 mm, 2.72 mm, and 3.76 mm, respectively. The displacements corresponding to the peak loads were 3.31 mm, 4.86 mm, and 7.07 mm, respectively. The greater the thickness of the steel tube was, the greater the yield displacement and peak displacement were. The thickness had no obvious influence on the yield load, peak load, or initial stiffness, and the difference between the specimens in groups 150-B and 150-C was less than 10%.
4. By comparative analysis, the yield displacements of 150-C1 and 200-A1 were 3.76 mm and 5.05 mm, respectively, and the peak displacements were 7.07 mm and 10.53 mm, respectively. The yield loads were 161.04 kN and 111.63 kN, respectively, and the peak loads were 191.43 kN and 131.00 kN, respectively. The initial stiffnesses were 93.31 kN/mm and 33.27 kN/mm, respectively. As the height of the steel tube increased, the yield displacement and peak displacement increased by 34.48% and 48.87%, respectively, and the yield load, peak load, and initial stiffness decreased by 30.68%, 31.57%, and 64.34% respectively.

5.   Comparing the welded specimen 150-C1 and the bolted specimen 150-D1, the yield displacements were 3.76 mm and 8.0 mm, respectively; the peak displacements were 7.07 mm and 17.49 mm, respectively; the yield loads were 161.04 kN and 53.37 kN, respectively; the peak loads were 191.43 kN and 64.10 kN, respectively; and the initial stiffnesses were 93.31 kN/mm and 13.57 kN/mm, respectively. Compared with the bolted specimen 150-D1, the yield displacement and peak displacement of the welded specimen 150-C1 were increased by 115.02% and 147.21%, respectively, and the yield load, peak load, and initial stiffness were decreased by 66.86%, 66.51%, and 86.45% respectively. The results showed that the welded damper had better yield load, peak load, and initial stiffness, but the bolted damper had greater yield displacement and peak displacement. Therefore, it is necessary to select a reasonably sized damper according to the actual engineering requirements.

3.3. Hysteretic Curves

The hysteretic curve, which is the relation curve between the load and displacement of a specimen, can reflect the deformation characteristics and energy dissipation capacity of the specimen. The hysteretic curves of each specimen are shown in Figure 9.

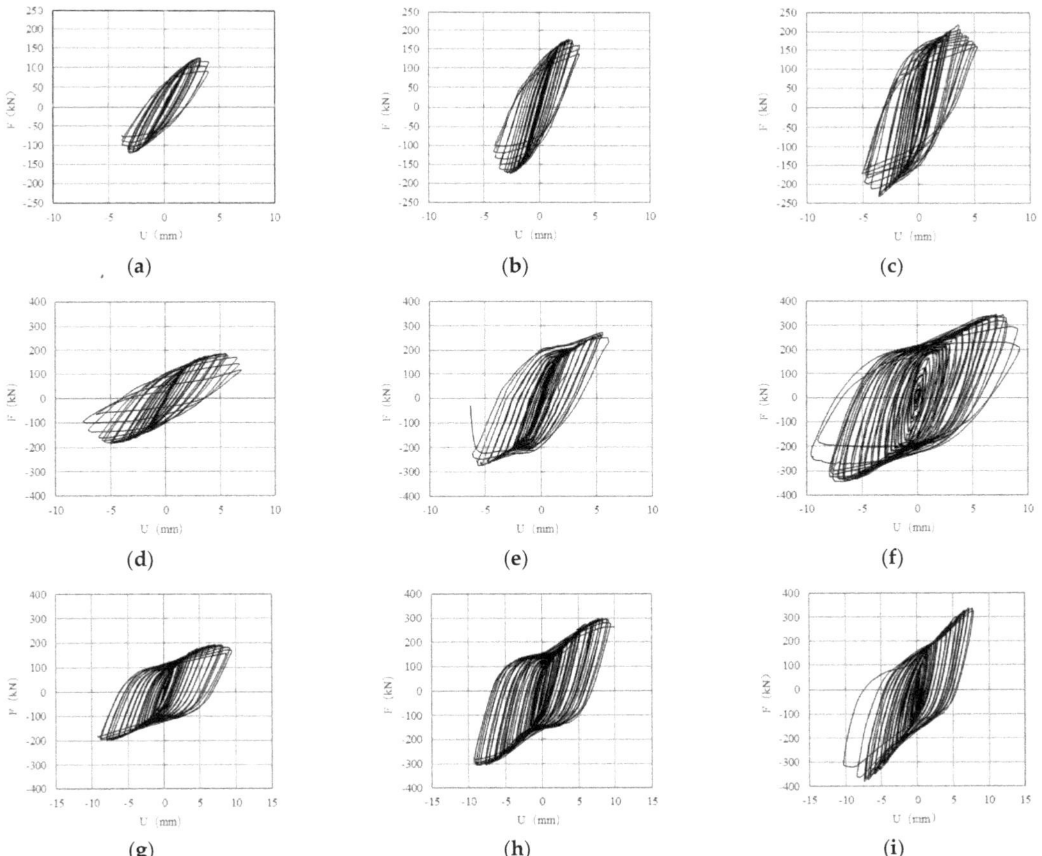

Figure 9. *Cont.*

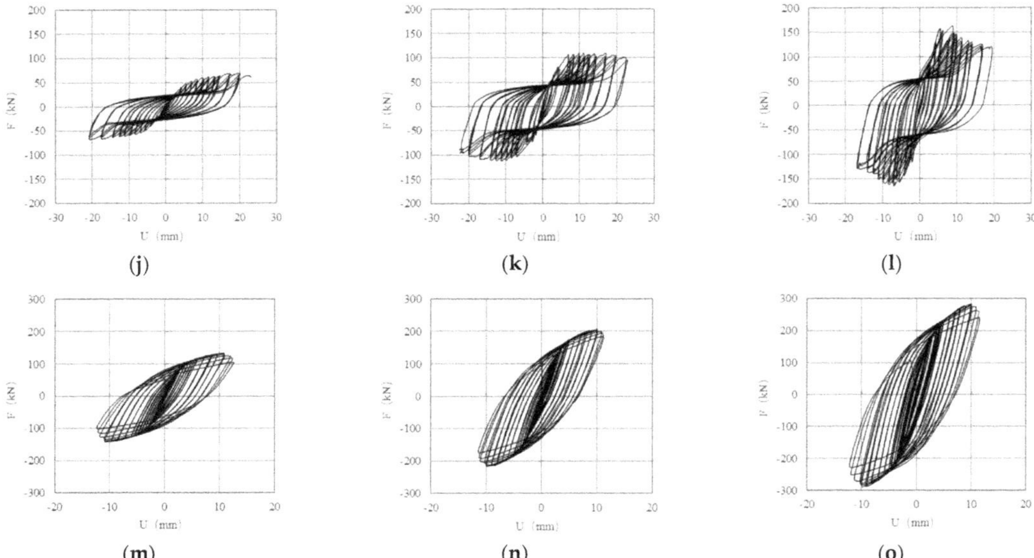

Figure 9. Hysteresis curves of the specimens. (**a**) 150−A1, (**b**) 150−A2, (**c**) 150−A3, (**d**) 150−B1, (**e**) 150−B2, (**f**) 150−B3, (**g**) 150−C1, (**h**) 150−C2, (**i**) 150−C3, (**j**) 150−D1, (**k**) 150−D2, (**l**) 150−D3, (**m**) 200−A1, (**n**) 200−A2, (**o**) 200−A3.

It can be observed from Figure 9 that:

1. The hysteretic curves of specimens 150-A1, 150-A2, and 150-A3 were fusiform, and with increasing width, the hysteretic curves of the shear square steel tube dampers gradually grew full, and the energy dissipation capacities of the dampers were enhanced. When the damper reached peak load, the load-bearing capacity gradually decreased. Before the load-bearing capacity dropped, the hysteretic curves of the dampers were relatively full, indicating that the dampers' energy dissipation capacity is strong.

2. The hysteretic curves of specimens 150-B1, 150-B2, and 150-B3 were in the shape of fusions, and the hysteretic curves of specimen 150-B3 appeared as a slight buckling phenomenon, which may be caused by the out-of-plane buckling of the steel tube due to the lack of out-of-plane constraints. With increasing width, the hysteretic curves of the specimens gradually grew full, and the energy dissipation capacities of the dampers were enhanced. Before the load-bearing capacity dropped, the hysteretic curves of the dampers were relatively full, indicating the dampers' strong energy dissipation capacity. After the load-bearing capacity reached the maximum, it decreased rapidly and soon lost its working capacity.

3. The hysteretic curves of specimens 150-C1, 150-C2, and 150-C3 were full, with a shape close to the parallelogram, and reflected good energy dissipation capacity. With increasing width, the hysteretic curves of the specimens gradually became full, and the energy dissipation capacities of the specimens increased. The positive and negative loads of the damper were not the same, which may have been due to an error in installation. All three specimens showed slight pinching, which was caused by out-of-plane buckling due to the lack of out-of-plane constraints on the steel tube. Comparing the specimens 150-A, 150-B, and 150-C, when other parameters were the same, the greater the steel tube thickness was, the greater the load-bearing capacity of the dampers was, and the greater the hysteresis loop area was.

4. The hysteretic curves of specimens 150-D1, 150-D2, and 150-D3 were Z-shaped. After the damper yielded, it could remain at high load for a period. Because the corner of the steel tube of the bolted damper warped at late loading, there was no obvious growth of the horizontal peak load. In the middle of the hysteretic curve, displacement grew without obvious growth of the horizontal load.

5. The hysteretic curves of specimens 200-A1, 200-A2, and 200-A3 were fusiform and reflected good energy dissipation capacity. The wider the curves were, the fuller they were, and the more energy was dissipated from the corresponding specimens. Compared with the specimen 150-C, the specimens with heights of 200 mm had greater ultimate displacement, but their load-bearing capacity was lower.

3.4. Skeleton Curves and Stiffness Degradation Curves

The skeleton curve of each specimen is shown in Figure 10.

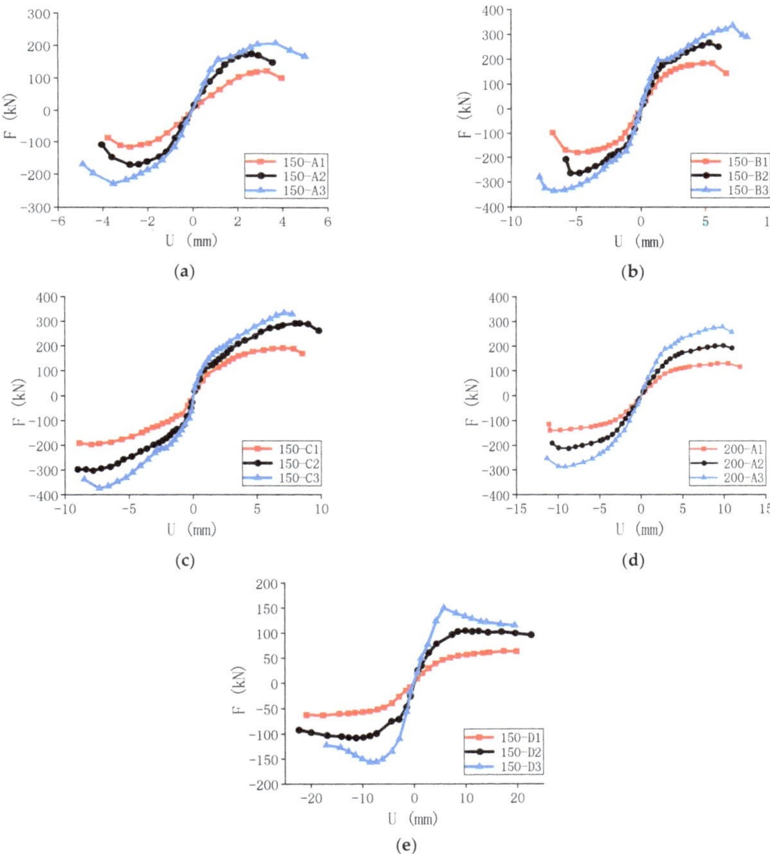

Figure 10. Skeleton curves of the specimens (**a**) 150-A, 150-A2, and 150-A3; (**b**) 150-B1, 150-B2, and 150-B3; (**c**) 150-C1, 150-C2, and 150-C3; (**d**) 200-A1, 200-A2, and 200-A3; (**e**) 150-D1, 150-D2, and 150-D3.

Equivalent stiffness K is the slope of the line between the origin and the load-bearing capacity peak point of each circle of hysteretic curves, which can reflect the change in stiffness. The formula for equivalent stiffness is

$$K = \frac{|+P_\mathrm{i}| + |-P_\mathrm{i}|}{|+\Delta_\mathrm{i}| + |-\Delta_\mathrm{i}|}$$

where P_i is the peak load of the circle I and Δ_i is the displacement corresponding to the load P_i. The stiffness degradation curve of each specimen is shown in Figure 11.

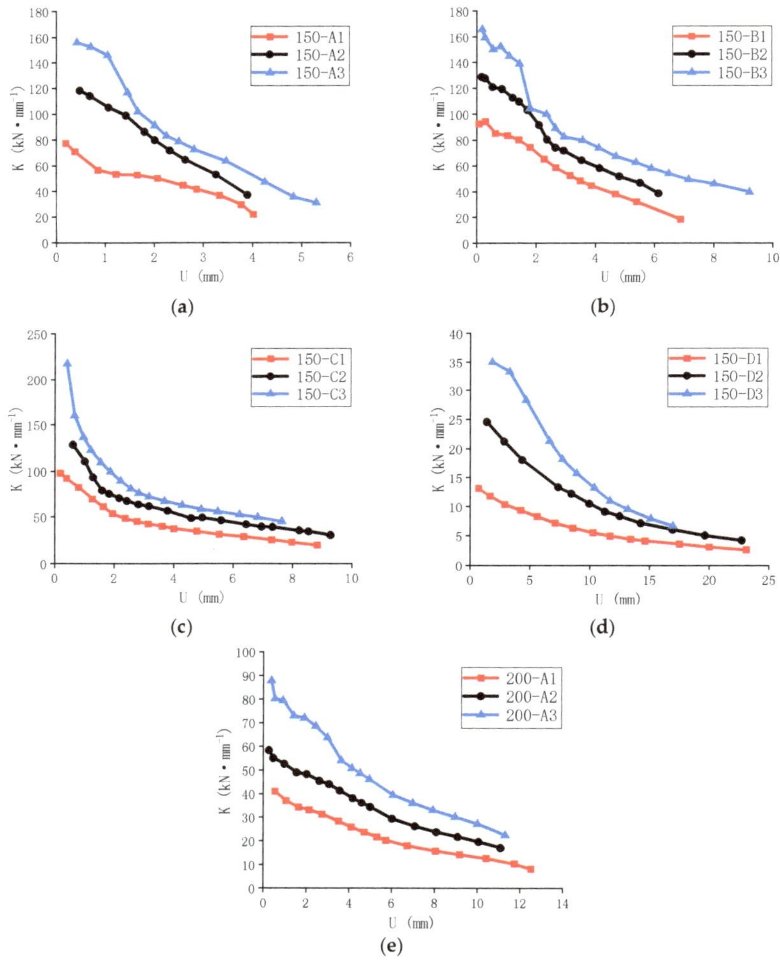

Figure 11. Skeleton curves of the specimens (**a**) 150–A, 150–A2, and 150–A3; (**b**) 150–B1, 150–B2, and 150–B3; (**c**) 150–C1, 150–C2, and 150–C3; (**d**) 150–D1, 150–D2, and 150–D3; (**e**) 200–A1, 200–A2, and 200–A3.

It can be seen from Figures 10 and 11 that:

1. Comparing all the specimens, the wider the steel tube was, the greater the equivalent stiffness was. With increasing displacement, the equivalent stiffness of all three specimens showed a downward trend.

2. The equivalent stiffness degradation curve of 150−A2 was almost linear throughout the whole process, and the stiffness degradation range of 150-A1 and 150-A3 at the initial stage was slightly higher than that at the later stage of loading.

3. For specimens 150−C1, 150−C2, and 150−C3, the stiffness degradation at the initial stage of loading was significantly higher than that at the later stage of loading. Moreover, the greater the initial stiffness was, the more obvious the stiffness degradation was.

4. Comparing specimens in group 150-C (with welded connections) with the specimens of the same size in group 150−D (with bolted connections), the stiffness of the welded dampers was always greater than that of the bolted dampers under the same displacement.

3.5. Energy Dissipation

The energy dissipation of a specimen can reflect its ability to dissipate seismic energy in an earthquake to reduce the seismic response of the main structure. The area enclosed by the hysteretic curves of the damper can reflect the absorbed and dissipated energy, so the energy dissipation capacity of the damper can be reflected by the area of the hysteretic loops. The sum of the hysteresis loop area and the equivalent viscous damping coefficients of specimens from the beginning of loading to the final failure mode are shown in Table 4. The hysteresis loop area was calculated by Origin software. The force and displacement data were imported into the software to draw the hysteresis curve, and the hysteresis loop area was calculated by the software. The equivalent viscous damping coefficient is equal to the area of the hysteretic curve divided by the nominal elastic potential energy, which is equal to the product of the yield displacement and the yield load. Generally, in order to simplify the calculation, the Chinese specifications "Specification for seismic test of buildings" JCJ/T 101–2015 were used for the calculation, schematically as follows:

$$\xi = \frac{1}{2\pi} \cdot \frac{S_{ABCD}}{S_{(OBE+ODF)}}$$

where ξ is the equivalent viscous damping coefficient, S_{ABCD} represents the area of the hysteresis loop, and $S_{(OBE+ODF)}$ represents the area of the triangles S_{OBE} and S_{ODF} as shown in Figure 12.

Table 4. Hysteresis loop area of welded seamless steel tube dampers.

Specimens	Energy Dissipation/kN·mm	S_{ABCD}/kN·mm	$S_{(OBE+ODF)}$/kN·mm	Equivalent Viscous Damping Coefficient
150-A1	3905.40	596.75	433.24	0.22
150-A2	7723.53	977.12	553.36	0.28
150-A3	15,795.12	1396.19	854.52	0.26
150-B1	13,978.32	1562.05	934.93	0.27
150-B2	30,195.37	3004.94	1483.52	0.32
150-B3	64,488.01	5406.19	2600.12	0.33
150-C1	48,224.81	3286.22	1686.38	0.31
150-C2	67,790.33	5097.96	2579.84	0.31
150-C3	97,713.07	4169.23	2815.41	0.24
200-A1	36,869.10	2206.40	1248.55	0.28
200-A2	51,116.49	3595.52	2128.33	0.27
200-A3	78,015.56	4939.12	2722.30	0.29
150-D1	22,752.19	1505.28	1155.87	0.21
150-D2	51,914.42	3523.44	2126.96	0.26
150-D3	58,807.43	3907.24	2185.04	0.28

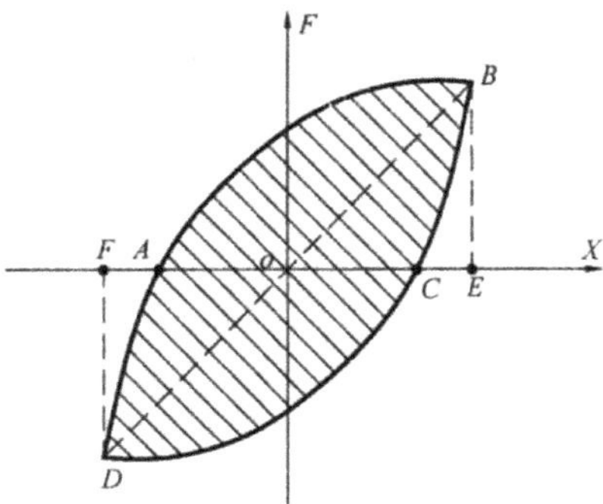

Figure 12. Calculation diagram of equivalent viscous damping coefficient.

It can be observed from Table 4 that:

1. Specimens 150-C1, 150-C2, and 150-C3 had energy dissipation values of 48,224.81 kN·mm, 67,790.33 kN·mm, and 97,713.07 kN·mm, respectively. With increasing steel tube width, the area of the hysteretic loops and the energy dissipation capacity of the dampers gradually increased.

2. Specimens 150-A1, 150-B1, and 150-C1 had energy dissipation values of 3905.40 kN·mm, 13,978.32 kN·mm, and 48,224.81 kN·mm, respectively. This indicates that as the thickness of the steel tube increased, both the area of the hysteretic loops and the energy dissipation of the damper increased.

3. The energy dissipation values for specimens 150-C1 and 200-A1 were 48,224.81 kN·mm and 36,869.10 kN·mm, respectively, showing a reduction of 23.5%. This shows that the longer and the higher the steel tube was, the worse the energy dissipation capacity was.

4. The energy dissipation values for specimens 150-C1 and 150-D1 were 48,224.81 kN·mm and 22,752.19 kN·mm, respectively. The bolted connection mode's energy dissipation value was lower than that of the welded connection mode by 53.82%, indicating that the welded damper could dissipate much more energy.

4. Conclusions

In this paper, the effects of different design parameters on the energy dissipation capacity of the proposed double-sided shear low-cost steel tube damper were explored through experiments. The test phenomena and data were compared and analyzed, and the following conclusions were obtained:

1. The thickness of the steel tube damper and the width of the shear surface had a positive effect on the initial stiffness, and the influence of shear surface width was more significant. Increasing steel tube height had two effects: to increase the yield displacement and peak displacement and to reduce the yield load, peak load, and initial stiffness.

2. The hysteretic curve of each specimen was full, and the energy dissipation capacity is good. After the load-bearing capacity reached the peak load, the load-bearing capacity of each damper decreased rapidly. Comparing the test results of each group, the greater the thickness of the steel tube was, the greater the load-bearing capacity of the damper was, and the larger the hysteresis loop area was. The ultimate displacement of the specimens with high steel tube height was greater, but the load-bearing capacity was reduced.

3. The greater the width of the steel tube was, the greater the equivalent stiffness was. With increasing displacement amplitude, the equivalent stiffness of the specimen showed a downward trend. The initial decline amplitude was significantly higher than that in the later stage of loading. The greater the initial stiffness was, the more

obvious the stiffness degradation was, and the stiffness degradation tended to be gentle in the later stage of the test.

4. With increasing steel tube width and thickness, the hysteretic loop area and the energy dissipation capacity of the specimens gradually increased, while with increasing steel tube height, the energy dissipation and the seismic effect decreased.

5. The welded damper had better yield load, peak load, and initial stiffness, but the bolted damper had greater yield displacement and peak displacement. The bolt connected damper had good ductility and still maintained a good energy dissipation capacity for a period of time after reaching yield and plastic deformation, but the corner of the steel tube tilted up in the later stage of loading, and the peak value of horizontal force did not increase significantly in the later stage. Under the same displacement, the stiffness of the welded damper was always greater than that of the bolted damper, and the welded damper could dissipate more energy. Therefore, a damper with reasonable connection form would need to be selected according to actual engineering needs.

Author Contributions: Conceptualization, L.X. and C.H.; methodology, Y.L.; software, Y.L.; validation, L.X. and C.H.; formal analysis, Z.Z.; investigation, F.D.; resources, Z.Z.; data curation, Z.Z.; writing—original draft preparation, Y.L.; writing—review and editing, C.H.; visualization, Y.L.; supervision, Z.Z.; project administration, Z.Z.; funding acquisition, Z.Z. All authors have read and agreed to the published version of the manuscript.

Funding: This study was funded by the National Key Research and Development Program of China (2019YFC1509304), the Training Plan for Young Key Teachers of the Institution of Higher Education in Henan Province (2019GGJS147), and the Key Scientific Research Project of the Institution of Higher Education in Henan Province (20A560026).

Institutional Review Board Statement: Not applicable.

Informed Consent Statement: Not applicable.

Data Availability Statement: Data are contained within the article.

Conflicts of Interest: The authors declare that they have no known competing financial interests or personal relationships that could have appeared to influence the work reported in this paper.

References

1. Huang, Z.; Li, R.Q.; Yang, L.Z. Study on the Performance-Based Aseismic Design for RC Frame with Mild Steel Shear Damper. *Adv. Mater. Res.* **2014**, *1065–1069*, 1513–1517. [CrossRef]
2. Mirzai, N.M.; Attarnejad, R.; Hu, J.W. Experimental investigation of smart shear dampers with re-centering and friction devices. *J. Build. Eng.* **2021**, *35*, 102018. [CrossRef]
3. Mirzai, N.M.; Attarnejad, R.; Hu, J.W. Analytical investigation of the behavior of a new smart recentering shear damper under cyclic loading. *J. Intell. Mater. Syst. Struct.* **2019**, *31*, 550–569. [CrossRef]
4. Zhu, B.; Wang, T.; Zhang, L. Quasi-static test of assembled steel shear panel dampers with optimized shapes. *Eng. Struct.* **2018**, *172*, 346–357. [CrossRef]
5. Zhao, Y.; Zhang, L. Damage Quantification of Frame-Shear Wall Structure with Metal Rubber Dampers under Seismic Load. *Revue des Composites et des Matériaux Avancés* **2020**, *30*, 227–234. [CrossRef]
6. Wang, B.; Yan, W.; He, H. Mechanical Performance and Design Method of Improved Lead Shear Damper with Long Stroke. *Shock Vib.* **2018**, *2018*, 1–18. [CrossRef]
7. Chen, R.; Xing, G. SEISMIC Analysis of high-rise buildings with composite metal damper. In *MATEC Web of Conferences; Da lian, China*; EDP Sciences: Dalian, China, 2015; Volume 31. [CrossRef]
8. Bakhshayesh, Y.; Shayanfar, M.; Ghamari, A. Improving the performance of concentrically braced frame utilizing an innovative shear damper. *J. Constr. Steel Res.* **2021**, *182*, 106672. [CrossRef]
9. Ghamari, A.; Haeri, H.; Khaloo, A.; Zhu, Z. Improving the hysteretic behavior of Concentrically Braced Frame (CBF) by a proposed shear damper. *Steel Compos. Struct.* **2019**, *30*, 383–392. [CrossRef]
10. Ghamari, A.; Kim, Y.-J.; Bae, J. Utilizing an I-shaped shear link as a damper to improve the behaviour of a concentrically braced frame. *J. Constr. Steel Res.* **2021**, *186*, 106915. [CrossRef]
11. Lin, X.; Wu, K.; Skalomenos, K.A.; Lu, L.; Zhao, S. Development of a buckling-restrained shear panel damper with demountable steel-concrete composite restrainers. *Soil Dyn. Earthq. Eng.* **2019**, *118*, 221–230. [CrossRef]

12. Deng, K.; Pan, P.; Li, W.; Xue, Y. Development of a buckling restrained shear panel damper. *J. Constr. Steel Res.* **2015**, *106*, 311–321. [CrossRef]
13. Yao, Z.; Wang, W.; Zhu, Y. Experimental evaluation and numerical simulation of low-yield-point steel shear panel dampers. *Eng. Struct.* **2021**, *245*, 112860. [CrossRef]
14. Kamaludin, P.; Kassem, M.M.; Farsangi, E.N.; Nazri, F.M.; Yamaguchi, E. Seismic resilience evaluation of RC-MRFs equipped with passive damping devices. *Earthq. Struct.* **2020**, *18*, 391–405. [CrossRef]
15. Aydin, E.; Ozturk, B.; Bogdanovic, A.; Noroozinejad Farsangi, E. Influence of soil-structure interaction (SSI) on optimal design of passive damping devices. *Structures* **2020**, *28*, 847–862. [CrossRef]
16. Cruze, D.; Gladston, H.; Farsangi, E.N.; Loganathan, S.; Dharmaraj, T.; Solomon, S.M. Development of a Multiple Coil Magneto-Rheological Smart Damper to Improve the Seismic Resilience of Building Structures. *Open Civ. Eng. J.* **2020**, *14*, 78–93. [CrossRef]
17. Cruze, D.; Gladston, H.; Farsangi, E.N.; Banerjee, A.; Loganathan, S.; Solomon, S.M. Seismic Performance Evaluation of a Recently Developed Magnetorheological Damper: Experimental Investigation. *Pract. Period. Struct. Des. Constr.* **2021**, *26*, 04020061. [CrossRef]
18. Khalili, M.; Sivandi-Pour, A.; Noroozinejad Farsangi, E. Experimental and numerical investigations of a new hysteretic damper for seismic resilient steel moment connections. *J. Build. Eng.* **2021**, *43*, 102811. [CrossRef]
19. Fathizadeh, S.F.; Dehghani, S.; Yang, T.Y.; Noroozinejad Farsangi, E.; Vosoughi, A.R.; Hajirasouliha, I.; Takewaki, I.; Málaga-Chuquitaype, C.; Varum, H. Trade-off Pareto optimum design of an innovative curved damper truss moment frame considering structural and non-structural objectives. *Structures* **2020**, *28*, 1338–1353. [CrossRef]
20. Dehghani, S.; Fathizadeh, S.F.; Yang, T.Y.; Noroozinejad Farsangi, E.; Vosoughi, A.R.; Hajirasouliha, I.; Málaga-Chuquitaype, C.; Takewaki, I. Performance evaluation of curved damper truss moment frames designed using equivalent energy design procedure. *Eng. Struct.* **2021**, *226*, 111363. [CrossRef]
21. Barbagallo, F.; Bosco, M.; Marino, E.M.; Rossi, P.P.; Stramondo, P.R. A multi-performance design method for seismic upgrading of existing RC frames by BRBs. *Earthq. Eng. Struct. Dyn.* **2017**, *46*, 1099–1119. [CrossRef]
22. Nuzzo, I.; Losanno, D.; Caterino, N. Seismic design and retrofit of frame structures with hysteretic dampers: A simplified displacement-based procedure. *Bull. Earthq. Eng.* **2019**, *17*, 2787–2819. [CrossRef]
23. Mazza, F. A simplified retrofitting method based on seismic damage of a SDOF system equivalent to a damped braced building. *Eng. Struct.* **2019**, *200*, 109712. [CrossRef]
24. Zare Golmoghany, M.; Zahrai, S.M. Improving seismic behavior using a hybrid control system of friction damper and vertical shear panel in series. *Structures* **2021**, *31*, 369–379. [CrossRef]
25. Keykhosro Kiani, B.; Hosseini Hashemi, B. Development of a double-stage yielding damper with vertical shear links. *Eng. Struct.* **2021**, *246*, 112959. [CrossRef]
26. Lin, K.; Zhou, A.; Liu, H.; Liu, Y.; Huang, C. Shear thickening fluid damper and its application to vibration mitigation of stay cable. *Structures* **2020**, *26*, 214–223. [CrossRef]
27. Suzuki, T. Analytical Study on Web Imperfection Design of Shear Panel Damper for Mechanical Performance Control. *J. Struct. Constr. Eng. (Trans. AIJ)* **2019**, *84*, 1401–1409. [CrossRef]
28. Xu, L.-Y.; Nie, X.; Fan, J.-S. Cyclic behaviour of low-yield-point steel shear panel dampers. *Eng. Struct.* **2016**, *126*, 391–404. [CrossRef]

MDPI

St. Alban-Anlage 66

4052 Basel

Switzerland

Tel. +41 61 683 77 34

Fax +41 61 302 89 18

www.mdpi.com

Metals Editorial Office

E-mail: metals@mdpi.com

www.mdpi.com/journal/metals